MW01064567

Environmental Applications of Nanoscale and Microscale Reactive Metal Particles

ACS SYMPOSIUM SERIES **1027**

Environmental Applications of Nanoscale and Microscale Reactive Metal Particles

Cherie L. Geiger, Editor
University of Central Florida

Kathleen M. Carvalho-Knighton, Editor
University of South Florida St. Petersburg

Sponsored by the
ACS Division of Industrial and Engineering Chemistry

American Chemical Society, Washington DC

Library of Congress Cataloging-in-Publication Data

Environmental applications of nanoscale and microscale reactive metal particles / [edited by] Cherie L. Geiger, Kathleen M. Carvalho-Knighton ; sponsored by the ACS Division of Industrial and Engineering Chemistry.

p. cm. -- (ACS symposium series ; 1027)

Includes bibliographical references and index.

ISBN 978-0-8412-6992-7

1. Heavy metals--Environmental aspects--Congresses. 2. Nanoparticles--Environmental aspects--Congresses. 3. Environmental chemistry--Congresses. I. Geiger, Cherie L. II. Carvalho-Knighton, Kathleen M. III. American Chemical Society. Division of Industrial and Engineering Chemistry.

TD879.H4E586 2009

628.5'2--dc22

2009034965

The paper used in this publication meets the minimum requirements of American National Standard for Information Sciences—Permanence of Paper for Printed Library Materials, ANSI Z39.48–1984.

Distributed by Oxford University Press

PRINTED IN THE UNITED STATES OF AMERICA

Foreword

The ACS Symposium Series was first published in 1974 to provide a mechanism for publishing symposia quickly in book form. The purpose of the series is to publish timely, comprehensive books developed from the ACS sponsored symposia based on current scientific research. Occasionally, books are developed from symposia sponsored by other organizations when the topic is of keen interest to the chemistry audience.

Before agreeing to publish a book, the proposed table of contents is reviewed for appropriate and comprehensive coverage and for interest to the audience. Some papers may be excluded to better focus the book; others may be added to provide comprehensiveness. When appropriate, overview or introductory chapters are added. Drafts of chapters are peer-reviewed prior to final acceptance or rejection, and manuscripts are prepared in camera-ready format.

As a rule, only original research papers and original review papers are included in the volumes. Verbatim reproductions of previous published papers are not accepted.

ACS Books Department

Contents

Field Simulation Studies

Technology Demonstrations and Field Applications

Indexes

Preface

In the past decade, research focusing on the use of nano- and micro- scale particles for environmental application has been at the forefront of environmental science and engineering. The enhanced reactivity of these materials has yielded a scientific renaissance leading to research on systems that had previously proven nearly impossible to degrade safely. Materials that may have lacked reactivity at the macro-size have become useful remediation tools as nanoscale particles. Nobel metals that would have never been used due to economics are now becoming feasible alternatives for field applications because of the low mass required when incorporated with nanoparticle base metals. In addition, the complexities of matrices that are innate to environmental remediation make these small particles of particular interest for improved emplacement technologies.

The book is divided into three sections that illustrate laboratory and mechanistic studies, field simulation studies, and technology demonstrations and field applications. Laboratory and mechanistic studies include research in degradation mechanisms of compounds ranging from polybrominated di-ethers and polychlorinated biphenyls to energetic compounds such as trinitrotoluene and triacetone triperoxide when exposed to reactive particles. Field simulation topics include bench-scale application to prove viability of technologies before progressing to deployment and finally studies done at the pilot-scale and field application level.

This book provides an international perspective of the current research efforts and recent results of leaders in the science of microscale and nanoscale environmental applications. Authors who have contributed to this book represent six countries from four different continents. The involvement of leading scientists from around the world is indicative of the importance placed on improving the quality of our most valuble natural resources.

This book is composed primarily of topics that were presented at sessions of the symposium organized by Cherie L. Geiger and Kathy Carvalho-Knighton (editors of this volume) under the support of the Division of Industrial and Engineering Chemistry. In addition, several bonus papers have been added to provide a complete overview of the current research.

As such, the book should appeal to chemists and engineers from academia, industry, and government who are involved in fundamental research, methods development and application. Our hope is that this book will provide a wealth of information and serve as an essential resource for a global perspective of the environmental application of nanoscale and microscale reactive particles.

Acknowledgments

We thank the many people who contributed their time and efforts toward making this volume possible. Candice Simmons, Simone Novaes-Card and Phillip Maloney freely gave of their time to help with formatting and proofreading manuscripts. Thanks to ACS Books Department Staff, including Jessica Rucker and Bob Hauserman. In addition, we would like to thank the following reviewers for their comments, suggestions and participation:

Mary Buthelezi, Ph.D. – Wheaton College
Isabel C. Escobar, Ph.D. – University of Toledo
Young-Shin Jun, Ph.D. – Washington University in St. Louis
Thomas J. Manning, Ph.D. – Valdosta State University
Diana Phillips, Ph. D. – Kettering University
Christian A. Clausen, Ph.D. – University of Central Florida

And most of all, thanks to all of the authors whose exceptional contributions have made this volume so worthwhile.

Overview

Chapter 1

A Review of Environmental Applications of Nanoscale and Microscale Reactive Metal Particles

Cherie L. Geiger[1], Kathleen Carvalho-Knighton[2], Simone Novaes-Card[1], Phillip Maloney[1] and Robert DeVor[1]

[1] Department of Chemistry, University of Central Florida, Orlando, FL 32816
[2] Environmental Science, Policy, and Geography, University of South Florida St. Petersburg, St. Petersburg, FL 33701

The problem of environmental contamination has seen increased interest in recent years as the full impact of various contaminants on flora, fauna, and human beings is being uncovered. In light of this, a great deal of research has been conducted to develop remediation technologies capable of combating this problem. A large variety of pollutants are currently found in the environment, including organic halides, energetic materials, and heavy metals. One of the most promising avenues of research involves the use of microscale and nanoscale metal particles in the remediation process, due to the vast array of useful properties possessed by these particles. Different metals have been tested for activity against several different classes of environmental pollutants, and a synopsis of the results for many of these tests will be provided in this chapter.

Introduction

Traditional Remediation Options

The past century has introduced many technological and scientific advances which at the time appeared to be only beneficial to humanity. Time, along with further advances in instrumentation and analysis techniques, has shown that this has not always been the case. Many of these advances have turned out to have unanticipated and unpredictable consequences, some of which are now only beginning to be addressed. Many chemical compounds created in mankind's past fall into this category.

The scientific community has proposed a variety of methods for the remediation and/or containment of these environmental hazards, with varying degrees of success. Environmental contaminants such as those discussed in this chapter are often subject to stringent legislation which limits the detectable amount in the environment before action must be taken to reduce that level (or limit access). Therefore, a great deal of time has been invested in developing techniques to remediate such contaminated zones. Depending on the type of contamination present (halogenated organic compounds, heavy metals, energetic materials) and where that contamination is located (sediments/soil, groundwater, atmospheric, enclosed systems), different remediation options exist. Techniques currently in use or under investigation include incineration, dredging, landfilling, soil washing/extraction, microbial degradation, capping, monitored natural recovery (MNR), chemical reduction, chemical oxidation, photolytic/radiolytic degradation, and others. However, many of these techniques are hampered by specific limitations or drawbacks that make them impractical for large-scale field use. Some of the more commonly utilized options will now be discussed.

Halogenated Organic Compounds (HOCs)

One of the most commonly used remediation technologies at present is the incineration of contaminated wastes. This technology can be applied to most any type of contamination or hazardous waste, as long as the waste has been isolated (e.g. extracted from soils or sediments via solvent washing). This is an *ex situ* technique which thermally dechlorinates the contaminant into harmless products. However, incomplete combustion is a serious concern with this option. This can lead to the production of even more toxic byproducts, such as the production of polychlorinated dibenzo-*p*-dioxins and polychlorinated dibenzo-furans (PCDFs), commonly referred to as dioxins, from the low-temperature combustion of polychlorinated biphenyls (PCBs) (*1-3*). Incomplete combustion occurs when incineration is carried out at low temperatures, which is why the incineration of PCBs must be performed at temperatures of 1200°C or greater. This calls for a large increase in fuel consumption, which leads to increased costs for this remediation technique (*3*). An additional concern associated with incineration (and all other off-site *ex situ* processes) is the transport of the hazardous materials to the incineration site (*4*). Costs for this technique have been on the rise due to the limited facilities capable of performing high-temperature incineration (*5*).

A more recent development in the remediation of chlorinated wastes is degradation using microbial agents. Two distinct types of microbial degradation exist, utilizing either anaerobic or aerobic microbes to reductively dechlorinate or oxidize the halogenated compounds, respectively. Aerobic degradation causes the oxidation of waste through a series of intermediates, ultimately destroying the contaminant. Unfortunately, research has shown that aerobic microorganisms are not capable of degrading highly chlorinated compounds (*6*) and will preferentially degrade low-chlorinated congeners, which are not those primarily used in the mixtures found as environmental contaminants. In addition, aerobic micro-organisms are only found in the top few millimeters of soils and sediments, limiting their effectiveness as an *in situ* remediation option. Anaerobic microbes, on the other hand, degrade HOCs through reductive dechlorination, carried out via the removal of chlorine atoms as halogen ions. In the case of PCBs, a larger fraction of more highly chlorinated congeners can thus be degraded than when utilizing aerobic microbes, though this is normally limited to the dechlorination of *meta-* and *para*-substituted congeners. *Ortho*-substituted chlorines are much more resistant to this type of attack, which can lead to incomplete degradation of the contaminants (*2*, *7*). Anaerobic treatment followed by aerobic treatment has been used successfully in the degradation of PCE and benzene, where neither treatment alone was capable of degrading both contaminants (*6*).

Sediment contamination, though prevalent in the United States, can be difficult to deal with because of the difficulty involved in accessing the sites of pollution. Several options exist, although the most commonly used is dredging of the contaminated sediments, followed by a second *ex situ* step such as landfilling. Dredging has a major drawback in that disturbing the contaminated sediments can lead to re-release of the contaminants into the surrounding water supply. This allows for the mobilization of the hazardous materials, and can spread the problem into surrounding areas that were previously free of contaminants. Some contaminants, such as PCBs, can also volatilize and enter the atmosphere while the dredged sediment is being de-watered (*8*). Even if the contaminants do not spread to new uncontaminated areas, release into the water system can cause increased contamination of local biota (*9*, *10*). Another serious limitation to dredging is that it does not eradicate the existing problem; it simply moves it from one location to another (in the case of landfilling) or requires an additional remediation technique (i.e. incineration) to complete the degradation.

Another common technique for remediation of sediments is called "capping", and is often then followed by monitored natural recovery. Capping refers to the placement of an inert chemical substance above contaminated sediments in order to prevent the pollutants from interacting with or mobilizing into the surrounding water system. As in the case of dredging, this does not act to degrade the wastes themselves; it is a means to immobilize the immediate threat presented by the contaminant. Another problem associated with sediment capping is the possibility of breaches occurring in the "cap", allowing release of the hazardous materials into the surrounding environment.

Monitored natural recovery (also called monitored natural attenuation, MNA) refers to allowing natural processes to reduce the magnitude and bioavailability of hazardous contamination over time. A variety of biological (biotransformation, biodegradation), chemical (sorption, oxidation, reduction) and physical (volatilization, dispersion, dilution) processes are considered pertinent to MNR. The time frame for MNR is much longer than that of other current remediation options, but it does have the advantage of being more cost-effective than other techniques, since an active *in situ* technology is not utilized. MNR is also less disruptive to the surrounding area, so in this way it is suitable for sensitive environments that might be destroyed by human activity (*11*). However, one risk is that nature can upset the recovery process in the form of sediment disturbances from phenomena like flooding, hurricanes, or earthquakes.

Heavy Metals

Remediation of heavy metals is generally classified into three categories: containment or capping, *ex situ treatment*, and *in situ* treatment. Containment involves isolating the contaminant and preventing its migration via the application of physical barriers made of steel, cement, bentonite or grout. This method can be employed either *ex situ* or *in situ*; however, commercial availability presently favors *ex situ* usage. Containment following isolation (often dredging) can be performed by means of solidification, which enables the contaminant itself to be trapped within a solid matrix or stabilized; this containment strategy requires chemical reactions to reduce contaminant mobility. Vitrification is another stabilizing method in which electrodes are inserted into the contaminated material, allowing the application of electrical current. Physical and chemical changes that take place during vitrification are driven by the applied current, although this requires large amounts of energy. As the material cools, the contaminant will be solidified in a glasslike matrix (*12*).

The *ex situ* methods involve the removal of soil or sediments from the environment and then the use of various treatment options such as soil washing, physical separation, or hydrometallurgical recovery (for highly contaminated soils). *Ex situ* methods for sediments are performed after dredging. Each year, three hundred million cubic yards of sediments are dredged to maintain the navigability of US waterways. Of those, three to twelve million cubic yards are so heavily contaminated that they require special handling and remediation (*12*). Re-suspension of the offending contaminant into the surrounding water environment during sediment removal is of particular concern with the dredging of sediments for ex situ treatment, as it can exacerbate the problem (*13*). Another complication inherent to sediment remediation is the high level of silt, clay and organic matter present, which can impede effective remediation. Sometimes the problems engendered by remediation can be worse than the initial presence of the contaminant. In these cases, the best option may be leaving the sediment as is and implementing continuous monitoring (*14*).

Common *in situ* methods for heavy metal remediation include soil flushing, electrokinetic techniques, and phytoremediation. Soil flushing consists of infiltrating the affected soils with extracting solutions. These solutions may contain chemical additives such as organic or inorganic acids, and complexing agents such as ethylenediaminetetraacetic acid, EDTA (*15*). While not as disruptive as *ex situ* methods like dredging, soil flushing can prove problematic because strict control must be kept over the mobilized contaminants to prevent their migration to other areas. Electrokinetic techniques require the insertion of electrodes and the passage of a low current which promotes the migration of metals and anions to the appropriate electrodes, where they can be removed by electroplating or precipitation (*16*). Phytoremediation involves certain plant species and their propensity to accumulate metals such as cadmium, copper, lead, nickel and zinc. When cultivated in a contaminated area, these plants can help to extract metals from the surrounding soil and sequester them in their tissues (*12*, *14*).

Energetic Materials

Both *ex situ* and *in situ* techniques have been employed for the remediation of highly energetic materials such as hexahydro-1,3,5-trinitro-1,3,5-triazine (RDX) and 2,4,6-trinitrotoluene (TNT), to varying degrees of success. *Ex situ* techniques include the use of granulated carbon, anaerobic bioreactors, UV-oxidation reactors, and electrochemical cells. All of these techniques suffer from high implementation costs stemming from re-injection and pumping, which limits the overall usefulness and applicability of these techniques (*17*).

Support for Remediation by Microscale and Nanoscale Metallic Particles

A variety of metals have been investigated for possible use as part of an environmental remediation technology. Some of the earliest research in this vein concentrated on the use of zero-valent metals for the reductive dehalogenation of commonly used industrial solvents and wastes. Early success was seen in the use of granular iron turnings for the remediation of contaminants including chlorinated ethylenes (*18*), carbon tetrachloride (*19*, *20*), and halogenated aromatics (*20*). Early applications of this technology included the installation of permeable reactive barriers (PRBs) to reduce contaminants found in groundwater flow (*21-23*). While PRBs did prove capable of remediating the contaminants found in underground plumes, they were unable to treat the source zone of the contamination (*23*). This was due to the difference in the inherent natures of the zero-valent iron being used and the contaminant itself. Many of these halogenated groundwater contaminants fell under the classification of dense non-aqueous phase liquids (DNAPLs) (*24-26*). As such, they had a hydrophobic nature, while the zero-valent iron surface was more hydrophilic. The metal was thus only capable of treating the dissolved phase contaminant, not the liquid DNAPL zone itself, due to the lack of interaction between the metal and the contaminant (*25, 27*).

In spite of this, metal particles have proven to be effective in the remediation of several important halogenated organics (*18-20*, *28-33*). Additional research has shown that granular zero-valent iron also is capable of reducing and precipitating heavy metals such as hexavalent chromium (*34*, *35*), and degrading nitroaromatics such as nitrobenzene (*36*), pentaerythritol tetranitrate (*37*), and nitroglycerin (*38*).

The initial work using metal particles such as zero-valent iron focused on the macroscale, using easily obtainable and less-costly iron turnings as a basis for the chemical remediation. Since then, work has focused on using smaller particles (from several microns to submicron in diameter) which provide increased reactivity. These small metallic particles possess many inherent advantages in terms of environmental remediation. One of the most obvious is the enhanced reactivity of the particles due to the increase in surface area available for reaction (*39-42*).

There are three common effects that are known to occur when using microscale and nanoscale particles.

1. Reaction rates are increased when compared to larger particles with smaller surface areas (*39*, *41*).
2. Contaminants that do not react readily with larger particles (such as PCBs and hexachlorobenzene) show enhanced degradation with microscale/nanoscale particles (*39*, *41*).
3. More completely reacted byproducts are formed from contaminants (such as the production of methane rather than chloroform from carbon tetrachloride) than when larger bulk particles are used (*40*).

While some of these advantages of are primarily due to the increased surface area of the smaller micro/nanoscale particles, this is not the only effect that decreased size has on their properties. It is well documented that micro- and nanoscale particles have different properties than those exhibited by bulk media comprised of the same elements. In the range of 1 to 100 nanometers, particles are in an intermediate phase between atomic and bulk states. In this size range, their electronic properties can be quite different than those of bulk metal turnings. This is primarily due to the fact that in the microscale and nanoscale range, both physical and chemical properties mainly rely upon the atoms located at the surface of the particles rather than those deeper in the bulk material. This often leads to a more active metal particle (*43*).

Contaminants of Interest

Micro/nanoscale particles have been proven to be effective in the remediation of a large number of environmental contaminants. These different compounds can vary widely in physical and chemical characteristics, demonstrating the versatility of these particles for use in remediation efforts. There are three main categories into which these compounds of interest can be broken down: halogenated organics, heavy metals, and energetic materials. Each category of contaminants has its own unique challenges and obstacles to overcome, yet microscale and nanoscale metal particles have been shown to successfully degrade or sequester analytes from each category.

Halogenated Organic Compounds

The most prevalent types of contaminants discussed in the literature are halogenated organic compounds. Aliphatic chlorinated compounds have been used for a variety of purposes over the years, such as trichloroethylene (TCE) which was used an industrial solvent with a wide variety of applications including as an extraction solvent, a dry cleaning solvent, and a metal degreaser. Chemicals of these types have been implicated in a variety of health effects such as liver or kidney damage and spontaneous abortions, and many are considered carcinogenic (*44*). Aliphatic HOCs which have proven susceptible to remediation using metal particles include carbon tetrachloride (*45*, *46*), chloroform, dichloromethane, chloromethane, bromoform (*47*), dibromochloromethane, dichlorobromomethane, tetrachloroethene (*48*), trichloroethene (*49-52*), *cis*-dichloroethene, *trans*-dichloroethene, 1,1-dichloroethene, vinyl chloride (*48*), and others (*53*). Aromatic halogenated compounds pose a more significant challenge, due to the inherent stability afforded by the aromatic nature of these molecules. These include halogenated benzenes (*41*, *54*), halogenated phenols (*55*), polychlorinated biphenyls (*39*), polybrominated diphenylethers, and polychlorinated dibenzo-*p*-dioxins (*33*, *56*). Limited success has been reported using single-metal micro and nanoscale particles alone; remediation for these compounds often requires the addition of a catalyst to create a bimetallic particle which is capable of dehalogenating aromatics (*57*).

The resistance of these compounds to degradation is both logical and ironic. PCBs, for example, were chosen for industrial applications due to the fact they are resistant to both chemical and thermal degradation. Now that their toxicity has been discovered, this recalcitrance has become a serious environmental concern.

The mechanism for degradation of HOCs using zero-valent metal particles is thought to occur primarily through reductive dechlorination. Aliphatic compounds such as carbon tetrachloride are readily capable of accepting electrons which are supplied by the oxidation of the zero-valent metal particles. The commonly accepted stoichiometric equation for the degradation of carbon tetrachloride with zero-valent iron is shown in Equation 1.

$$CCl_4 + 2Fe^0 + 4H^+ \rightarrow CH_4 + 2Fe^{2+} + 4Cl^- \qquad (1)$$

There are three proposed mechanisms for how this degradation takes place (*58*).

1. Reduction of the contaminant at the surface of the zero-valent metal particle.
2. Generation of oxidized form of the metal particle (e.g. ferrous ion) and subsequent reduction of contaminant by the oxidized form.
3. Generation of hydrogen at the surface of the metal particle, followed by reduction of the contaminant by hydrogen.

Heavy Metals

Heavy metals are a prevalent source of contamination found in the environment, and present several risks to the local biota. Contaminants of these types are capable of bioaccumulation, and therefore present a risk to organisms found higher up the aquatic food chain, including human beings (*59*). Certain heavy metals have priority status on the Environmental Protection Agency's National Priority (Superfund) List, including cadmium, copper, lead, mercury, nickel and zinc. Exposure to heavy metals (such as lead) can lead to a variety of ill-effects, including but not limited to neurological impairment, organ damage and immune system depression. Contamination of this type is caused by domestic and industrial effluents, storm water runoff, and leaching from minerals. Anthropogenic sources of lead include lead-zinc smelters and the production of ammunition, solder, glass, piping, insecticides, paints, and lead storage batteries (*60*, *61*).

The use of zero-valent metal technologies for the remediation of heavy metals has been explored for several years. Studies have shown that nano-scale and microscale zero-valent metals can reduce or remove heavy metals more efficiently than larger-scale metallic particles or conventional absorptive media. Nanoscale and microscale iron particles have shown success in remediating heavy metal contaminants such as cadmium, lead, silver, nickel, arsenic, and zinc (*53*, *62*). Two primary mechanisms have been documented for the remediation of these contaminants: reduction of the contaminants to a more inoffensive form, or the surface-mediated sorption of the contaminant (*53*).

An example of a heavy metal which undergoes the first remediation mechanism (reduction) is chromium in the presence of zero-valent iron. Chromium exists in two primary forms in the environment, hexavalent and trivalent. The adverse affects of chromium in the environment primarily depend upon its valence state, due to the difference in solubility that this creates. Cr^{6+} is considered the more toxic and immediately dangerous species, as it is water-soluble and capable of transport throughout an ecosystem once it has been released. Conversely, Cr^{3+} is considered to be a much less toxic species than Cr^{6+} because it so much less soluble and more immobile within the environment. Remediation of chromium contamination is considered successful once reduction from Cr^{6+} to Cr^{3+} has been achieved. The reaction pathway involves the reduction of dissolved chromate (CrO_4^{2-}) to insoluble Cr^{3+} (*63*). Precipitation then occurs with iron oxide and iron hydroxides, incorporating the contaminant within the surface oxide layer (*34*). Research has shown this reaction to be pseudo-first-order with respect to the normalized surface area of the zero-valent iron used. Use of nanoscale iron has shown rate constant increases of up to 30 times when compared to degradation utilizing conventional iron filings (*64*). Similar but less impressive increases have been seen with microscale iron. Uranium is also removed from water by zero-valent iron via reduction. Uranium(VI) in the form of uranyl ions (UO_2^{2+}) is directly reduced to uranium(IV) in the form of UO_2, which is a solid and will precipitate or adsorb to the iron oxides and hydroxides as in the case of chromium (*65*). Other heavy metal contaminants that are primarily reduced by zero-valent metals include copper, silver, and mercury (*53*).

Remediation of Ni^{2+} has been shown to be controlled by both reduction and surface complexation. Initially, the contaminant is adsorbed to the surface of zero-valent iron primarily as a nickel hydroxide, followed by reduction to a zero-valent form (*53*). The concentration of the zero-valent form increases until equilibrium is achieved, at which point XPS analysis demonstrates that 50% of the nickel has been reduced to the zero-valent form and remaining 50% exists as surface adsorbed hydroxide. Other metals with a standard reduction potential slightly above that of the zero-valent metal being used, such as lead(II), are also remediated by a combination of reduction and sorption to the surface of the metal particle (*53*).

Energetic Materials

Nanoscale/microscale metallic particles have recently been used as a remediation technique for the cleanup of energetic materials. Residual energetic materials are often found in sites where the production and use of munitions has occurred over the years, such as military installations. Testing has shown that partial detonation of munitions at these sites has caused the dispersal of these contaminants onto the soil surface, where transport to ground water may then occur (*66*). Research has demonstrated that several of the more commonly used energetic materials (TNT, RDX, TATP) are susceptible to reductive degradation upon exposure to zero-valent metals such as magnesium and iron (*67-69*). Other nitrated materials such as nitroglycerin, nitrobenzene, and PETN have also been successfully denitrated by metallic particles (*36-38*).

Specific Metal Particle Applications

A variety of different metallic particles and compounds have been investigated for use in the remediation of environmental contaminants over the past twenty years. These have included microscale/nanoscale zero-valent metal particles, enhancement via metal catalysts; bimetallic reductive compounds, transition metals, and others. It is difficult to create a completely comprehensive list of all metals used in remediation technologies; however the following is a list of selected metals that have undergone testing for use in environmental technologies. It is by no means a complete list, nor does it attempt to delineate every documented use of metals in remediation processes.

Alkaline and Alkali Earth Metals

Alkaline and alkali earth metals have been studied extensively in the scientific community and have been identified as viable electron sources in reductive methods of remediation.

Alkaline Metals

Alkaline metals such as sodium have been documented as capable of acting as electron sources for the reductive dehalogenation of contaminants. A specific example of remediation using sodium metal has been demonstrated in the reductive dehalogenation of brominated flame retardants and solvolysis oils. A Na/NH_3 mixture is capable of complete debromination of printed circuit board material, and an optimum debromination temperature of 100°C-120°C was eventually determined. In this case, the sodium acts as a source of solvated electrons capable of reducing the contaminants (*70*).

Similarly, potassium has been used as an electron donor source in the remediation of halogenated organic contaminants. Certain studies have shown an enhanced capability for dehalogenation when potassium is alloyed with sodium, as in the work done by Miyoshi *et al* (*71*). Complete dechlorination of 2,2',4,4',5,5'-hexachlorobiphenyl was demonstrated, and this technique also did not produce any toxic byproducts as are often seen during degradation of polychlorinated biphenyls.

Magnesium

Magnesium has a high reduction potential of -2.2V vs. standard hydrogen electrode (SHE), compared to -0.44V vs. SHE for iron (*72*). Combined with the fact that magnesium forms a self-limiting oxide layer rather than being completely consumed when exposed to an oxic environment, this makes it a good candidate for field environmental applications (*73*).

Magnesium zero-valent metal has been employed for the degradation of halogenated organic molecules under a variety of conditions. In a flash vacuum pyrolysis process, it was found to completely dehalogenate a variety of benzylic halides into the corresponding dibenzyls and toluenes (*74*). Elemental magnesium has also been used to debrominate cyclopropyl bromide in the presence of diethyl ether and gentle heat under an argon atmosphere. This reaction was found to yield 25-30% cyclopropane and 25% cyclopropyl magnesium bromide, a Grignard reagent. Cyclopropyl radical disproportionation on the surface of the magnesium was found to be responsible for 85% of cyclopropane production, providing an insight towards the mechanism (*75*). While this is true in the ground state, the reaction of triplet state magnesium with benzyl halides was found to proceed via the abstraction of the halogen by the magnesium, which is present as a biradical (*76*).

Zero-valent magnesium has also been used to reduce chromium(VI), as an alternative to iron which is often used for this purpose. Free magnesium metal is less toxic than free iron, and its oxides are more soluble than iron oxides, which could help to forestall the clogging of permeable reactive barriers. While chromium(VI) was successfully reduced by magnesium, the free metal was found to solubilize in water to the extent that the pH could not be mediated over time (*77*).

Calcium

Metallic calcium has been used to degrade polychlorinated dibenzo-p-dioxins, coplanar polychlorinated biphenyls, and dibenzofurans by stirring in ethanol under mild conditions (*56*). Calcium is similar to magnesium in that particles are stable in air due to the presence of an oxidized outer layer that prevents further reaction with the environment, making it safer to handle than sodium.

Transition Metals

Iron

Zero-valent iron has met with great success in the dehalogenation of chlorinated organic molecules. Matheson and Tratnyek (*19*) have demonstrated the ability of fine iron powder (43μm to 149μm) to reduce carbon tetrachloride to methylene chloride under anaerobic conditions. Studies conducted by Gillham and O'Hannesin (Gillham 1994) showed that 149μm iron powder was able to dechlorinate (though not always completely) 13 of the 14 chlorinated compounds tested with half-lives several orders of magnitude lower than those reported for natural abiotic processes. Further work by Orth and Gillham (*78*) focused on identifying the products of the TCE and iron reaction. Their work indicated that only a small percentage of chlorinated byproducts were produced, and the majority of products consisted of various small hydrocarbons. While chlorinated aromatic compounds are more difficult to degrade than chlorinated aliphatics, iron nanoparticles with an average diameter of approximately 90 nm were shown to partially dechlorinate hexachlorobenzene, with the lowest chlorinated byproduct being dichlorobenzene (*41*).

Other experiments have demonstrated the ability of micro- and nanoscale iron to dechlorinate PCBs. In a study by Yak *et al* (*79*), iron was used in subcritical water at 250 °C under 10 MPa to successfully degrade the highly chlorinated congeners in Arochlor 1260 to lower-chlorinated ones.

Zero-valent iron has also been used in several cases as a means of removing heavy metals. Chen *et al* (*62*) illustrated that iron can be used for heavy metal removal. However, the removal efficacy was shown to be pH-dependent, with the best removal occurring in solutions with a pH of two.

Nitroaromatic compounds have also been successfully degraded using zero-valent iron. Under anaerobic conditions, iron powder was shown to nitrobenzene to nitrosobenzene, then to aniline, with the reaction rate controlled by mass transfer to the iron surface (*36*). The necessity of surface access, together with the negligible effect of pH changes on reaction rate, support a mechanism of direct reduction by the iron metal. High-surface-area zero-valent iron was also demonstrated to be capable of a stepwise reduction of nitroglycerin to glycerol and ammonium ions (*38*), and granular iron was used to reduce PETN to pentaerythritol and ammonium via a similar mechanism (*37*).

Nickel

Unsupported nickel powder was found to provide 12-14% dechlorination of 4-chlorophenol to phenol at room temperature in the presence of sodium borohydride, but not with molecular hydrogen, suggesting that it is operating via a different mechanism from zero-valent iron. Nickel supported on alumina was shown to convert 81% of 4-chlorophenol to phenol, and a further 5% to cyclohexanol in the presence of sodium borohydride. Under the same conditions, high-surface-area Raney nickel was capable of complete dechlorination of the starting material (*80*). Nickel has also been used in conjunction with iron to produce bimetals with an eye towards use in permeable reactive barriers (*52*).

Zinc

The complete dechlorination of various polychlorinated dibenzo-dioxins has been achieved using zero-valent zinc powder under aqueous conditions (*81*). Degradation was shown to be stepwise, with more highly chlorinated compounds being degraded the most quickly.

Zero-valent zinc has also been employed in the degradation of 1,1,1-trichloroethane using 2-butyne as the solvent (*49*). This reaction produced primarily ethane, with a secondary product of 1,1-dichloroethane (1,1-DCA), which is degraded very slowly and does not participate in ethane formation. When zero-valent iron was used under the same conditions, the primary product was 1,1-DCA, with a smaller fraction of completely dechlorinated ethane. A study using a new analysis technique by Chen *et al* (*82*) monitored PCE degradation by zero valent zinc which showed approximately 75% conversion to TCE in 25 hours. Another study indicated that granular ZnO/Al_2O_3 has the capacity to degrade gaseous TCE however the experimental setup would be more applicable to an industrial application than one meant for *in situ* remediation.

Zinc is also an attractive substitute for iron because its oxides are much more soluble than iron oxides, preventing them from passivating the metal surface as quickly (*83*). However, the release of Zn^{2+} as a dissolution product is problematic. Excessive zinc intake has been known to cause copper deficiency and anemia in humans by increasing the production of metallothionein, a protein that has a high affinity for copper ions and may also play a part in the removal of iron (*84*). To minimize these risks, studies have been conducted using hydroxyapatite ($Ca_5(PO_4)_3OH$) as a Zn^{2+} scavenger. Hydroxyapatite was found to remove Zn^{2+} by simultaneous co-precipitation, adsorption, and substitution, substantially reducing the amount remaining in solution (*83*).

Vanadium

V_2O_5/TiO_2 nanoparticles have been used to catalytically oxidize PCBs (*85*), monochlorobenzene (*86*) and 1,2-dichlorobenzene (*87*) Work conducted by Ide *et al* (*88*) used a $TiO_2/V_2O_5/WO_3$ catalyst to degrade dioxins, chlorobenzenes, chlorophenols and coplanar PCB congeners in flue gas with various decomposition percentages.

Other Metals

Indium, Silicon, and Tin

Research has indicated that microscale indium metal is a viable reductant for functionalized nitrobenzenes in the presence of ethanol (*89*).

Silicon has been shown to be capable of reductive dehalogenation of chlorinated compounds. Carbon tetrachloride and tetrachloroethylene are degraded into chloroform and trichloroethylene by zero-valent silicon with an efficiency surpassing that of treatment with zero-valent iron (*46*). The silicon is released as $HSiO_3^-$ ion, which is both water-soluble and environmentally benign. Another study by Doong *et al* (*90*) demonstrated that zero-valent silicon showed promise for Cr(IV) removal and the reduction of Cu(II), Pb(II) and Ni(II) ions.

Zero-valent tin has been used to dechlorinate trichloroethylene via a similar pathway to that followed by elemental iron. The main degradation product is *cis*-dichloroethane, and others include ethene, ethane, and 1,1-dichloromethane. Di-chlorinated compounds constituted a higher percentage of product in reactions with tin than with iron (*51*). Tin has also been shown to degrade carbon tetrachloride in the presence of water, producing an intermediate species Cl_3CSnCl. The final reaction products, chloroform and CO_2, are theorized to come from protonation of this species by water (yielding chloroform) or elimination of CCl_2, which will yield HCl and CO_2 in an aqueous environment (*45*).

Bimetallic Systems

There are many combinations of metals which have been explored for environmental remediation. Most of these follow the formula of using one metal as the "base", which will be consumed in the same way that a lone metal would, and another as the "promoter" or catalytic partner, which is not consumed (*31, 91*) in order to maximize the rate and efficiency of the reaction. The incorporation of a thin coating of discontinuous bumps or plaques of catalytic metal greatly increases the already large surface area of micro- or nano-scale zero-valent metal, contributing to increased reactivity (*57*).

Fe/Pd

Palladium is known to be capable of catalyzing hydrodehalogenation of organic compounds (*54, 92, 93*), although when used alone it is not necessarily effective under the same conditions as iron (*91*). After the initial success of zero-valent iron, palladium was a logical choice for experiments in adding a catalyst.

While iron alone was found to be capable of hydrodechlorinating PCE, TCE, and their chlorinated byproducts to ethane within several hours, bimetallic Fe/Pd was found to accomplish the same in minutes. Fe/Pd was also able to dechlorinate the chloromethanes to methane within hours (*31*). When the bimetal was gravity-fed into an aquifer containing TCE, the concentration of TCE in the groundwater fell by 96% over a four-week monitoring period (*57*). Fe/Pd has also been used to completely dechlorinate the parasiticide dichlorophen (2,2'-methylenevis(4-chlorophenol)) within 90 minutes, while an iron/ruthenium bimetal was only 40% effective and iron/silver not effective at all (*94*).

Studies have found that the reactivity constant (normalized for surface area) of Fe/Pd is 100 times that of zero-valent iron alone, contributing to the marked decrease in persistent chlorinated byproducts in the presence of the bimetallic system (*91*).

Fe/Cu

Deposition of copper from solution onto microscale iron powder was shown by x-ray spectroscopy techniques to produce particles coated with a heterogeneous layer of copper. These were successfully used to dechlorinate 1,1,1-trichloroethane, with the observed rate constant highly dependent on copper loading up to one monolayer equivalent (*95*). A rise in temperature was also seen to increase the rate of reduction, which promoted the production of more fully dechlorinated species by increasing the concentration of available hydrogen near the metal surface (*96*).

Fe/Ni

Degradation of trichloroethylene by iron/nickel nanoparticles has been shown to occur 50 times faster than with iron-only nanoparticles (*52*). A physical mixture of iron and nickel nanoparticles was found to reduce trichloroethylene more quickly than nanoscale zero-valent iron alone, but not as quickly as bimetallic particles generated by joint reduction of the two elements. This underscores the importance of electronic contact for catalytic activity. As dehalogenation occurs, released hydroxide ions will react to form iron oxides and passivate the iron surface. The active lifetime of the Fe/Ni nanoparticles in TCE has been estimated by Schrick, et al (*52*) to be approximately 300 days, making these catalytic particles more suitable for shorter-term remediation efforts.

Fe/Si

The addition of iron to zero-valent silicon in an aqueous environment utilizes properties of both elements in order to remediate chlorinated hydrocarbons. In an iron-only system, as hydrogen from the water is consumed by dechlorination, the pH of the solution rises, leading to the oxidation of the outer layer of the zero-valent iron particles (*24*). When silicon is also present, hydroxide ions from the water will reactivate the surface of the silicon (SiO_2 +

$OH^- \rightarrow HSiO_3^-$), and the pH is reduced (*46*). The combination of these processes prevents the deactivation of the iron particles over time. Degradation of tetrachloroethylene in aqueous solution by Fe/Si is complete, yielding primarily ethane and ethane (*46*).

A study by Zheng *et al* (*97*) demonstrated that iron nanoparticles incorporated into a silica matrix of nano to microscale size enhances the efficiency of TCE degradation by the iron by preventing iron aggregation and facilitating the adsorption of TCE in the vicinity of the reactive iron.

Mg/Pd

Due to the formation of a self-limiting oxide layer by magnesium, dechlorination of organic compounds by magnesium/palladium bimetals can take place in the presence of oxygen (*72*). This provides a significant advantage over the anaerobic conditions and surface pre-activation required by treatments containing zero-valent iron (*31*, *49*). At ambient temperatures, Mg/Pd was shown to be capable of complete dechlorination of DDT (1,1-bis(4-chlorophenyl)-2,2,2-trichloroethane) to its organic skeleton, 1,1-diphenylethane, more quickly than Fe/Pd (*72*). Mg/Pd was also shown to be capable of completely dechlorinating polychlorinated biphenyls to biphenyl, with which single zero-valent metals have shown little success (*73*).

Conclusion

Emerging technologies utilizing microscale and nanoscale metallic particles have demonstrated their effectiveness in the treatment of environmental contaminants, with zero-valent iron as their flagship metal. Further chapters in this book will describe the reaction mechanisms and uses of zero-valent single-metal and bimetallic particles, and the effects of structure, surface properties, and level of catalytic metal present on remediation efficiency.

Many factors contribute to the difficulties encountered in treatment of contaminants. Some are intrinsic to the pollutant being studied (such as aromaticity or solubility properties), and others are a function of the environment (such as the presence of natural organic matter and the variability between oxidizing and reducing conditions). Following chapters will address some of these issues, and also describe some combination technologies utilizing zero-valent metallic particles in conjunction with mechanical or electrical energy in hopes of surmounting these problems.

The examination of various treatment methods also helps to provide an understanding of how different classes of chemicals will respond to these technologies, perhaps guiding their future use. The overarching goal in the use of microscale and nanoscale metallic particles is to promote the utilization of more efficient and effective remediation techniques in an effort to help restore the environment to its original conditions.

References

1. Chuang, F. et al. *Environ. Sci. Technol.* 1995, 29, 2460-2463.
2. Erickson, M.D. et al. *Analytical Chemistry of PCBs*. 1992, Lewis, Boca Raton, pp.5-45.
3. Wu, W. et al. *Chemosphere*. 2005, 60(7), 944-950.
4. De Filippis, P. et al. *Ind. Eng. Chem. Res.* 1999, 38, 380-384.
5. Jones, C.G. et al. *Environ. Sci. Technol.* 2003, 37(24), 5773-5777.
6. Beeman, R.E.; Bleckmann, C.A. *J. Contam. Hydrol.* 2002, 57, 147-159.
7. Wiegel, J., Wu, Q. *FEMS Microbiol. Ecol.* 2000, 32, 1-15.
8. Chiarenzelli, J.R. et al. *Chemosphere*. 1997, 34(11), 2429-2436.
9. Rice, C.P., White, D.S., *Environmental Toxicology and Chemistry*. 1987, 6(4), 259-274.
10. Schmidt, T.S., et al. *Environmental Toxicology and Chemistry*. 2002, 21(10), 2233-2241.
11. Förstner, U; Apitz, S.E. *J. Soils Sediments*, 2007, 7(6) 351–358.
12. Mulligan, C.N. et al. *J. Haz. Mater.* 2001, 85, 145-163.
13. Degtiareva, A., Elektorowicz, M. *Water Quality Res. J. of Canada*. 2001, 36(1), 1-19.
14. Milum, K.M. et al. *Abstr. Papers Am. Chem. Soc.* 2005, 229(1), U936-U937.
15. Pichtel, J. et al. *Environ. Engineering Sci.* 2001, 18(2), 91-98.
16. Virkutyte, J. et al. *Science of the Total Environment*. 2002, 289(1-3), 97-121.
17. Ahmad, F. et al. *J. Contam. Hydrol.* 2007, 90, 1-20.
18. Gilham, R.W. ; O'Hannesin, S.F. *Ground Water*. 1994, 32(6), 958-967.
19. Matheson, L.J. ; Tratnyek, P.G. ; *Environ. Sci. Technol.* 1994, 28(12), 2045-2053.
20. Johnson, T.L. et al. *Environ. Sci. Technol.* 1996, 30(8), 2634-2640.
21. Tratnyek, P.G. *Chemistry and Industry*. 1996, 13, 499-454.
22. Tratnyek, P.G. et al. *Ground Water Monitoring and Remediation*. 1997, 17(4), 108-114.
23. USEPA (1999). Solid Waste and Emergency Response (5102G). EPA 542-R-99-002.
24. Ritter, K. et al. *J. Contam. Hydrol.* 2002, 55, 87-111.
25. VanStone, N. et al. *J. Contam. Hydrol.* 2005, 78, 313-325.
26. Baciocchi, R. et al. *Water, Air, and Soil Pollution*. 2003, 149, 211-226.
27. USEPA (1998). Permeable Reactive Barrier Technologies for Contaminant Remediation. EPA/600/R-98/125.
28. Arnold, W.A. ; Roberts, A.L. *Environ. Sci. Technol.*, 2000, 34 (9), 1794-1805.

29. Scheutz, C. et al. *Environ Sci Technol.* 2000, 34(12), 2557-2563.
30. Schlicker, O. et al. *Ground Water.* 2000, 38(3), 403-409.
31. Muftikian, R. et al. *Water Res.* 1995, 29(10), 2434-2439.
32. Ma, C.; Wu, Y. *Environ. Geol.* 2008, 55, 47-54.
33. Kluyev, N. et al. *Chemosphere.* 2002, 46, 1293-1296.
34. Wilkin, R.T. et al. *Environ. Sci. Technol.* 2005, 39(12), 4599-4605.
35. Jeen, S.-W. et al. *J. Contam. Hydrol.* 2008, 95, 76-91.
36. Agrawal, A. ; Tratnyek, P.G. *Environ. Sci. Technol.*, 1995, 30 (1), 153-160.
37. Zhuang, L. et al. *Environ. Sci. Technol.* 2008, 42(12), 4534-4539.
38. Oh, S.Y. et al. *Environ. Sci. Technol.* 2004, 38(13), 3723-3730.
39. Lowry, G.V. ; Johnson, K.M. *Environ. Sci. Technol.* 2004, 38(19), 5208-5216.
40. Nurmi, J.T. et al. *Environ. Sci. Technol.* 2005, 39(5), 1221-1230.
41. Shih, Y. et al. *Coll. Surf. A.* 2009, 332, 84-89.
42. Tratnyek, P.G. et al. *Abstr. Papers. Am. Chem. Soc.* 2005, 230, U1536.
43. Akamatsu, K., Deki, S. *NanoSructured Materials.* 1997, 8(8), 1121-1129.
44. Moran, M. J. et al. *Environ. Sci. Technol.* 2007, 41, 74-81.
45. Boronina, T. et al. *Environ. Sci. Technol.* 1995, 29(6), 1511-1517.
46. Doong, R. et al. *Environ. Sci. Technol.* 2003, 37(11), 2575-2581.
47. Lim, T.-T. et al. *Water Res.* 2007, 41, 875-883.
48. Lien, H.-L.; Zhang, W. *Coll. Surf. A.* 2001, 191, 97-105.
49. Fennelly, J.P. ; Roberts, A.L. *Environ. Sci. Technol.* 1998, 32(13), 1980-1988.
50. Li, W. ; Klabunde, K.J. *Croatica Chemica Acta.* 1998, 71(4), 853-872.
51. Su, C. ; Puls, R. *Environ. Sci. Technol.* 1999, 33(1), 163-168.
52. Schrick, B. et al. *Chem. Mater.* 2002, 14(12), 5140-5147.
53. Li, X.; Zhang, W. *J. Phys. Chem. C.* 2007, 111(19), 6939-6946.
54. Liu, M. et al. *Environ. Sci. Technol.* Article ASAP, 06 March 2009.
55. Menini, C. et al. *Catalysis Today.* 2000, 62, 355-366.
56. Mitoma, Y. et al. *Environ. Sci. Technol.* 2004, 38(4), 1216-1220.
57. Elliott, D.W. ; Zhang, W. *Environ. Sci. Technol.* 2001, 35(24), 4922-4926.
58. Nyer, E.K. ; Vance, D.B. *Ground Water Monitoring and Remediation.* 2001, 21(2), 41-46.
59. USEPA,(2003). Methodology for Deriving Ambient Water Quality for the Protection of Human Health, EPA-822-R-03-030.
60. Mulligan, C.N. et al. *J. Soil. Contam.* 1999, 8(2), 231-254.
61. Dantas, T.N.C. et al. *Water Research.* 2003, 37(11), 2709-2717.
62. Chen, S.Y. et al. *Water Science and Technology.* 2008, 58(10), 1947-1954.
63. Ponder, S.P. et al. *Environ. Sci. Technol.* 2000, 34(12), 2564-2569.
64. Manning, B.A. et al. *Environ. Sci. Technol.* 2007, 41(2), 586-592.
65. Simon, F.G. et al. *The Science of the Total Environment.* 2003, 307, 231-238.
66. Fuller, M.E. et al. *Chemosphere.* 2007, 67(3), 419-427.
67. Oh, S.Y. et al. *Water Sci. & Technol.* 2006, 54, 47-53.
68. Bandstra, J.Z. et al. *Environ. Sci. Technol.* 2005, 39, 230-238.
69. Welch, R. et al. *Environ. Eng. Sci.* 2008, 25(9), 1255-1262.
70. Mackenzie, K., Kopinke, F.-D. *Chemosphere.* 1996, 33, 2423.
71. Miyoshi, K. et al. *Chemosphere.* 2000, 41, 819.

72. Engelmann, M.D. et al. *Chemosphere*. 2001, 43, 195-198.
73. DeVor, R. et al. *Chemosphere*. 2008, 73, 896-900.
74. Aitken R.A. et al. *Chem. Commun*. 1997, 1163-1164.
75. Walborsky, H.M., Zimmermann, C. *J. Am. Chem. Soc.* 1992, 114, 4996-5000.
76. Egorov, A.M. et al., *Russ. Chem. Bull.* 1999, 48(1), 147-151.
77. Park, J.-S., Lee, G. *Geochimica et Cosmochimica Acta*. 2008, 71(12), A723.
78. Orth, W.S., Gillham, R.W. *Environ. Sci. Technol.* 1996, 30, 66-71.
79. Yak,H.K. et al. *Environ. Sci. Technol.* 1999, 33(8), 1307-1310.
80. Roy, H.M. et al. *Applied Catalysis A*. 2004, 271, 137-143.
81. Wang, Z. et al. *Chemosphere*. 2008, 71, 360-368.
82. Chen, Z.L. et al. *J Liquid Chromatography & Related Technol.* 2004, 27(5), 885-896.
83. Song, H. et al. *Chemosphere*. 2008, 73, 1420-1427.
84. Fosmire, G.J. *Am. J. Clin. Nutr.* 1990, 51, 225-227.
85. Varanasi, P. et al. *Chemosphere*. 2007, 66(6), 1031-1038.
86. Graham, J.L. et al. *Catalysis Today*. 2003, 88, 73-82.
87. Krishnamoorthy, S. et al. *Catalysis Today*. 1998, 40, 39-46.
88. Ide, Y. et al. *Chemosphere*. 1996, 32, 189-198.
89. Pitts, M.R. et al. *J. Chem. Soc., Perkin Trans. 1*. 2001, 955.
90. Doong, R.A. et al. *Water Science and Technology*. 2004, 50(8), 89-96.
91. Zhang, W. et al. *Catalysis Today*. 1998, 40, 387-395.
92. Kovenklioglu, S. et al. *AIChE J.* 1992, 38, 1003.
93. Cheng, F. et al. *Environ. Sci. Technol.* 1997, 31, 1074.
94. Ghauch, A., Tuqan, A. *J. Haz. Mater.* 2009, 164(2), 665.
95. Bransfield, S.J. et al. *Environ. Sci. Technol.* 2006, 40(5), 1485-1490.
96. Bransfield, S.J. et al. *Applied Catalysis B*. 2007, 76, 348-356.
97. Zheng, T. et al. *Environ. Sci. Technol.* 2008, 42, 4494-4499.

Laboratory and Mechanistic Studies

Chapter 2

Use of Nanoparticles for Degradation of Water Contaminants in Oxidative and Reductive Reactions

Ishai Dror, Tal Ben Moshe and Brian Berkowitz

Dept. of Environmental Sciences & Energy Research
Weizmann Institute of Science Rehovot, Israel

Nanomaterials have received extensive attention recently as the next generators of scientific revolution. As such, nanomaterials are expected to be implemented in a wide range of applications. In the environmental field, nanomaterials hold promise for providing elegant solutions to numerous problems, from implementation of green chemistry processes for industrial and agrochemical uses, to production of novel materials for treatment of various contaminants. In this context the elimination of hazardous materials from the water environment is a major challenge facing environmental scientists today. In this chapter we present some results of our recent studies towards degradation of water contaminants. We exemplify both oxidative and reductive pathways for water remediation. In both cases we show the transformation of persistent contaminants through the use of nanomaterials as catalysts under ambient conditions.

Introduction

Engineered nanomaterials are usually described as inorganic materials of high uniformity, with at least one critical dimension below 100 nm. In this group we also consider naturally-occurring ultrafine particles with similar dimensions. These substances, especially engineered nanomaterials, have been the focus of many recent studies, and have been heralded as the next generators of scientific revolution. As such, nanosized materials are expected to be implemented in a wide range of applications, from medicine and cosmetics to new construction materials and industrial processes. Based on current estimates, nanotechnology is projected to become a $1 trillion market by 2015, and these materials are thus expected to be spread around the globe rapidly due to massive production and use.

Moreover, nanosized materials hold promise for providing elegant solutions to numerous environmental problems, from implementation of green chemistry processes for industrial and agrochemical uses (e.g., *1-2*) to production of novel material for treatment of various contaminants (e.g., *3-6*). In this context the elimination of hazardous materials from the water environment is a major challenge facing environmental scientists today. Several analytical, biological and physical methods have been developed for this purpose (*7*).

In this chapter we discuss two major pathways to utilize engineered nanomaterials for the treatment of water contaminants. One pathway is the reductive transformation of aqueous contaminants in high oxidation states, mainly through the use of nano zero valent iron (nZVI) in combination with metalloporphyrin catalysts. The second pathway utilizes nanomaterials for the complete oxidation of organic compounds. Both pathways employ nanomaterials in aqueous environments under conditions pertinent to real life settings, and are shown to transform a wide range of water contaminants.

The reductive transformation pathway has been studied extensively over the last two decades. In this case, nanosized elemental metals, mainly nZVI but also several other zero valent metals (e.g., Zn, Al, Mn), demonstrated activity as efficient reducing agents for water contaminants (e.g., *4, 8-10*). In a review paper on the use of nanosized iron for environmental remediation, Zhang (*10*) indicates that such particles may hold the potential for cost-effective remediation for a large spectrum of contaminants. Some examples of treatable compounds include chlorinated aliphatic and aromatic compounds (e.g., trichloroethylene (TCE), tetrachloroethylene (PCE) and carbon tetrachloride (CT)), pesticides (e.g., DDT and lindane), polychlorinated biphenyls (PCBs), heavy metals (e.g., Cr^{VI}, Pb^{II}, Cd^{II} and Hg^{II}), and inorganic anions (e.g., dichromate, perchlorate and arsenate). In most cases, complete reduction of chloro-organic compounds produces environmentally innocuous compounds, while reduction of heavy metals renders them insoluble and immobile. The main limitation of this technique is the increase in pH during the dechlorination reaction, which causes oxidation of other ions in the aqueous solution and hinders access to the zero valent metal surface. To overcome this problem the use of metalloporphyrins and their derivatives as electron transfer mediators in the reduction processes is suggested.

Metalloporphyrins are naturally-occurring, organic tetrapyrrole macrocycles composed of four pyrrole-type rings joined by methane bridges. Metalloporphyrins have several properties that make them very appealing for the treatment of persistent organic pollutants: (1) they are effective as redox catalysts for many reactions, and have a long range of redox activity; (2) they are electrochemically active with almost any core metal; (3) they are active catalysts in aqueous solution under conditions pertinent to groundwater environment, and (4) they have high stability which enables reactions under severe conditions that do not allow other ways of treatment.

Reductive dechlorination involves electron transfer from a bulk electron donor, through an electron mediator (catalyst – metalloporphyrin), to an electron acceptor (organic pollutant). The addition of electron carriers or mediators to the laboratory systems was found to greatly accelerate the measured reduction rates (e.g., *11-12*). An electron shuttle system has been postulated to account for the enhanced reactivity observed in these laboratory studies and presumably in reducing natural systems. A bulk electron donor (reductant) reduces an electron carrier (i.e., it transfers electrons to an intermediate molecule), which transfers the electrons to the pollutant of interest. The oxidized electron mediator is then reduced again by the bulk reductant, which enables the redox cycle to continue. Figure 1 illustrates the catalytic cycle occurring in natural and laboratory systems.

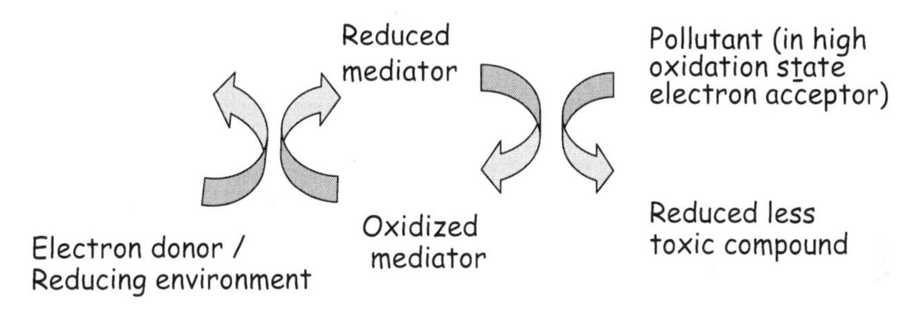

Figure 1. The reductive transformation catalytic cycle.

The oxidative pathway utilizing nanomaterials is related to a family of methods called Advanced Oxidation Processes (AOPs) (*13-14*). These methods involve in-situ generation of active radical species such as hydroxyl or superoxide radicals. This can be achieved by addition of chemical oxidants such as oxygen, ozone or Fenton's reagent (hydrogen peroxide and iron salts), or by employment of an energy source, usually UV irradiation, directly on the contaminants, or in conjugation with an oxidizing agent such as UV/O_3 or UV/H_2O_2.

One such mechanism of degradation is complete oxidative mineralization by heterogeneous photocatalysis (*15-17*). In this process, a suspension of metal oxide particles (or other semiconductors) acts as a catalyst for the reaction. The process is activated by UV/visible irradiation. These catalysts can efficiently absorb large fractions of the spectrum. The technique is considered particularly attractive due to its ability to quickly and completely oxidize different types of pollutants to water and carbon dioxide, without the formation of hazardous byproducts, according to the formula:

$$pollutant + O_2 / H_2O_2 \xrightarrow{catalyst/UV} CO_2^{\cdot} + H_2O + mineral\ acids$$

The mechanism of the reaction is a photocatalytically-induced radical oxidation (*18-19*). A photon is absorbed by the semiconductor particle, resulting in the transfer of electrons from the valence band into the conduction band, leaving behind a hole. The electron-hole pair may then recombine, releasing energy as heat, or it may migrate to the surface of the particle and initiate a reductive (electron) or oxidizing (hole) pathway for degradation on the particle surface or in its vicinity.

In recent years, nanosized particles have been considered particularly attractive catalysts because of their high reactivity, attributed to their enhanced surface area (*20-21*). The most common catalyst is titanium dioxide. This compound, in combination with UV irradiation at 300-400 nm, was shown to be an efficient catalyst for the degradation of many pollutants (e.g., *19,22-23*). Other metal oxide nanocatalysts include ZnO (*24-25*) and Fe_2O_3 (*26*). In addition, semiconductor nanocatalysts such as CdS (*27*) and ZnS are sometimes used.

In this chapter we present an example of the catalytic activity of copper oxide (CuO) nanoparticles for the degradation of alachlor and phenanthrene in aqueous solutions. Copper oxide is a semiconductor with an energy band gap of 1.21-1.5 eV (*28*). It is known as an efficient catalyst for reactions such as oxidation of carbon monoxide (*29-30*), degradation of nitrophenols (*31*), production of H_2O_2 (*32*) and E. Coli inactivation (*33*).

Experimental:

Reductive transformation systems:

Detailed procedures used in the experiments described here are given elsewhere (*3,34*). A brief dscription of the experimenal procedures is given below.

Preparation of Immobilized Metalloporphyrins in Sol-Gel Matrices

Trimethylorthosilane, methanol and metalloporphyrins dissolved in aqueous solution (2 mM) were mixed in a 1:5:8 molar ratio, respectively. The solution was mixed until gelation occurred and then dried in a hood for about one week until constant weight was achieved and catalyst concentration determined by calculating the amount of catalyst inserted per weight of matrix obtained.

Dechlorination Reaction Systems

Batch reactor systems were prepared in an anaerobic chamber. Each aliquot was prepared in an 8 mL clear sealed bottle containing 50 μL of 2 mM aqueous metalloporphyrin solution (in the homogeneous catalyzed reaction systems) or an equivalent molar amount of immobilized metalloporphyrins in sol-gel (in the heterogeneous catalyzed reaction systems). Zero-valent iron (nZVI) powder (30 mg) was added to the bottles, followed by an addition of 7.5 mL substrate stock solution of 50 mg/L, giving a final substrate concentration of 47 mg/L. The last step initiated the reaction and each reactor was sealed immediately. Control homogeneous systems consisted of nZVI and carbon tetrachloride stock solution. Control heterogeneous systems consisted of sol-gel without metalloporphyrins, nZVI and carbon tetrachloride stock solution. The terms "heterogeneous system" and "homogeneous system" relate hereafter to the catalysts. Thus, for example, a system with solid nZVI and dissolved metalloporphyrins is considered an homogeneous system. The samples were then mixed in a shaker (250 rpm).

For the atrazine transformation studies, 25 mL of atrazine aqueous stock solution of 12 mgL^{-1} was introduced into a 40 mL glass vial, followed by addition of porphyrin (TMPyP-Ni) aqueous solution (1.5×10^{-4} mmol). The reaction was initiated by addition of 100 mg nZVI; the samples were then mixed in a shaker (250 rpm) at room temperature for a desired period of time. The pH of the aqueous system was set to 6.6. The reaction was stopped by exposing the solution to air and extracting the aqueous solution with 5 mL of dichloromethane. The obtained products were identified by GC/MS.

In all cases, control experiments were performed to identify possible activity of each of the single components. It was found that metalloporphyrins were unable to catalyze reductive dechlorination of atrazine in the absence of the electron donors. Also, no reaction was observed under our experimental conditions when atrazine aqueous stock solution and nZVI were used without the presence of metalloporphyrins. It should be noted that ZVI alone is known to reduce atrazine (e.g. *35-38*) but in this case higher amounts of ZVI were used for lower concentrations of atrazine, and acidic to neutral pH was applied.

Oxidative transformation systems:

Three types of CuO nanoparticles were used in the experiments, two samples were purchased commercially and one was synthesized in the laboratory (*39*). The catalytic activity of the nanoparticles was tested for the degradation of two organic compounds: alachlor, an acetanilide herbicide and phenanthrene, a polycyclic aromatic hydrocarbon. The nanoparticles were suspended in aqueous solutions containing the pollutants. Hydrogen peroxide was used as the oxidizing agent in all experiments. The reaction mixture was stirred for 30 min at room temperature under ambient fluorescent lighting. Samples of the reaction mixture were taken at different times and the concentration of the organic pollutants was determined by GC/MS. Three control experiments were also performed, consisting of reaction mixtures without: (1) catalyst, (2) oxidizing agent, and (3) both catalyst and oxidizing agent. To check whether the catalysis is photo-induced, the experiment was repeated in complete darkness. To study the effect of ionic strength on the reaction, different concentrations of NaCl, a common groundwater salt, were added to the reaction mixture.

The detailed procedure used in the experiments is described in (*40*).

Results and Discussion

Reductive Transformation Pathways

The reductive transformation of many aqueous contaminants has been demonstrated mainly through the use of nZVI. Here we present the enhanced reductive reactivity through the use of metalloporphyrins as electron mediators together with electron donor – in our case nZVI. The first example is shown in Figure 2, which presents the reduction reaction of carbon tetrachloride (CT) and chloroform (CF) for systems with and without cyanocobalamin (vitamin B_{12}) as the electron mediator, and nZVI as the electron donor. The reaction was studied when the cyanocobalamin was dissolved in the contaminant solution (homogeneous system) and when the catalyst was trapped in a porous sol-gel matrix (heterogeneous system).

For these experiments it was expected that the reduction of CT would proceed by both the direct reduction activity of nZVI and by catalysis with the cyanocobalamin. As a result, the control presented in Figure 2 (for controls the same reaction setup was used but without the metalloporphyrins) also showed reduced concentrations with time. For the reductive dechlorination of nZVI, three pathways were proposed (*41*): "1. direct electron transfer from iron metal at the metal surface; 2. reduction by Fe^{2+}, which results from corrosion of the metal; 3. catalyzed hydrogenolysis by the H_2 that is formed by reduction of H_2O during anaerobic corrosion". The first pathway requires movement of the nZVI particles in the matrix to reduce the metalloporphyrins, which may be possible but probably has a steric limitation that slows the reaction. The other two pathways result in species that can readily move in solution and through the matrix to reduce the catalyst.

The total concentration of chlorinated compounds in the heterogeneous catalyst system decreased to 33.7 μM (i.e., 5.18 mg/L, ~10% of initial concentration), while the control sample decreased to 135.9 μM (i.e., 20.9 mg/L, ~40% of initial concentration). Similar (and even somewhat better results) were obtained for the homogeneous catalyst systems. In both experiments (the homogeneous and heterogeneous catalyst with nZVI), the addition of metalloporphyrins improved both the rate of CT reduction and the final CT degradation levels. Supporting results showing similar behavior in homogeneous catalysts systems (only) were presented by Morra et al. (*42*); they found that for some chlorinated organic compounds a combination of nZVI and catalyst may provide a remediation solution which is not attainable by nZVI alone.

The effect of the sol-gel matrix on the reduction of chlorinated compounds (mainly CF and dichloromethane) for the heterogeneous catalyst with nZVI (Figure 2) was shown to be more pronounced when the total concentration is reduced to about 10% of the initial concentration. This pattern is similar to the behavior found for the parallel reaction with titanium citrate as electron donor (*3*). Beyond this stage (~10%) the reduction reaction is essentially stopped. Because the only difference between the homogeneous and the heterogeneous systems is the sol-gel matrix, it is suggested that to achieve complete dechlorination of the CT, a longer residence time is needed to allow the contaminant to react with the reduced catalyst.

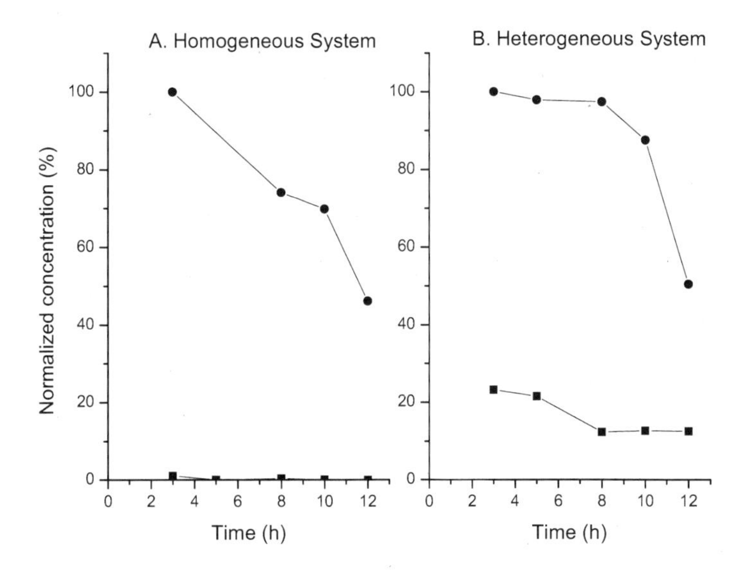

Figure 2. Concentration of total chlorinated compounds (CT + CF) vs. time for homogeneous and heterogeneous systems, catalyzed by cyanocobalamin using ZVI as an electron donor (30 mg ZVI for each reactor of 7.5 mL solution of 50 mg/L CT); ■ – electron donor plus catalyst; ● – control electron donor only (no catalysts). After Dror et al. (3).

A second example of reductive transformation of chlorinated organic compounds is the reductive dechlorination of atrazine. Atrazine, 2-chloro-4-(ethylamine)-6-(isopropylamine)-s-triazine, is a widely used herbicide that is known to be relatively persistent in soil. As such, it has a strong potential to contaminate surface water and groundwater, despite its moderate aqueous solubility (*43-45*). Atrazine is known to be toxic, and is suspected to be carcinogenic and endocrine-disrupting even at very low concentrations (*46-47*). Many efforts have been invested in the development of effective methods to decrease atrazine contamination in groundwater (e.g., *48-60*). Most of these processes were shown to involve alkyl-oxidation followed by dealkylation and/or dechlorination-hydroxylation as the major pathways of atrazine decay. In general, primary, secondary, chlorinated- and dechlorinated-tertiary derivatives have been identified. In most cases, cyanuric acid was formed as an end-product during oxidation processes of atrazine. Often, transforming atrazine into less toxic compounds involves its biodegradation, which was shown to proceed via enzyme-catalyzed hydrolysis reactions (*61-65*). However, microbial remediation of atrazine in the soil environment has been found to be a relatively slow process because of low biodegradability.

Only one reaction product was found for the reductive transformation of atrazine in the presence metallopophyrins and reducing agent, as shown in Figure 3. This methylated product is rarely mentioned in the literature. Gong et al. (*48*) obtained a large spectrum of byproducts resulting from atrazine photocatalysis (irradiation of soil samples containing atrazine by medium pressure mercury lamp in aerobic environment), one of the minor products of that reaction was the 2,4-bis(ethylamine)-6-methyl-s-triazine. In the current study the structure of the reaction product was verified by MS, NMR spectra and comparison of the GC retention time of the reaction product to a separately synthesized compound (by a procedure described by Highfill et al. (*66*)). In contrast to other dechlorination reactions of atrazine, where the chlorine atom is often replaced by an hydroxide group, in the current reaction a methyl group migrates from the isopropyl chain to form a symmetric molecule. The results of GC/MS analysis indicated the reaction to be very selective, with only one product, namely 2,4-bis(ethylamine)-6-methyl-s-triazine (methylated s-triazine) (Figure 3). Therefore, the reaction is assumed to be typical of reductive dechlorination of atrazine catalyzed by metalloporphyrins.

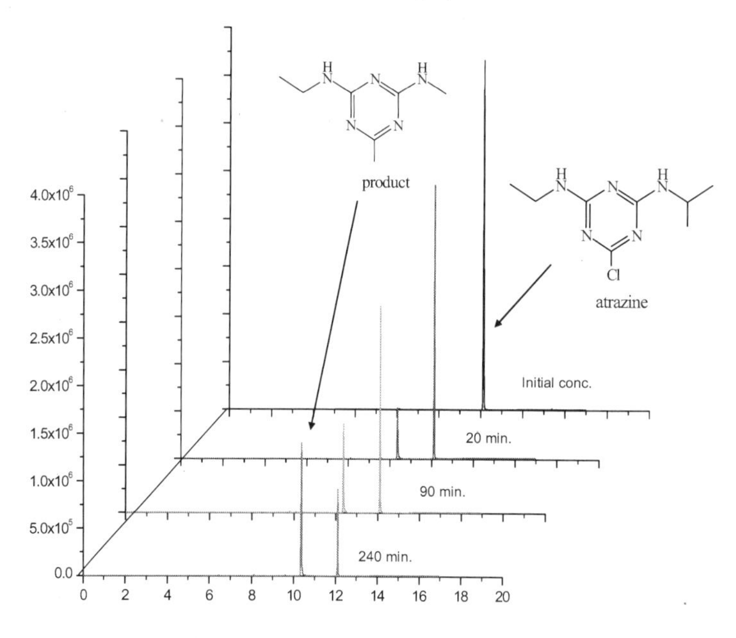

Figure 3. Gas chromatograms of samples collected during atrazine degradation at various times. After Nelkenbaum et al. (34).

Using the catalyst TMPyP-Ni, reactivity towards atrazine was found only when nZVI was used as a reducing agent (in contrast to cobalt metalloporphyrins, which were found active only when titanium citrate was used). Complete transformation of atrazine was achieved only after 140 h of reaction (Figure 4), with a rate constant of $1.24\times10^{-4}\ min^{-1}$. The dependence of the catalysts on the reducing agent indicates that even strong reducing agents such as nZVI or Ti citrate do not automatically reduce catalysts to their active form. It is also noted that the reaction remains active over time, and that the nZVI continues to reduce the oxidized catalyst, which in turn transforms atrazine along the same pathway as the Co catalyst does when titanium citrate is used.

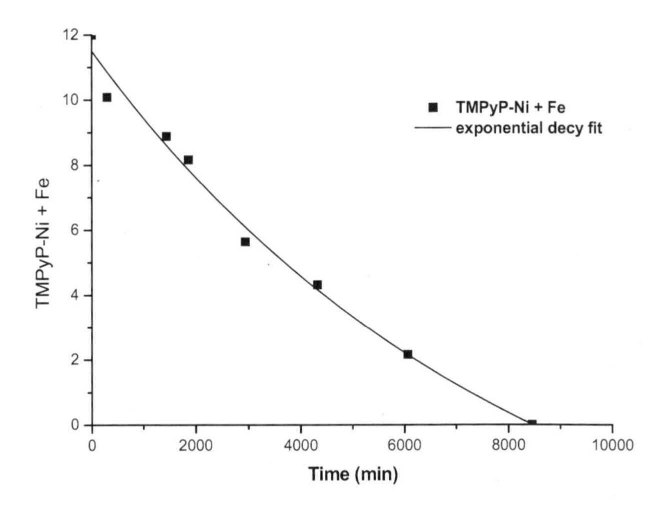

Figure 4. Atrazine degradation catalyzed by TMPyP-Ni using ZVI as an electron donor. Reaction conditions: initial atrazine concentration = 5.6×10^{-2} mM (12 ppm), [Ni] = 6×10^{-3} mM. After Nelkenbaum et al. (34).

The two examples described above demonstrate the powerful potential of combining strong reducing agents like nZVI with metalloporphyrins to achieve either faster reactions for the transformation of chloro-organic water pollutants and/or to enable new transformation pathways (such as in the case of the atrazine reaction).

Oxidative Degradation Pathways

The oxidative degradation pathway uses copper oxide nanoparticles and hydrogen peroxide to mineralize organic compounds. The results for the degradation of alachlor (an herbicide) and phenanthrene (a PAH substance) by commercial CuO nanoparticles are presented in Figures 5 and 6, respectively, with pollutant concentrations plotted vs. time. The nanoparticles demonstrated high catalytic activity, achieving 47.6% conversion after 3 min and complete conversion after 18 min for alachlor; and 39% conversion after 3 min and complete conversion after 21 min for phenanthrene. The plots decay exponentially, suggesting first-order kinetics with respect to the organic pollutant. The reaction rate constants were calculated to be 0.224 ± 0.008 min^{-1} for alachlor and 0.183 ± 0.010 min^{-1} for phenanthrene. In the control experiments, little or no change in concentration was seen, indicating that hydrogen peroxide alone or nano copper oxide alone will not degrade the studied contaminants. The small change in the control for the phenanthrene case (Figure 6) can be attributed to evaporation or sorption.

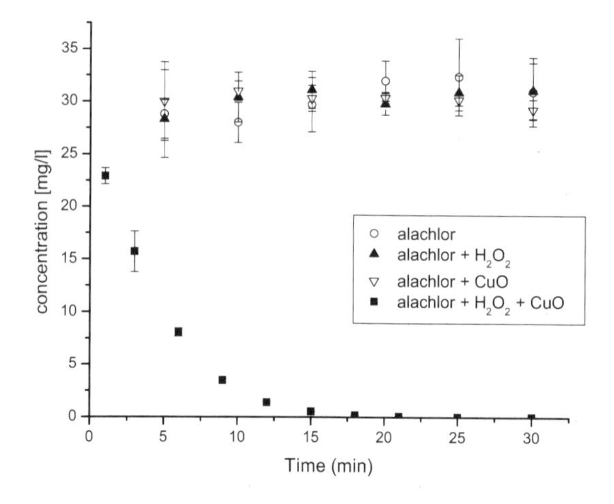

Figure 5. Degradation of alachlor in aqueous solution using H_2O_2 as oxidizing agent and CuO nanoparticles as catalyst. After Ben-Moshe et al. [40].

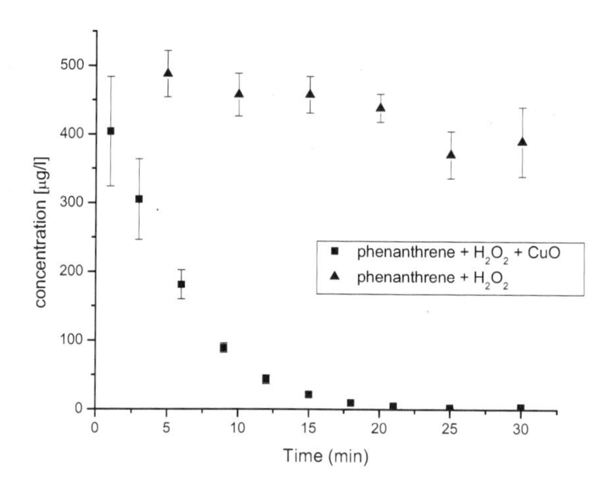

Figure 6. Degradation of phenanthrene in aqueous solution using H_2O_2 as oxidizing agent and CuO nanoparticles as catalyst. After Ben-Moshe et al. [40].

Furthermore, the catalytic activity of three types of CuO nanoparticles was compared for the degradation of alachlor. The results of the characterization of the nanoparticles and the catalysis experiments are summarized in Table I. The three samples had similar average particle sizes but different surface areas. This difference may be attributed to different porosity of the particles as a result of different preparation methods. The plot of the second sample appeared linear (not shown), suggesting zero-order kinetics with respect to alachlor. For the third sample, the shape of the plot is linear at the beginning but changes to exponential decay after about 15 minutes (not shown), indicating a transfer from zero-order to first-order kinetics. This change may be the result of increased numbers of active radical species and/or catalyst active sites available for reaction. The results were also normalized to surface area. The second sample showed the mildest catalytic activity despite having the highest surface area. The surface area was measured for powder while the experiments took place in a solution. This sample was seen to aggregate strongly in aqueous solution; therefore the actual surface area is smaller than measured, leading to a decrease in the number of accessible active sites for catalysis. The third sample had the lowest surface area and therefore showed the strongest catalytic ability after normalization.

Table I. CuO Samples Used for the Experiments.

Sample	Surface area [m^2g^{-1}]	Average particle size [nm]	% removal after reaction		Change in concentration normalized to surface area [$mgL^{-1}m^{-2}$]	
			15 min	30 min	15 min	30 min
Commercial CuO sample (Aldrich)	24	29	98.22 ±9.88	99.96 ±17.81	6.14 ±0.62	6.24 ±1.11
Commercial CuO sample (NaBond)	82	28	6.67 ±0.20	25.00 ±3.18	0.12 ±<0.01	0.46 ±0.06
Synthesis of CuO sample	7	30	64.67 ±3.89	98.43 ±23.20	13.86 ±0.83	21.09 ±4.97

To test whether the catalysis is photo-induced, the same experiment was repeated for alachlor in complete darkness. In both cases the results were very similar, achieving complete degradation after 18 min, suggesting that unlike other metal oxide compounds (such as TiO_2) this catalyst is not photo-induced and does not require irradiation.

To test the effect of ionic strength on the reaction, different concentrations of NaCl, a common groundwater salt, were added to the reaction mixture. The experiments were carried out at a pH lower than the point of zero charge of CuO. At this pH, the particles are positively charged, resulting in a higher concentration of anions in their vicinity. As the reaction takes place in the immediate vicinity of the particles, it is expected that these anions will be the ones affecting the catalysis. The results are presented in Figure 7. For low concentrations (up to 1 M), the reaction was slow. This is due to radical scavenging by the ions and competitive adsorption on the catalyst active sites. For concentrations in the range 0.01-0.1 M, the plot is linear, indicating pseudo-zero-order kinetics with respect to alachlor. For high salt concentrations, a dramatic increase in reaction rate was observed. The reaction rate constants for 1 and 5 M NaCl are 0.896±0.008 min^{-1} and 0.327±0.032 min^{-1}, respectively, and are larger than the rate constant without addition of salt (0.224±0.008 min^{-1}). After 3 min, 32%, 8.6%, 9.3%, 93% and 66.7% conversion was achieved for 0.001, 0.01, 0.1, 1 and 5 M, respectively. Complete degradation was achieved after 30, 6 and 15 min for 0.001, 1 and 5 M, respectively.

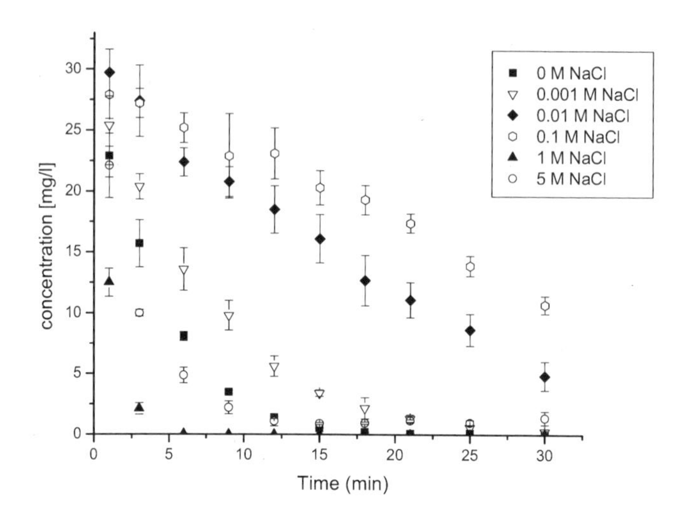

Figure 7. Effect of NaCl concentration on the catalytic degradation of alachlor using CuO nanoparticle catalysts in an aqueous solution. After Ben-Moshe et al. (40).

Conclusions

The use of nanomaterials for the treatment of contaminated aqueous solutions was presented. Following reductive and oxidative pathways, the use of nanomaterials holds promise for highly efficient remediation that can deal with a wide range of water contaminants under ambient conditions. In both cases, common persistent water contaminants (CCl_4, atrazine, alachlor, and phenanthrene) were demonstrated to be transformed quickly and easily to less toxic compounds, or completely mineralized, through the use of nanomaterials as catalysts.

References

1. Mohanty, A. K.; Drzal, L. T.; Misra, M. *Polymeric Mater. Sci. Eng.* **2003**, 88, 60-61.
2. Mckenzie, L. C.; Hutchison, J. E. *Chemistry Today* **2004**, 22, 30-33.
3. Dror, I.; Baram, D.; Berkowitz, B. *Environ. Sci. Technol.* **2005**, 39, 1283-1290.
4. Nurmi, J. T.; Tratnyek, P. G.; Sarathy, V.; Baer, D. R; Amonette, J. E.; Pecher, K; Wang, C. M; Linehan, J. C.; Matson, D. W.; Penn, R. L.; Driessen, M. D. *Environ. Sci. Technol.* **2005**, 39, 1221-1230.
5. Nagaveni, K.; Silalingam, G.; Hegde, M. S.; Madras, G. *Environ. Sci. Technol.* **2004,** 38, 1600-1604.
6. Kuhn, K. P.; Chaberny, I. F.; Massholder, K.; Stifkler, M.; Benz, V. W.; Sonntag, H. G.; Erdinger, L. *Chemosphere* **2003**, 53, 71-7.
7. Hoffmann, M. R.; Martin, S. T.; Choi, W.; Bahnemann, D.W. *Chem. Rev.* **1995**, 95, 69-96.
8. Gillham, R. W.; O'Hannesin, S. R. *Ground Water*, **1994**, 32, 958-967.
9. Boronina, T.; Klabunde, K. J.; Sergeev, G. *Environ. Sci. Technol.* **1995**, 29, 1511-1517.
10. Zhang, W. *J. Nanoparticle Res.* **2003**, 5, 323-332
11. Dror, I.; Schlautman, M. *Environ. Toxicol. Chem.* **2003**, 22, 525-533.
12. Dror, I.; Schlautman, M, *Chemosphere* **2004,** 57, 1505-1514.
13. Legrini, O.; Oliveros, E.; Braun, A.M., *Chem. Rev.* **1993**, 93, 671-698.
14. Ikehata, K.; El-Din, M. G.; *J. Environ. Eng. Sci.* **2006**, 5, 81-135.
15. Ollis, D. F. *Environ. Sci. Technol.* **1985** 19, 480.
16. Barbeni, M.; Pramauro, E.; Pelizzetti, E.; Borgarello, E.; Serpone, N.; Jamieson, M. A. *Chemosphere* **1986**, 15, 1913-1916.
17. Mills, A.; Davies R. H.; Worsley, D. *Chem. Soc. Rev.* **1993**, 22, 417-425.
18. Turchi, C. S.; Ollis, D. F.; *J. Catalysis* **1990**, 122, 178-192.
19. Mills, G.; Hoffmann, M. R. *Environ. Sci. Technol.* **1993**, 27, 1681-1689.
20. Bahnemann, D. W.; Kholuiskaya, S. N.; Dillert, R.; Kulak, A. I.; Kokorin, A. I. *App. Catal. B: Environ.* **2002**, 36, 161-169.
21. Huang, W. J.; Fang, G. C.; Wang, C. C. *Colloids and Surfaces A: Physicochem. Eng. Aspects* **2005**, 260, 45-51.
22. Matthews, R. *Water Res.* **1986**, 20, 569-578.
23. Tunesi, S.; Anderson, M. A. *Chemosphere* **1987**, 16, 1447-1456.
24. Hariharan, C. *Appl. Catal. A: General* **2006**, 304, 55-61.
25. Jung, H.; Choi, H., *Appl. Catal. B: Environ.* **2006**, 66, 288-294.
26. Khedr, M. H.; Abdel Halim, K. S.; Nasr, M. I.; El-Mansy, A. M. *Mater. Sci. Eng.* **2006**, A 430, 40-45.
27. Kapinu, E. I.; Viktorova, T. I.; Khalyavka, T. A. *Theor. Exp.Chem.* **2006**, 42, 282-286.
28. Marabelli, F.; Parravicini, G. B.; Salghetti-Drioli, F. *Phys. Rev. B* 52 **1995**, 1433-1436.
29. Jernaigan, G. G.; Somorjai, G. A. *J. Catal.* **1994**, 147, 567-577.
30. Huang, T. J.; Tsai, D. H.; *Catal. Lett.* **2003**, 87, 173-178.
31. Bandara, J.; Kiwi, J.; Pulgarin, C.; Peringer, P.; Pajonk, G. M.; Elaloui, A.; Albers P. *Environ. Sci. Technol.* **1996**, 30, 1261-1267.

32. Bandara, J.; Guasaquillo, I.; Bowen, P.; Soare, L.; Jardim, W. F.; Kiwi, J. *Langmuir* **2005**, 21, 8554-8559.
33. Paschoalino, M.; Guedes, N. C.; Jardim, W.; Mielczarski, E.; Mielczarski, J. A.; Bowen, P.; Kiwi, J. *J. Photochem. Photobiol. A: Chem.* **2008**, 199, 105-111.
34. Nelkenbaum, E.; Dror, I.; Berkowitz, B. *Chemosphere* **2009**, 75, 48-55.
35. Ghauch, A.; Rima, J.; Amine, C.; Martin-Bouyer, M. *Chemosphere* **1999**, 39, 1309-1315.
36. Ghauch, A.; Suptil, J. *Chemosphere* **2000**, 41, 1835-1843.
37. Dombek, T.; Dolan, E.; Schultz, J.; Klarup, D. *Environ. Pollut.* **2001**, 111, 21-27.
38. Dombek, T.; Davis, D.; Stine, J.; Klarup, D. *Environ. Pollut.* **2004**, 129, 267-275.
39. Fan, H.; Yang, L.; Hua, W.; Wu, X.; Wu, Z.; Xie, S.; Zou, B.; *Nanotechnology* **2004**, 15, 37-42.
40. Ben-Moshe, T.; Dror, I.; Berkowitz, B. *Appl. Catal. B: Environ.* **2009**, 85, 207-211.
41. Matheson, L. J.; Tartnyek, P. G. *Environ. Sci. Technol.* **1994**, *28*, 2045-2053.
42. Morra, M. J.; Borek, V.; Koolpe, J. *J. Environ. Qual.* **2000**, 29, 706-715.
43. Cohen, S.; Creeger, S.; Carsel, R.; Enfield, C. Treatment and disposal of pesticide waste. Am. Chem. Soc.: Washington DC. **1984**.
44. Briggs, S. A Basic Guide to Pesticides: Their Characteristics and Hazards, Hemisphere Publishing Corp.: Washington. **1992**.
45. Nelson, H.; Jones, R. *Weed Technol.* **1994**, 8, 852-862.
46. Hayes, T. B.; Collins, A.; Lee, M.; Mendoza, M.; Noriega, N.; Stuart, A. A.; Vonk, A. *Proc. Natl. Acad. Sci. USA* **2002**, 99, 5476-5480.
47. Mizota, K.; Ueda, H. *Toxicological Sci.* **2006**, 90, 362-368.
48. Gong, A.; Ye, C.; Wang, X.; Lei, Z.; Liu, J. *Pest Manag. Sci.* **2001**, 57, 380-385
49. Nélieu, S.; Kerhoas, L.; Einhorn, J. *Environ. Sci. Technol.* **2000**, 34, 430-437.
50. Chu, W.; Chan, K.H.; Graham, N. J. D. *Chemosphere* **2006**, 64, 931-936.
51. Parra, S.; Stanca, S. E.; Guasaquillo, I.; Thampi, K. R.; *Appl. Catal. B-Environ.* **2004**, 51, 107-116.
52. Bianchi, C. L.; Pirola, C.; Ragaini, V.; Selli, E. *Appl. Catal. B-Environ.* **2006**, 64, 131-138.
53. McMurray, T. A.; Dunlop, P. S. M.; Byrne, J. A.; *J. Photoch. Photobio. A: Chem.* **2006**, 182, 43-51.
54. Ma, J.; Graham, N. J. D. *Ozone Sci. Eng.* **1997,** 19, 227-240.
55. Ma, J.; Graham, N. J. D. *Water Res.* **2000**, 34, 3822-3828.
56. Héquet, V.; Le Cloirec, P.; Gonzalez, C.; Meunier, B. *Chemosphere* **2000**, 41, 379-386.
57. Krýsová, H.; Jirkovský, J.; Krýsa, J.; Mailhot, G.; Bolte, M. *Appl. Catal. B-Environ.* **2003**, 40, 1-12.
58. Chan, K. H.; Chu, W. *J. Hazard. Mater. B* **2005**, 118, 227-237.
59. Chu, W.; Chan, K. H.; Kwan, C.Y.; Choi, K.Y. *Chemosphere* **2007**, 67, 755-761.

60. Joo, S. H.; Zhao, D.Y. *Chemosphere* **2008**, 70, 418-425.
61. Wackett, L. P.; Lange, C.C.; Ornstein R.L.; In Biomacromolecules: From 3-D to Applic., Hanford Symp. Health and the Environ. 34th, Pasco: Wash., **1997**.
62. Ralebitso, T. K.; Senior, E.; van Verseveld, H.W. *Biodegradation* **2002**, 13, 11-19.
63. Neumann, G.; Teras, R.; Monson, L.; Kivisaar, M.; Schauer, F.; Heipieper, H. J. *Appl. Environ. Microbiol.* **2004,** 70, 1907-1912.
64. Radosevich, M.; Tuovinen, O. H. *Microbial degradation of atrazine in soils, sediments, and surface water. ACS Symp. Series 863,* **2004**, 129-139.
65. Pearson, R.; Godley, A.; Cartmell, E. *Pest Manag. Sci.* **2006**, 62, 299-306.
66. Highfill, M. L.; Chandrasekaran, A.; Lynch D. E.; Hamilton, D. G. *Cryst. Growth Des.* **2002**, 2, 15-20.

Chapter 3

Small Particle Size Magnesium in One-pot Grignard-Zerewitinoff-like Reactions Under Mechanochemical Conditions: On the Kinetics of Reductive Dechlorination of Persistent Organic Pollutants (POPs)

Volker Birke[1], Christine Schütt[1], Wolfgang Karl Ludwig Ruck[2]

[1]Faculty of Environmental Science and Technology, Campus Suderburg, Leuphana University of Lüneburg, 29556 Suderburg, Germany
[2]Faculty of Environmental Science and Technology, Institute of Ecology and Environmental Chemistry, Leuphana University of Lüneburg, 21335 Lüneburg, Germany

Small particle size base metals are produced in a ball mill mechanochemical (MC) reactor. During ongoing milling they react at room temperature simultaneously and in a single process step, to rapidly, and highly selectively reduce by dechlorination persistent polychlorinated organic pollutants, such as polychlorinated biphenyls (PCBs). In the presence of a low acidic hydrogen donor, reduction is to the parent hydrocarbons. Whether in a purified form or as mixtures, the polychlorinated organic pollutants can be destroyed both effectively and in an environmentally friendly manner. This *non-incineration and non-thermal* process extends even to contaminated materials, such as soils, filter dusts and the like. A kinetic study utilizing dichloro- and monochlorobenzene (DCB, MCB) as the model pollutants and excess magnesium and n-butylamine as reagents in a laboratory centrifugal ball mill at 25 °C, was carried out in order to investigate the reaction mechanisms for MC dehalogenation reactions employing magnesium and slightly acidic hydrogen donors. The focus of the study was on polyhalogenated *aromatic* POPs, such as PCBs or hexachlorobenzene which have chlorine-aromatic ring carbon bonds, with higher bond strength than chlorine-aliphatic carbon bonds. Analysis of the findings strongly suggested that one-pot, consecutive Grignard-Zerewitinoff-like reactions occur: first, formation of the Grignard intermediates from DCB or MCB, respectively, and then, in a stepwise manner, their protonation to monochlorobenzene or benzene, respectively, by the amine. Furthermore, a rationale is derived for the observed complete reductive dechlorinations to benzene, formed at approximately 100 % yield (mole/mole, based on DCB or MCB). The rate constants for the formation of intermediates 3-chloro-phenylmagnesium chloride and phenylmagnesium chloride were calculated. These correspond well to similar data previously reported elsewhere. The time-dependent development of the actual reactive surface of the small-sized magnesium particles during grinding was assessed.

Introduction

It has been demonstrated in recent years that persistent and recalcitrant polychlorinated compounds such as polychlorobenzenes, pentachlorophenol (PCP), or PCBs, can readily and highly selectively be dechlorinated within minutes to their parent hydrocarbons by a one-pot MC reaction inside a ball mill at room temperature (*1-3*). This approach, called DMCR ("Dehalogenation by Mechanochemical Reaction"), comprises the application of small particle size base metals such as iron, zinc, aluminum, alkaline earth or alkali metals, in combination with slightly acidic hydrogen donors, frequently bearing hydroxyl, amino, amido, polyoxyalkylic and/or ether groups as reagents. The presence of both reagents is required. For instance, mechanochemical dechlorination of polychlorinated biphenyls (PCBs) with small particle size iron, aluminum, magnesium or sodium in the presence of alcohols, ethers, or amines, in a ball mill to yield harmless chloride and biphenyl, a biodegradable compound, occurs even in complex matter, such as soils or filter dusts. Side-reactions such as formation of polychlorinated dibenzodioxins (PCDD) or related hazardous compounds could be ruled out by utilizing high resolution gaschromatographic (GC) analyses of the reaction products (i.e., gaschromatographs equipped with electron capture detectors, ECD, as well as a mass spectrometers, MSD) (*3*). Other MC approaches to dehalogenate polyhalogenated compounds, for instance, by application solely of metals, metal hydrides, or metal oxides such as calcium oxide (lime), silicon dioxide (quartz) etc and no other reagents, may give rise to numerous and complex degradation reactions of the polyhalogenated compounds. Inorganic chloride is found, often quantitatively, hence indicating any dehalogenating mechanisms, but numerous organic intermediates and degradation products or elemental carbon (not quantitatively) may be generated as well, therefore making it difficult or even impossible to identify and define the actually involved single mechanisms. Those degradation products which are still partly halogenated may have an unknown, but evidently potential toxicological risk associated with them (*2,4-13*). An example of this would be the milling of 1,1,1-trichloro-2,2-bis(4-chlorophenyl)ethane (DDT) with calcium oxide (*6*). In the case of metals alone being used to dehalogenate POPs in a ball mill and with no added hydrogen donor (*5*), parent hydrocarbons may be formed but only in very small amounts most probably due to self-protonation of the compound to be degraded. Main degradation products may be carbides, hydrides and/or elemental carbon. Thus, it may be inferred that organic molecules are entirely ruptured and as yet unidentified complex reaction mechanisms are occurring. Reaction rates are generally slow, on the order of hours. Highly sophisticated high energy mills (HEM) can accelerate these rates, but the mills are rather expensive and cannot readily be scaled up to industrial sizes for the destruction of pollutants in the environment.

It was also reported recently that individual Grignard reagents can be produced in a ball mill by grinding magnesium filings in the presence of organic halides dissolved in ether solvents (*14,15*). In addition, formation of a Grignard reagent from an organic halide and its subsequent protonation by alcohols or similar slightly acidic hydrogen donors in a *second* step, i.e., a Zerewitinoff-like reaction, is well known in conventional organometallic chemistry in solvents. But it has rarely been performed in a one-pot reaction and, when so, required elevated temperatures to obtain relatively high yields of the parent hydrocarbons (*16,17*). For instance, chlorobenzene underwent 89 % conversion to benzene using magnesium and propan-2-ol in decalin at 150 °C. Magnesium and *methanol* were not capable of reducing chlorobenzene or some of its derivatives at all however bromo- and iodobenzene were converted to benzene (*18-20*). Interestingly, magnesium in liquid ammonia reacted with iodobenzene to furnish some benzene in a one-pot reaction (*21*). Shaw (*22*) accomplished mechanical activation of magnesium for its successful application to one-pot consecutive Grignard formation and Grignard reactions by cutting magnesium chips and reacting them with organic halides in ethers using a specially designed cutting machine and reaction chamber at room temperature. Surprisingly, when water had also been added, bromobenzene yielded benzene, i.e., most probably from intermediary phenylmagnesium bromide, which must have been formed first (in the presence of water). However, that dehalogenation approach utilizing a specific kind of mechanical activation can be applied in liquid phase only, but would not work for treatment of contaminated liquid-solid or solid matter such as soils, sludges etc., due to its specific design. Furthermore, base metals other than magnesium or aluminum, such as iron, cannot readily be activated mechanically by cutting, only by ball milling, provided, in some cases, that specific grinding aids are also be employed.

By conducting DMCR in a ball mill, a highly selective one-pot MC reductive dehalogenation of polyhalogenated pollutants such as POPs can actually be accomplished both in liquid and solid-liquid and solid-matter phases. These reactions are always characterized by high degradation rates of the pollutants at room temperature (*1-3*). Moreover, mechanical stress/impact can continuously be exerted on the metal, hence, avoiding the poisoning of the reactive surface over the entire course of the reaction. This had been impossible in Shaw's approach.

The DMCR approach aims to boost the dehalogenation reaction primarily by the application of mechanical energy to the metal in three respects:

1. continuously producing small particle sized metal, with dimensions in the range of micro- or even nanometers, i.e., large, fresh supply of highly accessible metal surfaces,
2. continuously renewing surface layers, thereby making surface metal atoms accessible to reactants by removing impurities such as oxides, reaction intermediates, degradation products etc. adsorbed on the metal's surface,
3. activating accessible surface metal atoms for reaction by continuous impact.

Regarding DMCR, it has been assumed that the mechanically activated metal serves as the reducing agent that is 1) it delivers electrons for breaking down the carbon-halogen bond in the initiating mechanistic step *via* a single electron transfer, SET, or a double transfer, producing a radical plus inorganic halide or a carbanion plus inorganic halide, respectively, 2) this transfer occurs at the metal's surface while the halogenated molecule is adsorbed to the activated surface, and 3) afterwards, in a second mechanistic step, the hydrogen donor transfers hydrogen atoms or protons, to replace the stripped off halogen atom (*1,3*). It could be demonstrated that the addition of an appropriate, slightly acidic hydrogen donor is decisive regarding fast and defined overall degradation at room temperature (*1-3*).

A kinetic study was undertaken to identify the mechanisms operating in this one-pot reaction. There were two objectives of this study, targeted especially on magnesium metal: 1) whether or not there is initial formation of a Grignard intermediate followed, in a second step, by its protonation and 2) whether a quantitative estimate can be made of the effect of MC ball milling itself on the chemical reaction.

Materials and Methods

Centrifugal ball mills ("Retsch S 100", supplier "Retsch GmbH", Haan, Germany), each equipped with a 50 ml steel grinding jar (containing three steel balls, diameter 2 cm each), were utilized for bench-scale kinetic studies according to experimental procedures previously described (*1-3*).

Time-Dependent, Isothermal Dechlorination of 1,3-DCB in a Sand Slurry

The 50 ml steel grinding jar was loaded with 10 g sand, 1.5 g magnesium filings and 15 ml of a mixture of 1,3-DCB and excess n-butylamine (molar ratio DCB/amine appr. 1:40) dissolved in excess toluene (concentration of 1,3-DCB related to sand: 0.034 mol/kg). Toluene was employed as a diluting agent in order to intentionally lower reaction rates for better resolution of the data. Note that without the diluting inert liquid, DCB is degraded within minutes using magnesium and butylamine, making the timely collection of sufficient and reliable data points at such fast rates much more difficult. All compounds employed were supplied in reagent grade. The ball mills were always operated at a revolution of 580 rounds per minute (highest possible revolution regarding these devices) for the required time isothermally at 25 °C (+/- 1 °C) by deploying external cooling/heating jackets (coolant/heating liquid: water), fitted tightly to the steel grinding jars. After ceasing the milling process, the resulting dark grey slurry was immediately removed from the jar and filtered. An aliqot of the resulting liquid was treated with acidified water and washed with 10 ml n-hexane in a separation funnel. The organic phase was dried over sodium sulfate and afterwards adjusted to a defined volume by adding n-hexane. A small portion of it was injected into a gas chromatograph (GC) equipped with a flame ionization detector (FID). The gas chromatograph was a Perkin Elmer Autosystem XL equipped with the following devices and operated under appropiate conditions: column BPX-35 (SGE Analytical Science Pty. Ltd.), 30 m, ID 0,32 mm, 1,0 µm film thickness, carrier gas helium, flow rate 1,3 mL/min., on-column injector at 35 °C, autosampler, oven temperature initially at 35 °C for 1 min., ramp 12 °C/min. up to 240 °C, maintaining 240 °C for further 5 min., FID temperature a constant 280 °C. Calibration/quantitative analyses were performed both according to external and internal standard (1,2,4-trichlorobenzene) methods.

To get reliable data, the procedure was repeated in its entirety and at timed intervals for each data point. The procedure was also conducted in another, identical 50-ml steel grinding jar. The majority of the experiments were conducted with multiple runs for each data point. There was always a duplicate run, very frequently triplicate runs up to five identical runs. Regarding their kinetic evaluation, average values (percentiles) were used after a statistical analyses and error calculations had been conducted. Quality assurance was attained by checking the performances of the mills regularly by means of reference experiments as well as by means of having mechanical maintenance work done by the supplier Retsch GmbH. Due to a certain abrasion of steel in each experiment, grinding balls as well as grinding jars were replaced with new ones within certain periods of time. It was checked that the steel components (appr. 86 % iron, 12 % chromium, 2 % carbon) do not impact the course of the reaction by repeating single experiments with added small amounts of these components (adjusted to the average amount of abrasion in a single run after 60 minutes).

Computations for Kinetic Analyses

Data analyses were performed using the software package EASY-FIT 4.3 (*23,24*), which enables parameter estimation in explicit functions, steady state equations, Laplace transforms, ordinary differential- (ODE) and differential algebraic equations. EASY-FIT makes it also possible to fit several data sets *simultaneously* to an *entire set* of ODEs, e.g., ODEs representing a postulated mechanistic scenario of consecutive or other complex reactions.

Results and Discussion

Concentration measurements of the MC reductive dechlorination of 1,3-DCB (Figure 1) over time reveal several trends.

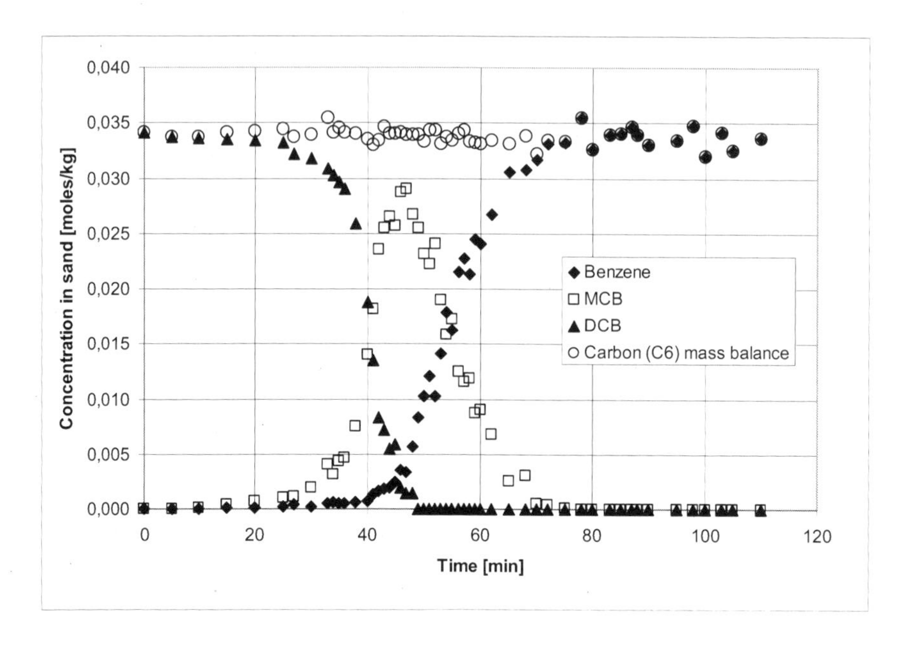

Figure 1. Time-dependent concentration data for the MC reductive dechlorination of DCB via MCB to benzene, employing magnesium and n-butylamine as reagents in the presence of sand and toluene (diluting agent) in a centrifugal ball mill (Retsch S 100 model, 50 ml steel grinding jar, 3 steel balls) at 25 °C.

1. According to Figure 1, a total disappearance of DCB is observed at 25 °C within appr. 50 minutes.
2. There is an induction period in the early stage of the reaction. Virtually no reaction takes place until after appr. 30 minutes of milling have elapsed. At around 30 min., DCB commences to undergo reaction. This induction period can be decreased by pre-milling of the metal in separate step or other pre-activation. It is also impacted by milling parameters, kind of metal, kind of hydrogen donor as well as other reaction parameters.
3. Once initiated, the degradation of DCB takes place rapidly, requiring only appr. 15-20 minutes for completion.
4. MCB is detected as the first and sole (intermediate) reaction product from DCB.
5. The sole and final product of the entire reaction is benzene, apparently formed from the intermediate MCB.

6. A full mass balance for this total dechlorination of DCB is observed: within the experimental error, DCB yields appr. 100 % benzene (mole/mole) *via* the intermediate MCB.
7. In conclusion, a total reductive dechlorination of DCB *via* MCB, as the one and only intermediate, to benzene is observed over appr. 70 minutes. There is an initial induction period of appr. 30 min.

According to the literature, the appearance of an induction period (Figure 1) may be typical for certain MC reactions involving metals (*25*). The reaction scheme in Figure 2 demonstrates the hypothesized mechanisms discussed above (*1-3*), and is useful in deriving a suitable kinetic model.

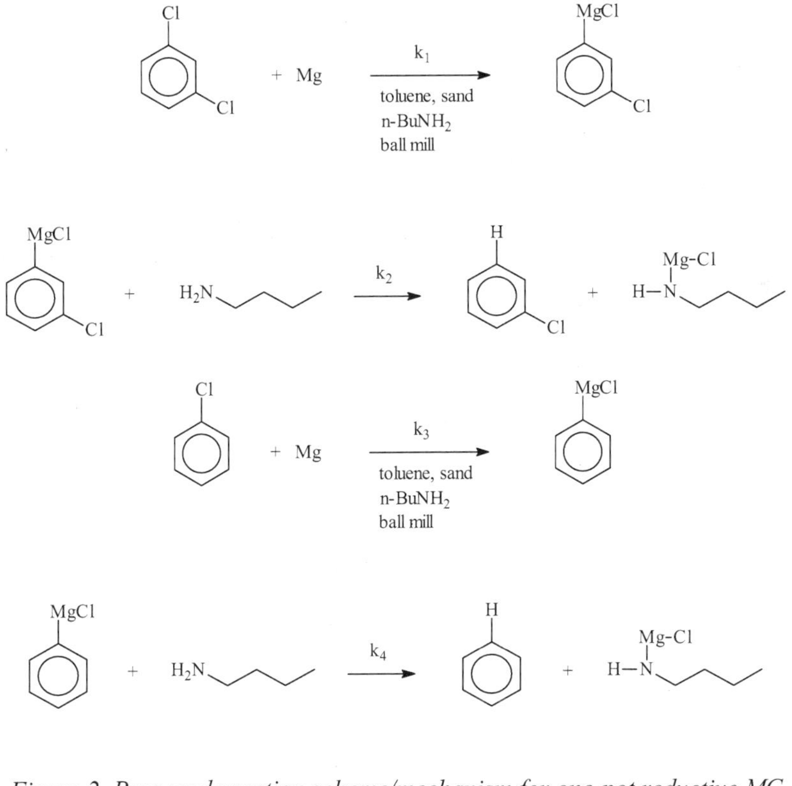

Cl
+ Mg
k_1
toluene, sand
n-$BuNH_2$
ball mill
MgCl
Cl

MgCl
Cl
+ H_2N
k_2
H
Cl
+ Mg-Cl
H—N

Cl
+ Mg
k_3
toluene, sand
n-$BuNH_2$
ball mill
MgCl

MgCl
+ H_2N
k_4
H
+ Mg-Cl
H—N

Figure 2. Proposed reaction scheme/mechanism for one-pot reductive MC dechlorination of DCB to benzene employing magnesium and n-butylamine (n-$BuNH_2$). It shows two consecutive Grignard-Zerewitinoff-like reactions.

In Figure 3, the assumed protonation step of the intermediary Grignard compound 3-chlorophenylmagnesium chloride by n-butylamine is shown in detail. Grignard compounds are known to be very strong bases and to readily undergo protonation by acids. This protonation occurs even with very weak acids such as amines, especially *primary* amines such as n-butylamine (*26,27*). Furthermore, it is known that amines can effectively form complexes with Grignard compounds such as shown in Figure 3, step 1, potentially promoting or catalyzing Grignard formations and/or Grignard reactions under certain conditions (*28*).

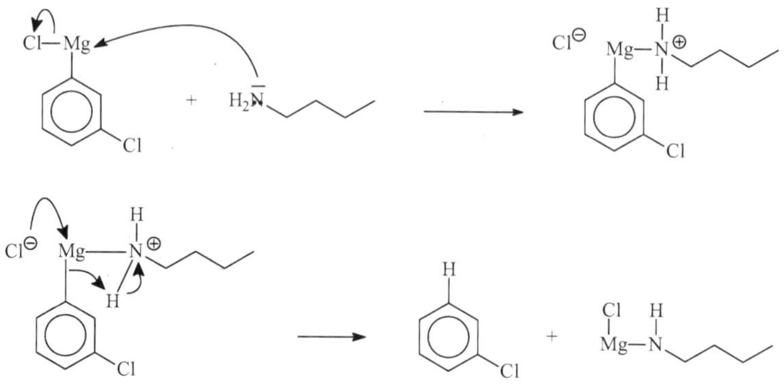

Figure 3. Postulated detailed reaction scheme for the protonation of intermediary 3-chlorophenylmagnesium chloride by n-butylamine yielding MCB (adapted from the literature (26-28)).

Rate equations corresponding to every single reaction step shown in Figure 2 are shown below. They represent a system of linear and first order ODEs and are based on the assumed mechanistic scenario shown in Figures 2 and 3.

$$\frac{dc_{DCB}}{dt} = -k_1 \cdot c_{DCB} \cdot S(t) \tag{1}$$

$$\frac{dc_{ClArMgCl}}{dt} = k_1 \cdot c_{DCB} \cdot S(t) - k_2 \cdot c_{ClArMgCl} \tag{2}$$

$$\frac{dc_{MCB}}{dt} = k_2 \cdot c_{ClArMgCl} - k_3 \cdot c_{MCB} \cdot S(t) \tag{3}$$

$$\frac{dc_{ArMgCl}}{dt} = k_3 \cdot c_{MCB} \cdot S(t) - k_4 \cdot c_{ArMgCl} \tag{4}$$

$$\frac{dc_{Benzene}}{dt} = k_4 \cdot c_{ArMgCl} \tag{5}$$

In Equations 1-5, c denotes the concentration of the relevant compound, k the rate constant of the detailed reaction step, *ClArMgCl* and *ArMgCl* stand for the intermediary Grignard compounds 3-chlorophenylmagnesium chloride and phenylmagnesium chloride (Ar denotes an aryl moiety), respectively, $S(t)$ for the actually reactive surface of the magnesium particles (magnesium surface atoms which are in an activated status for reaction with the organic halide), and t for time. Hence, rate constants k defined by Equations 1-5 actually represent surface area-related values, i.e., the products $k*S_{max}(t)$ represent an observed pseudo first order rate constant k_{obs}, according to common definitions for heterogeneous reactions *(29)* such as the formation of Grignard reagents (*30*). $S_{max}(t)$ is the maximum surface area comprising activated magnesium atoms which can be produced after a certain time of milling/permanent impact has elapsed.

The following assumptions were made for the purpose of simplification of equations 1-5:

1. The formation of intermediary Grignard compounds is dependent on the concentration of the organic halides as their precursors in solution and the particular part of the surface area S of the magnesium which is actually reactive, i.e., in an activated status.
2. The formations of both intermediary Grignard compounds are not diffusion-controlled at the magnesium surface. Because of continuous and intense mechanical impact on the surface, creating various, readily accessible and also activated magnesium surface atoms throughout the reaction, there are, at no time, protective solvent layers or reaction product layers and the like. Hence, individual reaction orders are regarded to be of first order.
3. Steady-state conditions (the "Bodenstein hypothesis") can be applied to the formation and consumption of the intermediary Grignard compounds (*31-33*), that is, the concentration of those compounds may be low at any time and the overall velocity ought to be virtually equal to zero.
4. The protonations of the intermediary Grignard compounds obey a pseudo-first order law in terms of the concentration of the Grignard, because the proton donor amine is applied in excess, hence its concentration is regarded constant.
5. The actual reactive surface $S(t)$ of the small size magnesium particles is assumed to be equal to zero at the beginning of the reaction/ball milling process and develops (increases) over time according to a sigmoidally-shaped (logistic) function, according to the literature (*34*), wherein time-dependent surface properties of ball-milled metals are discussed, in particular adsorption isotherms for ball milled iron metal.

According to these assumptions, the ODE system represented by Equations 1-5 can be reduced to 3 equations by simple rearrangements:

$$\frac{dc_{DCB}}{dt} = -k_1 \cdot c_{DCB} \cdot S(t) \tag{6}$$

$$\frac{dc_{MCB}}{dt} = k_1 \cdot c_{DCB} \cdot S(t) - k_3 \cdot c_{MCB} \cdot S(t) \tag{7}$$

$$\frac{dc_{Benzene}}{dt} = k_3 \cdot c_{MCB} \cdot S(t) \tag{8}$$

In Equations 6-8, *S(t)* is now defined as a modified logistic growth function (*35*), see Equation 9:

$$S(t) = k_5 \cdot \frac{1}{1+\exp(-k_6(t-k_7))} \tag{9}$$

It can be seen that the rearranged and reduced ODE system (Equations 6-9) takes into account the previously assumed steady-state conditions for the intermediary Grignard compounds. Hence, their corresponding rate equations are omitted, and the reduced and simplified ODE system thus generated contains the rate constants for the Grignard formations only and contains no rate constants for the protonation steps.

In order to ascertain whether the assumed mechanistic scenario was operating in this reaction and to see if the Grignard formation rate constants k_1 and k_3 as well the actually reactive magnesium surface function *S* could be determined, data points from two experiments employing differing initial concentrations of DCB were plotted.

Figure 4 shows the degradation concentrations over time of DCB *via* MCB to benzene at two differing initial concentrations of DCB. Six data sets are plotted using the reduced/simplified ODE system with EASY-FIT. Striking is the smoothness of the fit. This suggests, for the particular reaction investigated here, that the postulated mechanisms are in fact the very likely ones for DMCR process using magnesium and amines as dechlorinating agents.

Furthermore, EASY-FIT calculates the Grignard formation constants, but only their products $k_1 \cdot S(t)$ and $k_3 \cdot S(t)$ deliver explicit, defined values due to the mathematical structure of the associated ODEs. In other words, *S* cannot be calculated absolutely, but only relative to and dependent on its evolution in time (hence, $S(t) = S_{abs}(t)/S_{abs}(t_0)$). However, as $k_{1,obs} = k_1 \cdot S_{max}(t)$ and $k_{3,obs} = k_3 \cdot S_{max}(t)$, both k_{obs} values can readily be compared to reported values for similar Grignard formation reactions.

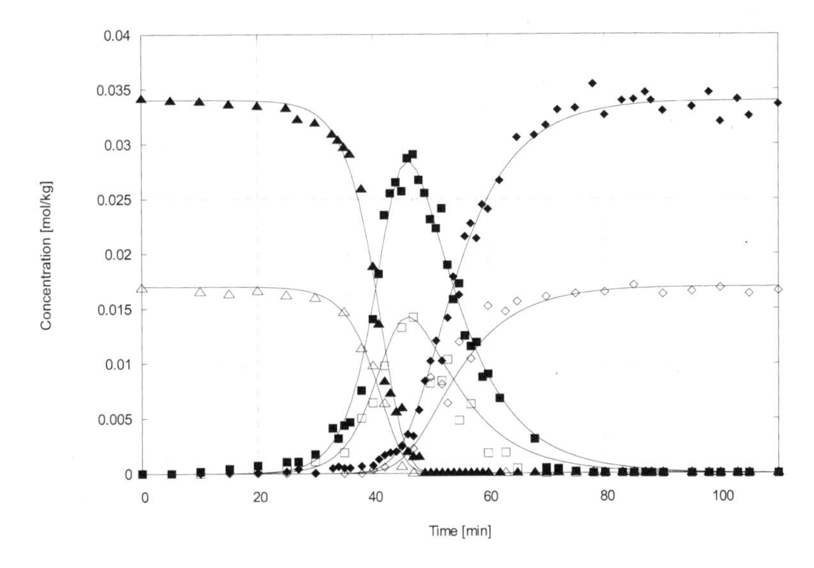

Figure 4. Simultaneous fit of 6 data sets to the ODE system (Equations 6-8), which was derived from the postulated mechanistic scenario (Figure 2 and 3), by using the least squares routine EASY-FIT. Triangles represent DCB, squares stand for MCB, and diamonds for benzene. Curves with empty symbols apply to the reaction, wherein 0.017 mole/kg DCB were employed (initial concentration), black symbols represent concentration/time values of DCB, MCB and benzene in the second reaction (0.034 mole/kg DCB initial concentration).

EASY-FIT calculates for the Grignard formation of phenylmagnesium chloride $k_{3,obs} = 2.0 \cdot 10^{-3}\ s^{-1}$ which is in good agreement with k_{obs} values reported for the formation of phenylmagnesium bromide in diethyl ether at 35 °C, in the range of $10^{-3}...10^{-2}\ s^{-1}$ *(30)* under varying reaction conditions (i.e., varying the magnesium and bromobenzene reactant ratio), and for the formation of ethylmagnesium bromide *(36)*: at 25 °C one obtains $k_{obs} = 1.74 \cdot 10^{-3}\ s^{-1}$ by recalculating data given in *(36)*.

Furthermore, EASY-FIT calculates for the Grignard formation of 3-chlorophenylmagnesium chloride $k_{1,obs} = 2.25 \cdot 10^{-2}\ s^{-1}$, a value one decimal power greater than the value for MCB. This can readily be understood, because of the activating influence of the second chlorine atom in 1,3-DCB for Grignard formation compared to MCB.

It must be emphasized that the proper selection of the surface function *S* is pivotal for the success of the overall calculations. Various kinds of other mathematical functions had been tested as *S* functions but did not provide satisfying results (fits). Note that various additional data sets could be recorded in numerous modified experimental runs and were also fitted successfully to the ODE system (Equation 6-9) providing nearly identical parameters. That is to say, the entire ODE system could be validated to a high degree in numerous additional experiments when Equation 9 was inserted, thereby successfully calibrating the model. (Those results are not shown here and are to be published elsewhere). The function employed here always shows a sigmoidal shape, i.e., S is equal to zero at the beginning of the milling process, then increases slightly and, after a certain point, increases exponentially over time, reaching a constant value in a later stage. In other words, the function reflects a hypothesized growth rate for reactive surface magnesium atoms (sites) which is proportional to the product of the present size and the future amount of growth of reactive sites. The parameter k_5 is some limiting growth parameter (Equation 9). Equation 9 represents the known logistic or autocatalytic growth function (*35*) in a slightly modified form. By applying Equation 9, one takes into account two major processes most probably generating the majority of reactive magnesium surface atoms. It is assumed that during continuous mechanical activation of magnesium particles during milling, reactive magnesium surface atoms are generated first by crushing/cracking the magnesium particles into smaller ones (production of fresh edges, cracks, dislocations etc.) and, second, by direct impact/collisions (between a ball and a metal particle, between two metal particles, and between a metal and a sand particle and so forth).

Hence, the proper selection of a semi-empirical S function plays a key role in the correct modeling this MC reaction. Currently ongoing investigations strive to set up and verify a theoretical model for predicting and adjusting this function regarding different reactions in a ball mill as well as to link its empirical parameters to milling parameters.

However, as a scanning electron microscope microscopy (SEM) examination coupled with an EDX analysis for magnesium reveals, highly irregularly shaped magnesium metal particles are most probably generated to a high degree during grinding (Figure 5). Clearly, this finding makes it hardly possible to readily estimate the development of the available magnesium surface during milling over time by simple geometrical calculations.

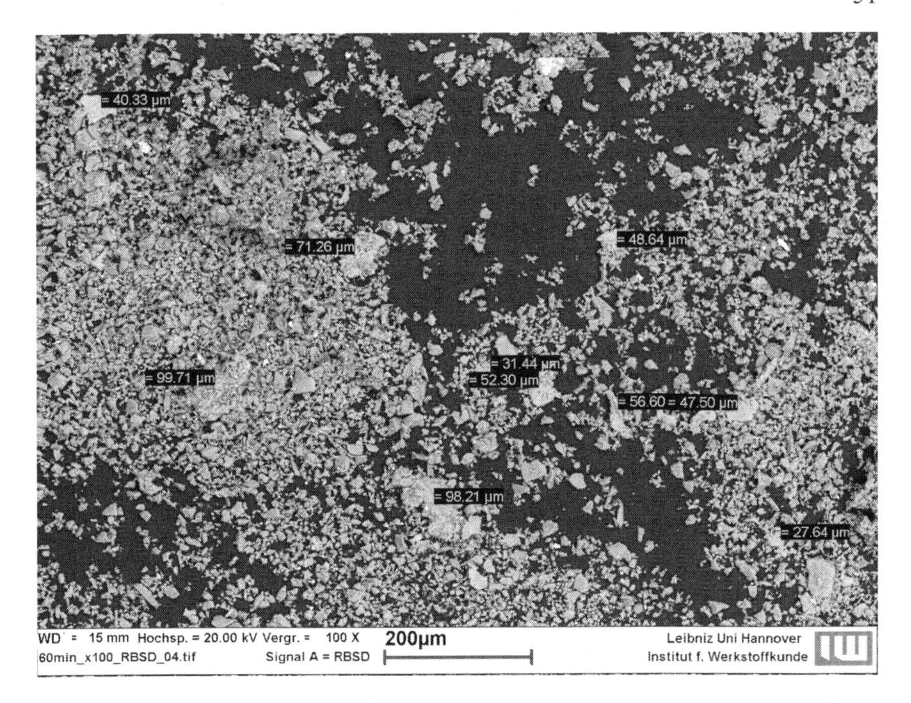

Figure 5. SEM image of a mixture containing sand and magnesium, after both were ground along with DCB, MCB, benzene in toluene and n-butylamine for 60 min. Areas showing high magnesium concentrations/magnesium particles are detected by EDX and reveal highly irregular shapes of presumed magnesium metal particles with diameters in the range of appr. 20-100 µm. Note that higher resolutions of single particles seem to show multiple cracks, dislocations etc. in these particles, thus indicating that the actual reactive surface area comprising the activated surface atoms may cover much smaller areas that is those having diameters below 10 µm (Leibniz University of Hanover, Institute of Materials Science, Germany, Dr. M. Jendras, Dipl.-Math. K. Kerber, and Prof. Dr.-Ing. F.-W. Bach)

Conclusions

MC reductive dechlorination of polychlorinated pollutants such as POPs in a ball mill or a similar device, when performed under DMCR conditions, that is with a base metal and hydrogen donor, is a promising approach to destroy those toxic compounds effectively and in an environmentally friendly manner:

1. The reaction takes place at room temperature. It is a *non-thermal* destruction process for POPs and can be completed in a *short time*.
2. The elimination of polychlorinated pollutants can be performed not only on pure substances but also on contaminated complex solid, solid-liquid, and liquid matter such as is found in soils, filter dusts, sludges or oils (e.g., transformer oils containing PCBs).

3. Only a few defined, and easily disposable degradation products are formed.
4. Moreover, full mass balances are found. No side reactions occur.
5. A model has been developed that explains and properly predicts the mechanisms involved and therefore the progress of the reaction/process over time. The chemical kinetics of the DMCR process has been elucidated and is well understood. In the case of magnesium, coupled, consecutive one-pot Grignard-Zerewitinoff-like reactions occur. This has been verified by fitting experimental data to a simple kinetic model. The selection of an appropriate semi-empirical surface function, which properly estimates the development of the reactive magnesium surface during milling, is pivotal. It was found that a sigmoidally shaped logistic growth function can successfully be applied, and the kinetic model derived here may be used to effectively design and scale up ball mill reactors for technical applications.
6. Simple and readily available reagents can be used. Note that although an amine as well as highly reactive magnesium metal were selected for this study as model reagents; similar reagents showing better handling capacities/lower toxicities etc. can as easily be chosen from a wide range of available reagents and adapted to numerous areas of application (*1-3*).
7. The recycling or economic and environmentally benign re-use of hazardous materials can be precisely predicted allowing flexibility in many areas of application.
8. Further mechanistic/kinetic studies are needed to provide precise data with which process engineers can design and operate full-scale applications.
9. Cost parameters have to be investigated to a higher degree: how the concentration/amount of pollutant(s), the kind and concentration of metal and promoter as well as water content of the waste may govern reaction kinetics and subsequently cost.

Acknowledgements

The authors are grateful to Dipl.-Ing. agr. Tobias Kinkel, Dipl.-Ing. Christian Holdack and Mrs. Sabine Henschel (CTA), Leuphana University of Lüneburg, Campus Suderburg, for performing numerous experimental and analytical works, to Prof. Harald Burmeier, Leuphana University of Lüneburg, Campus Suderburg, for organizational support of this work as well as to Dr. Wolf-Ulrich Palm, Leuphana University of Lüneburg, Institute of Ecology and Environmental Chemistry, for valuable advice. The authors also thank Dr. Michael Jendras and Dipl.-Math. Kai Kerber, Institute of Materials Science (Director: Prof. Dr.-Ing. F.-W. Bach), Leibniz University of Hanover, Germany, for conducting SEM and EDX examinations. One of the authors (VB) expresses his special thanks to Prof. Dr. Klaus Schittkowski, University of Bayreuth, for valuable technical support regarding application of EASY-FIT.

References

1. Birke, V.; Mattik, J.; Runne, D. *J. Mat. Sci.* **2004,** *39*, 5111.
2. Birke, V.; Runne, D.; Mattik, J.; Berger, A; Schütt, C.; Aresta A.; Dibenedetto, A. In *Remediation of Chlorinated and Recalcitrant Compounds —2004.* Proceedings of the Fourth International Conference on Remediation of Chlorinated and Recalcitrant Compounds, Monterey, CA, U.S.A., May 2004, Gavaskar, A. R., Chen, A. S. C., Eds.; Battelle Press: Columbus, 2004, ISBN 1-57477-145-0.
3. Birke, V.; Mattik, J.; Runne, D.; Benning, H.; Zlatovic, D. Dechlorination of Recalcitrant Polychlorinated Compounds Using Ball Milling In *Ecological Risks Associated with the Destruction of Chemical Weapons*; Kolodkin, V. M., Ruck, W., Eds.; NATO Security through Science Series – C: Environmental Security; Springer: Dordrecht, The Netherlands, 2006; p 111-127.
4. Rowlands, S. A.; Hall, A. K.; McCormick, P. G.; Street, R.; Hart, R. J.; Ebell, G. F.; Donecker, P. *Nature* **1994,** *367*, 223.
5. Donecker, P.; McCormick, P. G.; Street, R.; Rowlands, S. A. Patent PCT WO 94/14503, 1994.
6. Hall, A. K.; Harrowfield, J. M.; Hart, R. J.; McCormick, P. G. *Environ. Sci. Technol.* **1996,** *30*, 3401.
7. Loiselle, S.; Branca, M.; Mulas, G.; Cocco, G.; *Environ. Sci. Technol.* **1997,** *31*, 261.
8. Gock, E.; Opel, M.; Mayer, J. *Organohalogen Compounds* **2001,** *54*, 185.
9. Ikoma, T.; Zhang, Q.; Saito, F.; Akiyama, K.; Tero-Kubota, S.; Kato, T. *Bull. Chem. Soc. Jpn.* **2001,** *74*, 2303.
10. Zhang, Q.; Matsumoto, H.; Saito, F.; Baron, M. *Chemosphere* **2002,** *48*, 787.
11. Tanaka, Y.; Zhang, Q.; Saito, F. *J. Mat. Sci.* **2004,** *39*, 5497.
12. Nomura, Y.; Nakai, S.; Hosomi, M. *Environ. Sci. Technol.* **2005,** *39*, 3799.
13. Yan, J. H.; Peng, Z.; Lu, S. Y.; Li, X. D.; Ni, M. J.; Cen, K. F.; Dai, H. F. *J. Haz. Mat.* **2007,** *147*, 652.
14. Veit, M.; Hoffmann, U. *Chem. Ing. Tech.* **1996,** *68*, 1279.
15. Harrowfield, J. M.; Hart, R. J.; Whitaker, C. R. *Aust. J. Chem.* **2001,** *54*, 423.
16. March, J. *Advanced Organic Chemistry*, 4th ed.; Wiley: New York, 1992; p 441.
17. Bryce-Smith, D.; Wakefield, B. J.; Blues E. T. *Proc. Chem. Soc.* **1963,** 219.
18. Baeyer, A. *Ber. Dtsch. Chem. Ges.* **1905**, *38*, 2759.
19. Zechmeister, L.; Rom, P. *Liebigs Ann. Chem.* **1929**, *468*, 117.
20. Hutchins, R. O.; Suchismita; Zipkin, R. E.; Taffer, I. M. *Synth. Commun.* **1989**, *19*, 1519.
21. White, G. F.; Kraus, C. A. *J. Am. Chem. Soc.* **1923**, *45*, 768.
22. Shaw, M. C. *J. Appl. Mech.* **1948,** *15*, 37.
23. Schittkowski, K. *Numerical Data Fitting In Dynamical Systems*, 1st ed.; Kluwer Academic Publishers: Dordrecht/Boston/London, 2002.
24. Easy-Fit Model Design, Web site of Prof. Dr. Klaus Schittkowski, University of Bayreuth, Germany.

http://www.math.uni-bayreuth.de/~kschittkowski/easy_fit.htm (accessed Feb. 15th, 2009)

25. Heinicke, G. *Tribochemistry*, Akademie-Verlag: Berlin, 1984.
26. Silvermann, G. S.; Rakita, P. E. *Handbook of Grignard Reagents*, 1st ed.; Marcel Dekker: New York, 1996; pp 441-453.
27. *Metallorganische Verbindungen, Be, Mg, Ca, Sr, Ba, Zn, Cd*; Müller, E., Ed.; Methoden der Organischen Chemie (Houben-Weyl), Vol. XIII/2a; Georg Thieme: Stuttgart, 1973, pp 152-155.
28. Pällin, V.; Tuulmets, A. *J. Organomet. Chem.* **1999,** *584*, 185.
29. Levenspiel, O. *Chemical Reaction Engineering*, 2nd ed.; Wiley: New York, 1972.
30. Simuste, H.; Panov, D.; Tuulmets, A.; Nguyen, B. T. *J. Organomet. Chem.* **2005,** *690*, 3061.
31. Bodenstein, M. *Z. Physik. Chem.* **1913,** *85*, 329.
32. Mauser, H. *Formale Kinetik*, 1st ed.; Bertelsmann Universitätsverlag: Düsseldorf, 1974.
33. Hammett, L. P. *Physical Organic Chemistry, Reaction Rates, Equilibria, and Mechanisms*, 2nd ed.; New York: McGraw-Hill, 1970.
34. Husemann, R.; Bernhardt, C.; Heegn, H. *Powder Technology* **1976,** *14*, 41.
35. Draper, N.; Smith, H. *Applied Regression Analysis*; 2nd ed.; Wiley: New York, 1981; pp 507-513.
36. Kilpatrick, M.; Simons, H. P.; *J. Org. Chem.* **1937,** *2*, 459.

Chapter 4

Proposed Mechanisms for the Dechlorination of PCBs using Microscale Mg/Pd in Methanol

Robert DeVor[1], Phillip Maloney[2], Christian A. Clausen[2], Seth Elsheimer[2], Kathleen Carvalho-Knighton[3] and Cherie L. Geiger[2]

[1]ASRC Aerospace, Kennedy Space Center, Fl 32815
[2]Department of Chemistry, University of Central Florida, Orlando, FL 32816
[3]Department of Environmental Science, Policy, and Geography, University of South Florida St. Petersburg, St. Petersburg, FL 33701

The combination of zero-valent metals and hydrogenation catalysts, specifically mechanically alloyed, palladized magnesium (Mg/Pd) has been proven effective in the remediation of polychlorinated biphenyls (PCBs), although the exact mechanism of degradation has yet to be elucidated. Knowing the exact mechanism of dechlorination could prove useful in optimizing the bimetallic Mg/Pd for use in field-scale applications. A variety of experiments have been performed on individual PCB congeners in an attempt to determine the mechanism by which the degradation occurs. Studies have focused on experiments carried out in methanol, however, several studies have also been carried out in water to examine solvent specificity. Results of these studies have suggested three possible solvent specific mechanisms, all of which include the removal of the chlorine atom by hydrogen as the rate limiting step, varying only in the exact nature of the hydrogen species (radical, hydride, or "hydride-like" radical).

Introduction

Polychlorinated biphenyls (PCBs) are a family of 209 chemical compounds for which there are no known natural sources. They have a heavy, oil-like consistency (single congeners can exist as solids), high boiling points, a high degree of chemical stability, low flammability, low electrical conductivity, and specific gravities between 1.20 and 1.44. Because of the aforementioned characteristics, PCBs were used in a variety of applications such as: heat transfer and hydraulic fluids; dye carriers in carbonless copy paper; plasticizer in paints, adhesives, and caulking compounds; and fillers in investment casting wax (1). PCBs can volatilize from sources and are capable of resisting low temperature incineration. This makes atmospheric transport the primary mode of global distribution (2). PCBs are subject to reductive dechlorination, even though they are generally considered recalcitrant in the environment (3). The process of PCB reductive dechlorination replaces chlorines on the biphenyl ring with hydrogen, reducing the average number of chlorines per biphenyl in the resulting product mixture. This reduction is important because, in most cases, the less chlorinated products are less toxic (4, 5), have lower bioaccumulation factors, and are more susceptible to aerobic metabolism, including ring opening and mineralization (5, 6).

Currently, the most common remediation technique is incineration, but this procedure is not without its problems. Incineration requires a large amount of fuel and, if temperatures are not high enough, can lead to the formation of highly toxic by-products, including polychlorinated dibenzo-*p*-dioxins and polychlorinated dibenzo-furans (commonly referred to as dioxins) (7). Another traditional remediation technique for PCB contamination is dredging of contaminated soils and sediments followed by land filling of the resulting hazardous waste. Land filling is undesirable because the inherent stability of PCBs poses problems when leaching occurs. Microbial degradation, both aerobic and anaerobic, is another treatment option currently being investigated. Aerobic processes proceed via oxidative destruction of the PCBs, although dechlorination is limited to those congeners having five or less chlorines present on the biphenyl ring. Anaerobic microbial degradation occurs via a reductive dehalogenation pathway that can typically only remove chlorines from the *meta* or *para* position (8). Furthermore, slow reaction rates and incomplete degradation have limited the applicability of this approach in the field.

A more promising technique that has been studied in recent years is the use of zero-valent metals (including magnesium, zinc, and iron) for the *in situ* remediation of a variety of chlorinated compounds including PCBs. Dechlorination of PCBs by zero-valent iron has been demonstrated at high temperatures (9) however, little dechlorination occurred at temperatures less than 200°C. By using palladium, a known hydrodechlorination catalyst (10-14), as a coating on the zero-valent iron surface, rates of dechlorination by iron have been increased. Muftikian and co-workers demonstrated rapid degradation of perchloroethylene (PCE) with Fe/Pd (15) and Grittini showed that the Fe/Pd bimetallic system can degrade PCBs but did not quantify the degradation (16). While Fe/Pd has shown high levels of degradation in laboratory studies, the bimetal must be prepared under inert atmosphere after rigorous acid-washing of the iron (17). A report by Fernando and co-workers proposed that the enhanced reactivity of Pd/Fe might be due to the adsorption of hydrogen (H_2), generated by iron corrosion, on the palladium (16).

The disappearance of chlorinated organic compounds from aqueous solutions contacting ZVMs may be due to dechlorination reactions or sorption to ZVM-related surfaces. This investigation utilizes mechanically alloyed magnesium and palladium (1% on graphite), a bimetallic system that has proven effective in previous research in the degradation of PCBs (18-20). Magnesium has several advantages over other zero-valent metals. Due to the self-limiting oxide layer that forms on the surface of magnesium, it is capable of dechlorination even after exposure to oxygen (unlike iron, which will completely oxidize upon exposure to oxygen and become deactivated). Another advantage in using magnesium is that it has a greater thermodynamic driving force when compared to both iron and zinc, as shown below:

$$Mg^{2+} + 2e^- \rightarrow Mg^0 \quad E^0 = -2.37\ V$$
$$Fe^{2+} + 2e- \rightarrow Fe^0 \quad E^0 = -0.44\ V$$
$$Zn^{2+} + 2e^- \rightarrow Zn^0 \quad E^0 = -0.76\ V$$

An understanding of the underlying mechanism for the dechlorination is a necessary next step in order to fine-tune the bimetallic system for maximum possible effectiveness. Several studies are discussed within this paper to help elucidate and to propose a mechanism by which PCBs are degraded by palladized magnesium (Mg/Pd). Possible mechanistic pathways under investigation included degradation through a benzyne intermediate, nucleophilic aromatic substitution, and the use of hydrogen (in the form of a radical or hydride) in removing the chlorine atom. Experimental evidence suggests three different possible mechanistic pathways, all of which include the removal of the chlorine atom by a hydrogen atom as the rate-limiting step.

Methods

Materials and Chemicals

Neat PCB standards were obtained from Accustandard (New Haven, CT) and Optima© grade methanol and toluene were obtained from Fisher Scientific (Pittsburgh, PA). 99.0% Methanol-*d* was obtained from Acros Organics (Morris Plain, NJ). Magnesium (~ 4-µm) was obtained from Hart Metals, Inc (Tamaqua, PA). 1% palladium on graphite was obtained from Engelhard (Iselin, NJ), while 10% palladium on graphite was obtained from Acros Organics. All metals and catalysts listed above were used as received.

A ~0.08 wt% palladium-magnesium mixture was prepared by ball-milling 78 g Mg with 7 g of 1% palladium on graphite in a stainless steel canister (inner dimensions 5.5 cm by 17 cm) with 16 steel ball bearings (1.5 cm diameter, at a total mass of 261.15 g). The material was milled for 30 minutes using a Red

Devil 5400 series paint mixer. A ~0.8 wt% palladium-magnesium mixture was prepared in a similar fashion using 10% palladium on graphite.

Some experiments were conducted on bimetal material that had been milled up to three years earlier. In order to "reactivate" the bimetallic compound, the already prepared material (85 g) was milled for an additional 30 minutes on the Red Devil 5400 series paint shaker.

Individual PCB solutions were prepared by diluting the neat standards with Optima© grade methanol (or 99.0% methanol-*d*) to the desired concentration. Further dilution was done on several mono-chlorinated congener solutions with deionized water to bring the final solvent ratio to 9:1 water: methanol.

Experimental rate constants were normalized using a ρ constant (g of (Mg/Pd)/L of solution) to allow comparison between various studies.

Experimental Procedure

Vial studies using 0.25 g of Mg/Pd and 10 mL of individual PCB solution in 20-mL vials (with PTFE lined caps) were conducted. Samples were placed on a Cole-Parmer Series 57013 Reciprocating Shaker table (speed 7) until the appropriate extraction time. Samples were extracted using 10 mL of toluene (Optima©) which was added to the sample vial. The resulting mixture was then shaken by hand for two minutes. Next, 4 mL of this miscible solution was filtered with a Puradisc® 25-mm (0.45-μm pore size) nylon syringe filter attached to a glass syringe. This mixture was then placed into a centrifuge tube with a PTFE lined cap (to prevent evaporation) and approximately 2 mL of deionized water was added to separate the methanol from the toluene. The mixture was then shaken by hand for two minutes, followed by centrifugation for two minutes. The toluene layer of the extract was collected for further analysis. Studies in 9:1 water:methanol were conducted similarly, however the starting weight of Mg/Pd was 0.05 g and after 2 minutes of shaking, a five minute ultrasound treatment was performed, to ensure complete extraction of PCBs from the surface of the bimetal. The extraction efficiency for this separation technique was ~97% with a relative standard deviation of less than 4%.

Analysis of the extracted samples was performed on a Perkin Elmer Autosystem XL gas chromatograph equipped with an electron capture detector (GC-ECD), a Shimadzu GC-2014 w/TOF MS and a Thermo Finnigan Trace GC/DSQ, all using a RTX-5 column (30-m, 0.25-mm i.d., 0.25-μm df). Ultra high purity nitrogen was used as the ECD makeup gas at a flow of 30 mL/min. Helium was used as the carrier gas in both instruments, and was held at a constant flow of 1.3 mL/min. On the GC-ECD, the injector port temperature was held at 275°C and the detector was at 325°C. On the Shimadzu and the ThermoFinnigan mass spectrometers, the injector temperature was 220°C, and the ion source temperature was 250°C. All three instruments were equipped with autosamplers. An initial oven temperature of 100°C was used, and then ramped up to 270°C. Identification of each of the single congener PCBs was based upon the retention times and mass spectra of known standards.

Results and Discussion

Solvent Specific Reaction Kinetics

Monochlorinated PCB studies

Initial experiments using PCB-001, PCB-002 and PCB-003 were carried out in both methanol and water:methanol (9:1) in an attempt to determine solvent specificity.

Zero-valent magnesium mechanically alloyed with 1% palladium on graphite showed 50% dechlorination of a 10 mL 50ug/ml methanol solution of PCB-001 within 30 minutes while 50% dechlorination took just over two hours for PCB-002 and PCB-003 under identical reaction conditions. A summary of the normalized, pseudo-1st order rate constants of each congener from these studies can be found in Table I.

Table I. Solvent and Congener Specific Rate Constants

	Rate Constant, k ($L\ min^{-1}\ g^{-1}$)	
PCB Congener	*100% Methanol*	*90% Water/ 10% Methanol*
001	0.0011	0.00226
002	0.00045	0.00486
003	0.00052	0.00716

A similar set of experiments was conducted using a solvent system consisting of 90% water and 10% methanol (used to increase solubility). In these studies, 0.05 g of 1% Mg/Pd was used instead of 0.25 g, due to greater reactivity of the Mg/Pd in water. Also, the samples were exposed to a 5 minute period of ultrasound to assist in extraction of any adsorbed PCBs from the surface of the catalytic metal prior to syringe filtering. PCB starting concentrations were also lower (5 ug/ml-10 ug/ml) due to solubility issues of PCBs in water. Pseudo 1st order rate constants obtained from the experiments are also shown in Table I.

Kim *et al.* (2004) performed a similar study on monochlorinated biphenyls in water using Pd/Zn and Pd/Fe (21). They achieved results similar to what was observed in the experiments conducted in the water:methanol solvent system, but there were some differences when compared to the pure methanol system. Kim *et al* observed that the rate of dechlorination for monochlorinated congeners in water was PCB-003 > PCB-002 > PCB-001, which is in agreement with the experimental results presented above. The studies conducted in pure methanol solvent show a completely different preference for dechlorination, PCB-001 > PCB-003 > PCB-002. The differences presented above could be due to several factors. In the methanol study, the reversal in the dechlorination trend may indicate that bimetallic degradation is dependent upon the proton donor that is used. In addition, the previous study used both Pd/Zn and Pd/Fe, rather than Mg/Pd, which may indicate different bimetallics can undergo different mechanistic pathways. Finally, the bimetallics were prepared differently which may also affect the mechanism of degradation. Kim *et al.* (21) used electrodeposition to prepare the Pd/Zn and Pd/Fe whereas the Mg/Pd was prepared by mechanical alloying.

"Lag Time" investigation

One interesting data point that appeared in all three studies performed using methanol as the solvent system was that there appeared to an initial "lag time" before significant degradation would begin to occur. In all three of the monochlorinated studies performed in methanol that are presented in this paper, that lag time was approximately 30 minutes. Higher chlorinated congeners have exhibited similar "lag" times as well. One possible explanation for this is that a certain amount of time is required for enough molecular/atomic hydrogen to be generated before degradation can occur. To test this theory, a study was initiated in which pure methanol and ~0.25g of 1% Mg/Pd were allowed to react for ~30 minutes, at which point the samples were spiked with PCB-001 to a concentration consistent with prior studies. The samples were then extracted and analyzed as before. The data is included in Figure 1.

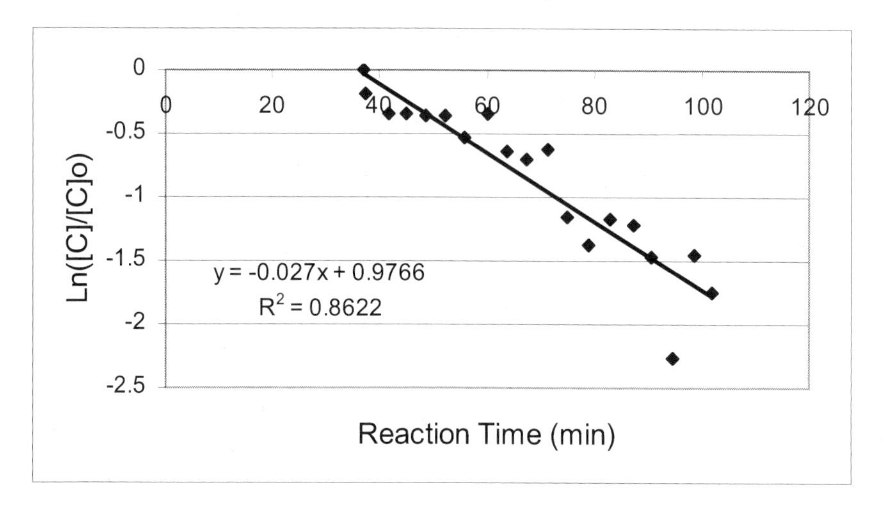

Figure 1. 1st Order kinetics plot of PCB-1 from "Lag" study. (Adapted from reference 19. Copyright 2008 Elsevier.)

This study produced a rate constant almost identical to previous study without the additional "lag" time. This indicates that the reason for the difference in kinetics is due to generation/adsorption of hydrogen rather than the adsorption of the contaminant itself, otherwise the same ~30 minute "lag" time would have been observed in this study. Interestingly enough, this trend is not seen in studies with water as the solvent, most likely due to water's greater ability to donate a proton, thus creating molecular hydrogen more quickly.

Another important question to answer is the final fate of the contaminant, whether or not biphenyl is the end product or if it is degraded further. Studies were conducted in both methanol and 9:1 water:methanol solvent systems using more active palladized magnesium (10% Mg/Pd vs. 1% Mg/Pd) in order to answer this question. The results are shown below in Figure 2. There was no significant degradation of biphenyl seen in studies using the pure methanol solvent even after 30 days. However, the studies with 9:1 water:methanol showed significant degradation within the first six hours, and near complete degradation within 3 days. This was in agreement with other published studies where the primary solvent is water (21).

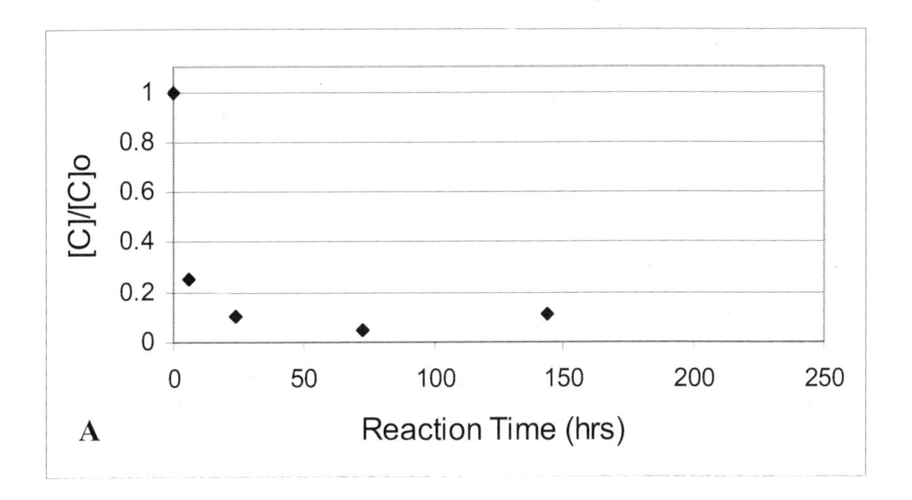

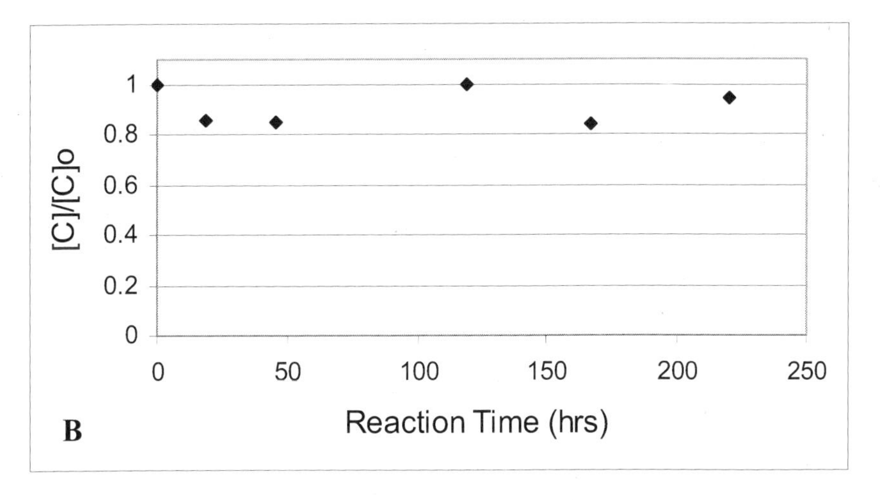

Figure2 . Relative degradation of Biphenyl in A) methanol and B) 9:1 water:methanol w/ 10% Mg/Pd. (Adapted from reference 19. Copyright 2008 Elsevier.)

Investigations of the Dechlorination Mechanism in Methanol

Isotopic Investigation of the Hydrogen Source

The source of hydrogen in this reaction was investigated to determine if it was coming from the alcohol or methyl moiety of the solvent. Gas-phase studies have shown that the methyl C-H bond requires less energy to break than the O-H bond in methanol (22). However, whether or not this is true in this reaction isn't clear, due to the role that solvent effects can play in bond strengths. In order to determine the source of the hydrogen, an isotopically labeled reaction study was performed using PCB-1. Methanol-D (CH_3OD) was used as the solvent, and the byproducts were determined by GC-MS. If the proton was coming from the alcohol group, then the *m/z* ratio of the biphenyl byproduct would be 1 amu higher than when the same study was run using pure methanol as the solvent. The mass spectra of both of these studies are given in Figure 3. As can be seen in the corresponding mass spectra, when pure methanol was used as the reacting solvent, the parent ion of the biphenyl byproduct was 154 *m/z*. When isotopically labeled methanol-d was used, however, the parent ion of biphenyl was 155 *m/z*. This conclusively demonstrated that the proton being donated in the reaction was coming from the alcohol group of the methanol, rather than from the methyl group (as gas-phase data suggests). Additionally, this information ruled out the possibility of the dechlorination going through a benzyne intermediate, which was being considered as a possible mechanistic pathway for Mg/Pd dechlorination. However, once the intermediate C-C triple bond is formed, two protons (or deuteriums) would have to add across the triple bond. The above data show only one additional deuterium was seen in the degradation of PCB-1 to biphenyl, eliminating the benzyne mechanism as a possible mechanistic pathway.

Kinetic Investigations of the Dechlorination Mechanism of a Polychlorinated Congener (PCB-151) in Methanol

Subsequent studies focused on the mechanism of dechlorination specifically in a methanol solvent using a hexachlorinated congener, 2, 2', 3, 5, 5',6-PCB (PCB-151). Zero-valent magnesium mechanically alloyed with 1% palladium on graphite was shown to be capable of degrading 85% of 10 mL of a 20-μg/mL PCB-151 in methanol solution within 72 hours (Figure 4).

In the preliminary PCB-151 study, after the initial rapid dechlorination, subsequent degradation of this hexachlorinated congener was halted. A possible explanation is the competition with the daughter compounds for the (relatively) few active palladium catalytic sites on the magnesium, which may occur if the contaminant is bound to the surface of the bimetal during the dehalogenation process. If this was the case, the newly created byproduct may not leave the surface once the first chlorine atom is removed. This would not allow non-degraded PCB-151 to interact with the active site. If there were not an excess of active sites in the reaction system (as is the case here), then competition would favor the dechlorination of the newly formed lower chlorinated byproduct, halting the degradation of the parent congener (Figure 5).

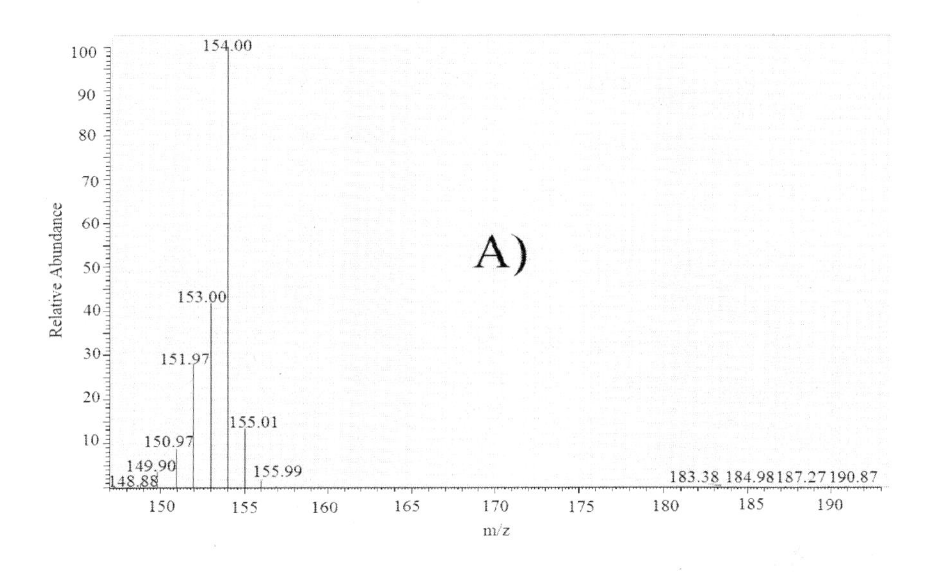

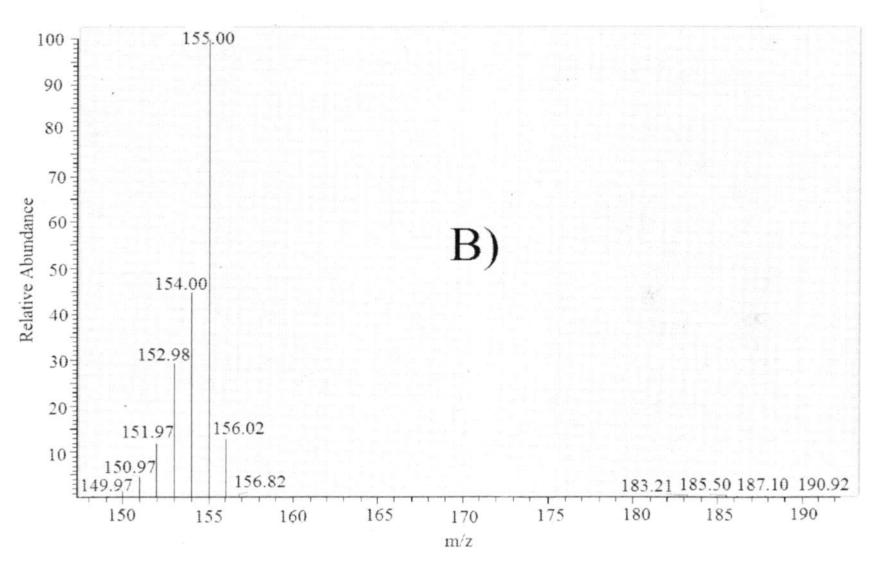

Figure 3. Mass spectra of PCB-1 degradation product (biphenyl and deuterated biphenyl) with Mg/Pd in A) MeOH and B) MeOD. (Adapted from reference 19. Copyright 2008 Elsevier.)

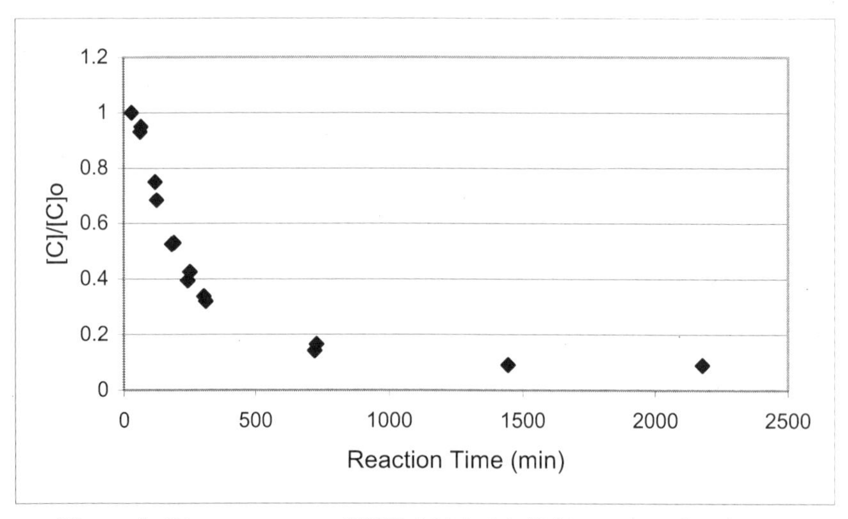

Figure 4. Disappearance of PCB-151 in Mg/Pd in methanol with time.

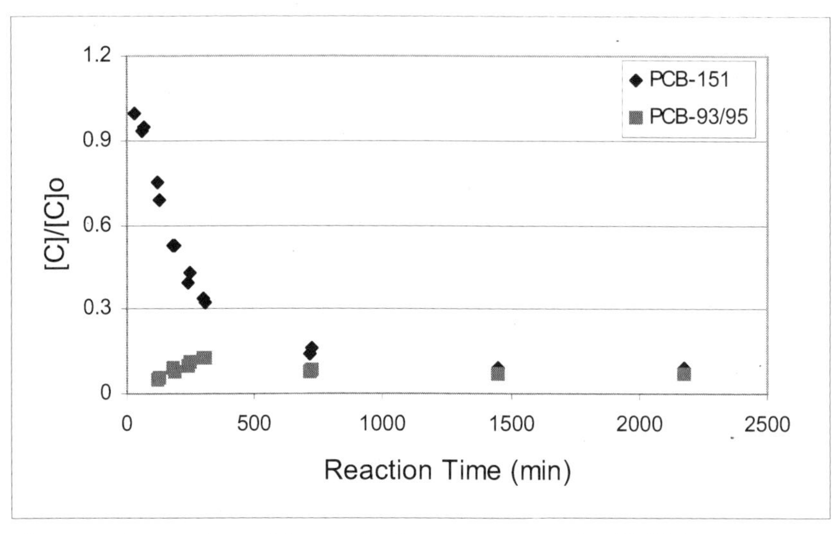

Figure 5. Comparison of PCB-151 disappearance and PCB-93/95appearance. (Adapted from reference 20. Copyright 2008 Elsevier.)

The disappearance of PCB-151 coincided with the production of PCB-93/PCB-95. The degradation of the parent congener slowed dramatically indicating that the active sites were occupied with the newly formed lower chlorinated byproducts. Subsequently, the degradation of the newly formed byproducts slowed down and halted as lower chlorinated byproducts were created and occupied the active sites.

The following experiment was performed to test this theory. A PCB-151 solution was spiked with biphenyl, so that the final concentrations of both analytes were equal (5-µg/mL). This solution was exposed to the Mg/Pd as was done in the previous experiments. When compared to a control study consisting of only PCB-151 exposed to Mg/Pd, the degradation of the parent congener was severely inhibited. This evidence supports the concept that competition is occurring at the active sites on the Mg/Pd surface.

The degradation of PCB-151 indicated a stepwise dechlorination due to the sequential production of degradation byproducts with the initial reaction pathway PCB-151 → PCB-93/PCB-95 (major products) + PCB-92 (minor products) → PCB-45 (Figure 6).

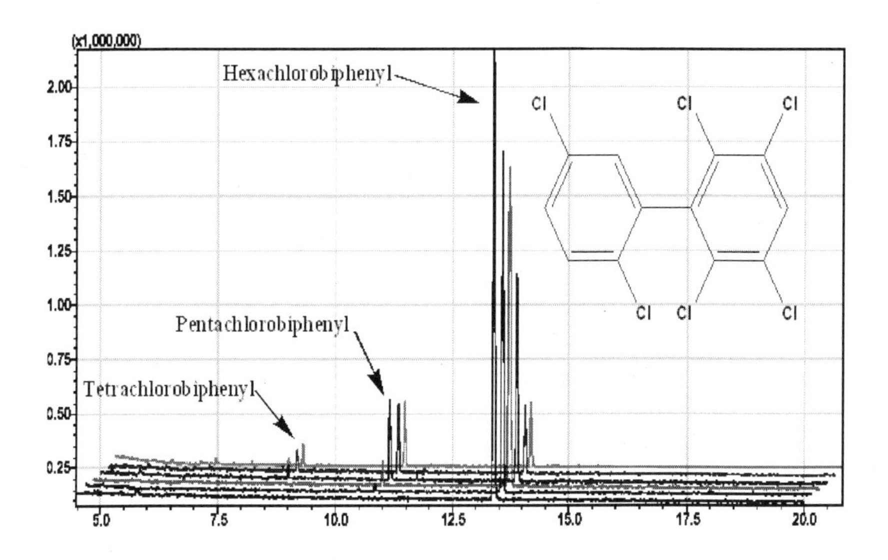

Figure 6. GC-MS analysis of the degradation of PCB-151 using Mg/Pd in methanol. Time 0 hr chromatogram is shown at the bottom of the figure with samples taken at longer time periods shown sequentially above. (Adapted from reference 20. Copyright 2008 Elsevier.)

Due to the change in the kinetics of the parent congener once there was competition for the active sites, a pseudo rate constant and reaction order were determined for the first six hours of the study. These data were fit to a pseudo 1st order plot, and is shown below in Figure 7.

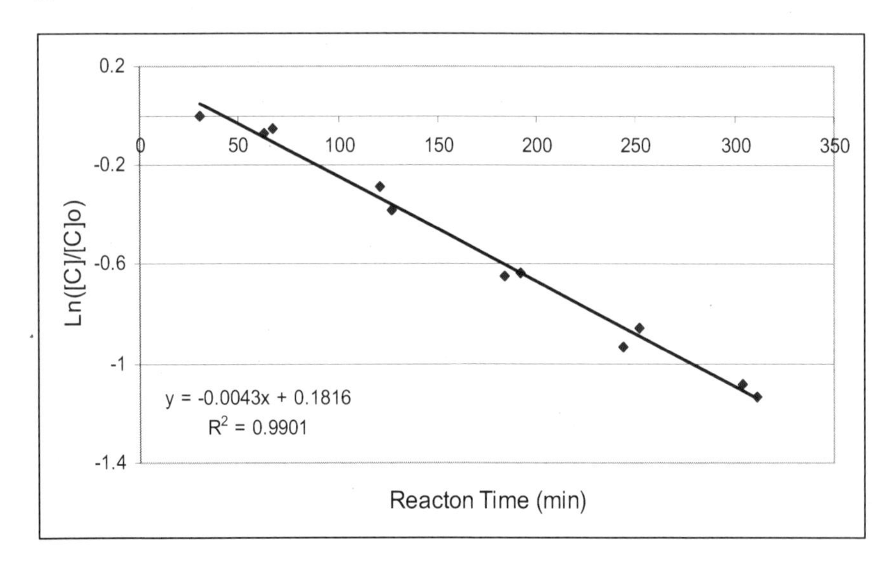

Figure 7. Pseudo 1st order kinetics plot of the degradation of PCB-151 with Mg/Pd in methanol. (Adapted from reference 20. Copyright 2008 Elsevier.)

From this plot, a pseudo 1st order rate constant of k = 1.72E^{-4} L min^{-1} g^{-1} (normalized by volume of solution and mass of Mg/Pd used) was obtained for PCB-151 degraded by Mg/Pd in methanol.

Change in Mg/Pd Activity Over Time and the Effect of Re-ballmilling

Subsequent testing of the original mechanically alloyed material had shown that the metal became less active after six months. In an attempt to reactivate the metal, the bimetal was subjected an additional 30 minutes of mechanical alloying (under identical conditions to the initial milling). The re-milled bimetal was tested using PCB-151 to determine the effectiveness of reactivation process. This study was done using identical conditions to the original PCB-151 degradation study, and is shown in Figure 8. A similar degradation profile was observed for the reactivated metal as seen with the original Mg/Pd bimetal, although the degradation of the parent congener appears to happen at a faster rate. A pseudo-first order kinetics plot is shown in Figure 9.

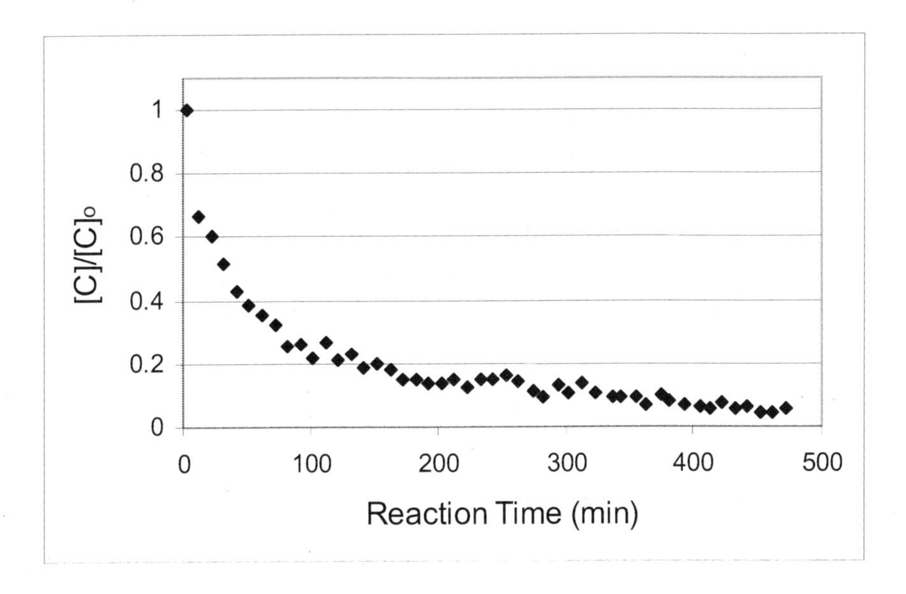

Figure 8. Degradation Plot of PCB-151 using reactivated Mg/Pd in methanol. (Adapted from reference 20. Copyright 2008 Elsevier.)

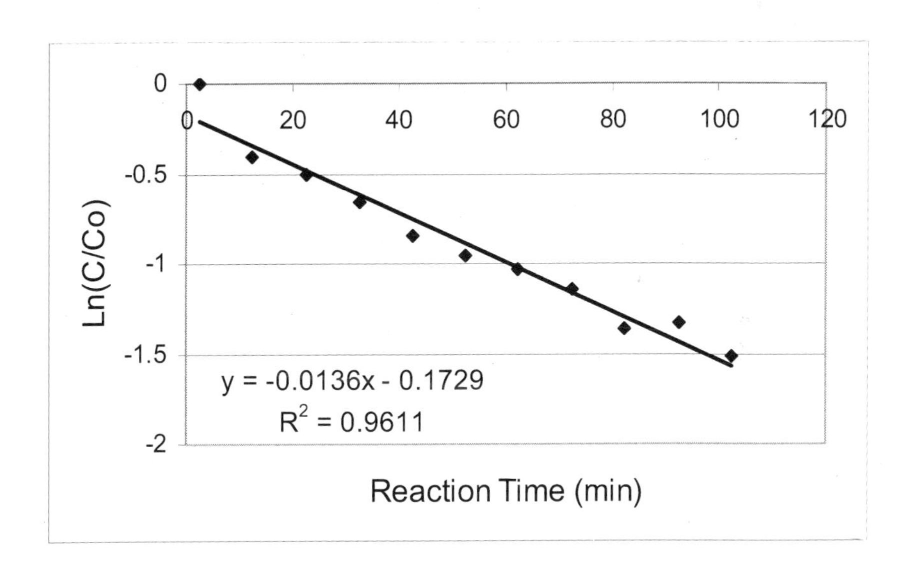

Figure 9. Pseudo 1st order kinetics plot of the degradation of PCB-151 using reactivated Mg/Pd in MeOH. (Adapted from reference 20. Copyright 2008 Elsevier.)

A normalized pseudo 1st order rate constant of 5.44E-4 L min^{-1} g^{-1} was obtained, which is more than 3x faster than the original bimetal. Not only was the reactivated metal more reactive, it appears to be less selective as well. Additional lower chlorinated byproducts were detected, including all possible penta-chlorinated congeners.

Kinetic Isotope Effect Data

As a mechanistic probe, kinetic isotope effects were explored using methanol-*d* (CH_3OD) with PCB-151. Samples were prepared in an inert atmosphere glove box to prevent hydrogen contamination in the solvent. A plot showing the pseudo 1st order kinetics is shown in Figure 10.

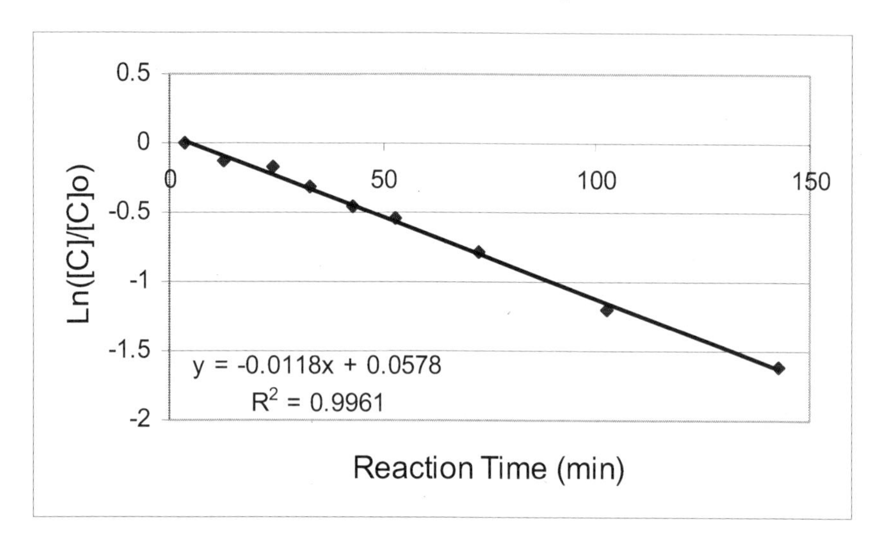

Figure 10. Pseudo 1st order kinetics plot of the degradation of PCB-151 using reactivated Mg/Pd in MeOD. (Adapted from reference 20. Copyright 2008 Elsevier.)

Applying mass normalization of the bimetal to the observed data yields a rate constant of 2.36E-4 L min^{-1} g^{-1}. Using this value and the rate constant from Figure 9 a ratio of k_H/k_D = 2.31 is calculated, indicative of a primary kinetic isotope effect.

Comparison of Possible Mechanisms

Several possible reaction mechanisms have been eliminated. It has been experimentally determined that PCBs are not passing through a benzyne intermediate during degradation. Deuterium labeled studies have shown only a single deuterium being added to the biphenyl structure upon the removal of a chlorine atom, rather than the two deuterium's expected if the degradation pathway did include a benzyne intermediate. Nucleophilic aromatic substitution is also unlikely, due to the fact that the rate constant decreases as the as the degree of chlorination increases. Nucleophilic aromatic substitution requires electron-withdrawing substituents to activate the aromatic system, which means a higher degree of chlorination should increase the rate constant of the reaction (depending on the pattern of the substituents, *ortho/para* and *meta*).

Having a k_H/k_D greater than 2 indicates that the hydrogen bond is being either broken or formed in the rate-limiting step of the reaction (23). It was unlikely that the rate-limiting step involves a hydrogen bond being broken, since it was already removed from the methanol to form molecular hydrogen. It is a more likely possibility that the formation of a new carbon-hydrogen bond was the rate-limiting step, occurring after the contaminant had been adsorbed on the bimetallic system. There are several mechanistic pathways that can account for the observed kinetic data presented in the paper, all of which contain aryl radical intermediates.

One possible mechanism (Mechanism A) is the formation of adsorbed atomic hydrogen (H·) at the surface of the palladium (formed from the dissociation of H_2 on the Pd surface), although this seems unlikely to occur due to the less favorable thermodynamics involved in atomic hydrogen desorption from the surface of the palladium (compared to other hydrogen/palladium species) (24). The chlorine atom would be abstracted in a homolytic bond cleavage, forming HCl and an aryl radical. The intermediate aryl radical would then react with a second H· to terminate the radical process.

A thermodynamically more reasonable approach would be that the hydrogen is absorbed within the first few sublayers of the palladium (Mechanism B). Electron density would be added to the hydrogen via the zero-valent magnesium, increasing the nucleophilicity of this species. Previous research has suggested that a "hydride like" species would then be capable of degrading halogenated species (25). The role of the magnesium is also clearer in this case. Not only does it produce the molecular hydrogen necessary for the reaction, it is also used as a means to help create and store this absorbed hydride-like species. This more activated nucleophilic species could then react with the contaminant and would then proceed in by the same mechanism as the atomic hydrogen.

The first two possible mechanisms discussed are similar to an $S_{RN}1$ type of reaction, in which a nucleophilic substitution occurs involving a radical intermediate. The proposed mechanisms differ from the $S_{RN}1$ type in that the initiation step proceeds via the generation of the aryl radical by the abstraction of the chlorine by a hydrogen radical or hydride-like radical in a homolytic bond cleavage, in which $HCl_{(aq)}$ is formed. Normally, $S_{RN}1$ mechanisms involving halides proceed by the unimolecular decomposition of a radical anion (which is obtained from the substrate) formed by an attacking nucleophile (23), which is seen in the third proposed pathway. In this case, the attacking nucleophile abstracts the leaving group as the aryl radical is formed, rather than imparting a negative charge to the aryl system (which would then be followed by the expulsion of the anion halide, leaving an aryl radical). A second nucleophile (atomic hydrogen or hydride-like radical) is then able to react with the aryl radical in a termination step. Subsequent dechlorination can continue as long as there are available chlorine atoms on the biphenyl ring to initiate the process. Dechlorination will stop once all chlorine atoms have been removed, leaving biphenyl as the final product for the degradation mechanism. Of these two, Mechanism B is more energetically favorable. Hydrogen is more easily able to bind to the subsurface sites than surface sites, the enthalpies of desorption are 32 kJ/mol and 80 kJ/mol, respectively (24).

A third pathway is the formation of hydride (H^-) moieties within bulk palladium, which could then react with the PCB substrate in a $S_{RN}1$ reaction. The hydride acts as a nucleophile which can transfer an electron to the contaminant substrate. This causes the expulsion of a chlorine atom, leaving an aryl radical that can quickly react with another H^-. The charged biphenyl species can transfer an electron to another PCB substrate, so that process can continue to propagate (23).

A balanced chemical equation is given in Figure 11, which holds true for the above possible mechanistic pathways.

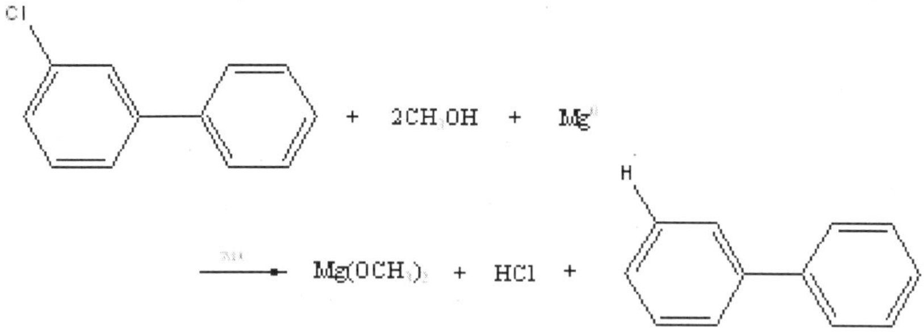

Figure 11. Balanced mechanism for the declorination of PCBs by Mg/Pd in methanol. (Adapted from reference 20. Copyright 2008 Elsevier.)

There are several reasons mechanisms of the proposed type are possible. Palladium catalysts are well known to produce both atomic hydrogen and hydride species when molecular hydrogen is present, as is the case here, so there would be no shortage of the initiating nucleophile. Also, this type of mechanism would explain the lack of dimerization and additional chlorinated byproducts. Radical chlorine is never produced as a reactive species in this reaction scheme; it is bound up with the atomic hydrogen in a covalent bond or as an anion species capable of abstracting a proton from the solvent as soon as it is removed from the biphenyl structure. This helps to explain the lack of chlorinated adducts. Additionally, the aryl radical would be unlikely to come into contact with a second aryl radical, which explains the lack of dimerization. It is unknown at this time if the reactive intermediate remains at the surface of the bimetal during the reaction, but it seems likely when looking at prior experiments involving chlorinated aromatics and Pd/C systems (11). These schemes require adsorption of the PCB onto the surface of the bimetallic compound, then reaction at the interface of the palladium and graphite, limiting the mobility of the aryl radical. This limited mobility and the overabundance of atomic hydrogen, hydride-like radicals, and hydrides on the surface of the catalyst almost certainly allows for the reaction of the aryl radical and second nucleophilic hydrogen, rather than two separate aryl radicals coming into contact. A visual representation these mechanistic schemes are shown in Figure 12 and Figure 13.

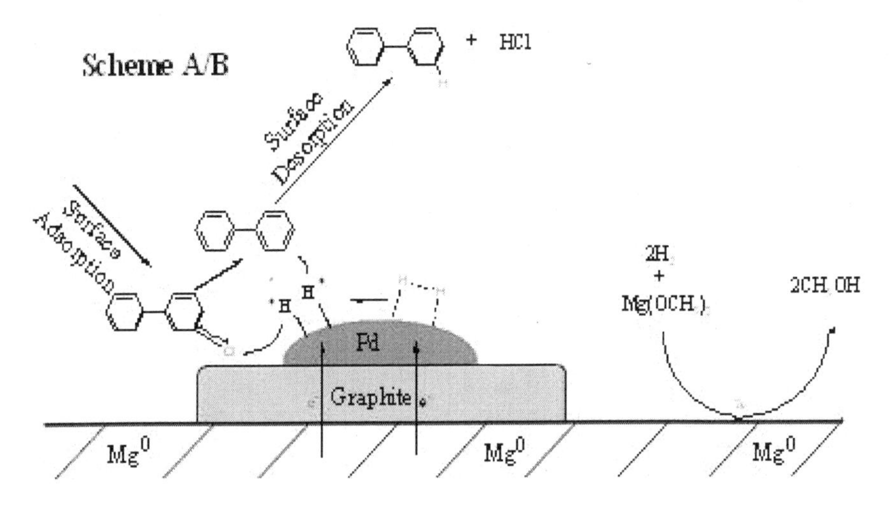

Figure 12. Proposed mechanism for the dechlorination of PCBs by Mg/Pd in methanol by (a) atomic hydrogen and (b) "hydride-like" radicals. (Adapted from reference 20. Copyright 2008 Elsevier.)

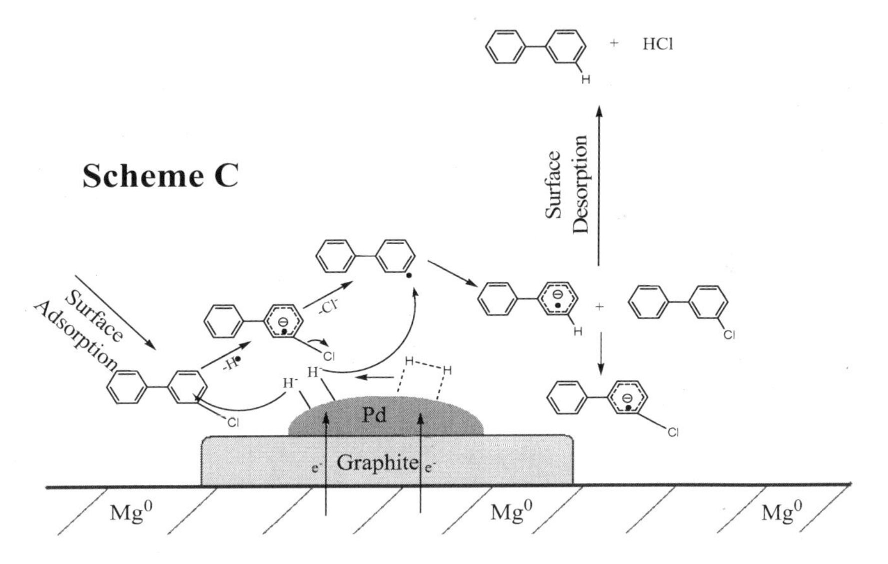

Figure 13. Proposed mechanism for the dechlorination of PCBs by Mg/Pd in methanol by hydride. (Adapted from reference 20. Copyright 2008 Elsevier.)

Conclusion

This work has confirmed mechanically alloyed Mg/Pd to be an effective remediation technique for the degradation of PCBs, and has indicated a distinct solvent specificity to the dechlorination reaction. The relative order of increasing rate constants for the dechlorination of the three mono-chlorinated congeners with palladized magnesium was found to be *ortho>para>meta* with pure methanol solvent systems. However, when the primary solvent is aqueous in nature, the order of increasing rate constants is *para>meta>ortho,* which is in agreement with previous published results on monochlorinated congeners in primarily aqueous systems (21). Additionally, it was found that biphenyl was capable of being degraded in an aqueous system, but was resistant to degradation in pure methanol solvents, another indication of a solvent specific mechanism. The investigation of a single polychlorinated biphenyl congener, PCB-151, in methanol was found to undergo a stepwise dechlorination substantiated by observance of sequential production of degradation byproducts, and exhibits pseudo 1st order kinetics. Based on these experimental results, three possible types of hydrogen species have been proposed as the intermediate reactant responsible for the abstraction of chlorine in the reductive dechlorination process.

Acknowledgements

This research has been supported by a grant from the U.S. Environmental Protection Agency's Science to Achieve Results (STAR) program. Although the research described in the article has been funded in part by the U.S. Environmental Protection Agency's STAR program through grant X832302, it has not been subjected to any EPA review and therefore does not necessarily reflect the views of the Agency, and no official endorsement should be inferred.

References

1. Toxic Information Series (1983) Polychlorinated biphenyls. U.S. EPA Toxics Information Series, United States Environmental Protection Agency, Office of Toxics Substances, TSCA Assistance Office (TS-799) Washington, District of Columbia.
2. Eisenreich S. J., Looney B. B. and Hollod G. J. *In Physical Behavior of PCBs in the Great Lakes*; Mackay, D.; Paterson, S., Eisenreich, S. J.; Simmons, M., Eds. Ann Arbor Science, Ann Arbor, Michigan, 1983, pp. 115-125.
3. Bedard, D. L.; Quensen, J. F., III. *In Microbial Transformation and Degradation of Toxic Organic Chemicals*; Young, L. Y.; Cerniglia, C., Eds.; John Wiley & Sons: New York, 1995; p 127.

4. Quensen, J. F., III; Mousa, M. A.; Boyd, S. A.; Sanderson, J. T.; Fruese, K. L.; Giesy, J. P. Reduction of Aryl Hydrocarbon Receptor-Mediated Activity of Polychlorinated Biphenyl Mixtures due to Anaerobic Microbial Degradation. *Environ. Toxicol. Chem.* 1998, 806.
5. Mousa, M. A.; Quensen, J. F., III; Chou, K.; Boyd, S. A. Microbial Dechlorination Alleviates Inhibitory Effects of PCBs on Mouse Gametes *in Vitro*. *Environ. Sci. Technol.* 1996, *30*, 2087.
6. Bedard, D. L.; Wagner, R. E.; Brennan, M. J.; Haberl, M. L.; Brown, J. F., Jr. Extensive degradation of Aroclors and environmentally transformed polychlorinated biphenyls by Alcaligenes eutrophus H850. *App. Environ. Microbiol.* 1987, *53*, 1094.
7. Wu Wenhai; Xu Jie; Zhao Hongmei; Zhang Qing; Liao Shijian. A practical approach to the degradation of polychlorinated biphenyls in transformer oil. *Chemosphere*. 2005, *60*(7), 944-50.
8. Wiegel, Juergen; Wu, Qingzhong. Microbial reductive dehalogenation of polychlorinated biphenyls. *FEMS Microbiology Ecology*. 2000, *32*, 1-15.
9. Chuang, F.; Larson, R. A.; Wessman, M. S. Zero-Valent Iron-Promoted Dechlorination of Polychlorinated Biphenyls. *Environ. Sci. Technol.* 1995, *29*, 2460-2463.
10. Wang, C.-B.; Zhang, W.-X. Synthesizing Nanoscale Iron Particles for Rapid and Complete Dechlorination of TCE and PCBs. *Environ. Sci. Technol.* 1997, *31*, 2154-2156
11. Cheng, I. F.; Fernando, Q.; Korte, N. Electrochemical Dechlorination of 4-Chlorophenol to Phenol. *Environ. Sci. Technol.* 1997, *31*, 1074-1078.
12. Li, T.; Farrell, J. Reductive Dechlorination of Trichloroethene and Carbon Tetrachloride Using and Palladized-Iron Cathodes. *Environ. Sci. Technol.* 2000, *34*, 173-179.
13. Kovenklioglu, S.; Cao, Z.; Shah, D.; Farrauto, R. J.; Balko, E. N. Direct Catalytic Hydrodechlorination of toxic organics in wastewater. *AIChE* J. 1992, *38*, 1003-1012.
14. Lowry, G. V.; Reinhard, M. Hydrodehalogenation of 1- to 3-Carbon Halogenated Organic Compounds in Water Using a Palladium Catalyst and Hydrogen Gas. *Environ. Sci. Technol.* 1999, *33*, 1905-1910.
15. Muftkian, R.; Fernando, Q.; Korte, N. A method for the rapid dechlorination of low molecular weight chlorinated hydrocarbons in water. *Water Res*. 1995, *29*, 2434-2439.
16. Grittini, C.; Malcomson, M.; Fernando, Q.; Korte, N. Rapid Dechlorination of Polychlorinated Biphenyls on the Surface of a Pd/Fe Bimetallic System. *Environ. Sci. Technol.* 1995, *29*, 2898-2900.
17. Doyle, John G.; Miles, Teri Ann; Parker, Erik; Cheng, I. Francis. Total Polychlorinated Biphenyl by Dechlorination to Biphenyl by Pd/Fe and Pd/Mg Bimetallic Particles. *Microchemical Journal*. 1998, *60*, 290-295.
18. Aitken, Brian; Geiger, C.L.; Clausen, C.A., Milum, K.M. and Brooks, K.. Variables Associated with Mechanical Alloying of Bimetals for PCB Remediation. In *Remediation of Chlorinated and Recalcitrant Compounds*, Proceedings of the 5th International Conference on Remediation of Chlorinated and Recalcitrant Compounds, Monterey, CA, May 22-25, 2006; Curran Associates, Inc: New York,.

19. DeVor, R; Carvalho-Knighton, K; Aitken, B; Maloney, P; Holland, E; Talalaj, L; Fidler, R; Elsheimer, S; Clausen, C; Geiger, C. Dechlorination Comparison of Monosubstituted PCBs with Mg/Pd in different solvent systems. Chemosphere. 2008, 73, 896-900.
20. DeVor, R; Carvalho-Knighton, K; Aitken, B; Maloney, P; Holland, E; Talalaj, L; Fidler, R; Elsheimer, S; Clausen, C; Geiger, C. 2009. Mechanism of the Degradation of Individual PCB Congeners Using Mechanically Alloyed Mg/Pd in Methanol. Chemosphere 2009, In Press.
21. Kim, Young-Hun; Shin, Won Sik; Ko, Seok-Oh. Reductive Dechlorination of Chlorinated Biphenyls by Palladized Zero-Valent Metals. *Environmental Sci. Health.* 2004, A39(5), 1177-1188.
22. Blanksby, Stephen J.; Ellison, G. Barney. Bond Dissociation Energies of Organic Molecules. *Acc. Chem. Res.* 2003, *36*, 255-263.
23. Carey, Francis A.; Sundberg, Richard J. Advanced Organic Chemistry Part A: Structure and Mechanisms. Kluwer Academic / Plenum Publishers: New York, 2000.
24. Cybulski, Andrzej; Moulijn, Jacob A. Structured Catalysts and Reactors. Taylor and Francis: Boca Raton, 2006, p 585.
25. Cwiertny, David M.; Bransfield, Stephen J.; Roberts, A. Lynn. Influence of the Oxidizing Species on the Reactivity of Iron-Based Bimetallic Reductants. *Environ Sci. Technol.* 2007, *41*, 3734-3740

Chapter 5

PBDE Degradation with Zero-valent Bimetallic Systems

Kathleen Carvalho-Knighton[1], Lukasz Talalaj[1] and Robert DeVor[2]

[1]Environmental Science, Policy, and Geography, University of South Florida St. Petersburg, St. Petersburg, FL 33701
[2]Department of Chemistry, University of Central Florida, Orlando, FL 32816

Polybrominated diphenyl ethers (PBDEs) are a group of widely used brominated flame retardants. Due to their extensive use, increasing levels of PBDEs have been found in humans, fish, birds, marine mammals, sediments, house dust, air, and supermarket foods. As a new environmental pollutant, a feasible in-situ remediation method is needed. In situ remediation methods for PCBs have been developed at UCF using palladium/magnesium bimetal created by mechanical alloying. The lessons learned from the PCB work have been applied to PBDEs in this chapter. Several bimetallic systems were examined to determine the rate of debromination of 2,2',4,4'- tetrabromodiphenyl ether (BDE-047). In addition, kinetic studies on BDE-047 were conducted with 99% degradation in five hours with 0.8% Mg/Pd. During the first 30 minutes, 80% of the BDE-047 is degraded with diphenyl ether detected as one of the byproducts.

Polybrominated diphenyl ethers (PBDEs) are a family of chemical compounds that share diphenyl ether as their back bone. Situated around the rings are bromine molecules that attach at 10 different sites. As a result, PBDEs can exist in 209 different variances called congeners and have adapted the same naming system as PCBs due to their structural similarities.

PBDEs thermal resistance has allowed them to be utilized as a flame retardant worldwide for the past 30 years. PBDEs contain the ability to slow down the ignition and allow for more escape time in the case of fires. During high heat scenarios and ignitions, PBDE starts to breakdown 50 degrees Celsius prior to its associated polymer (1). The flame retardant's decomposition allows for the bromines to take out the high energy radicals that are formed as a result of combustion (1). They are introduced into numerous consumer goods ranging from furniture to electronics. Out of 175 different flame retardants, PBDEs dominate the market because of their low cost and high performance efficiency (2).

Brominated diphenyl ethers (BDEs) usage is categorized into three commercial standards: penta-BDE, octa-BDE, and deca-BDE. Pent-BDE commercial standard mainly consist of BDE- 47 and BDE-99, are introduced into flexible polyurethane foam such as furniture and make up 10% of the products weight. Octa-BDE is primarily utilized in high impact plastic product, such as casings for computers, kitchen appliances, fax machines, phones, and car trimmings (3). Deca-BDE mainly consisting of BDE-209 and nonabrominated diphenyl ethers is used in plastics that act as adhesives and coatings for products that include television casings, wire coating, and other electronics (3). The world demand in 2001 for the three commercial PBDEs exceeded 67,390 tonnes, 49% of it being utilized in the United States as shown in Table I (4).

Table I: PBDE compositions and 2001 demand (Adapted from ref. 8).

Commercial PBDE Product	*% Composition of Commercial Mixtures*	*2001 Demand in American (Metric Tons)*	*Percentage World Demand in Americas*
PentaBDE (BDE-71)	TetraBDEs 24-38 PentaBDEs 50-60 HexaBDEs 4-8	7100	95
OctaBDE (BDE-79)	HexaBDEs 10-12 HeptaBDEs 44 OctaBDEs 31-35 NonaBDEs 10-11 DecaBDEs <1	1500	40
DecaBDE (BDE-83r) Saytex 102E	NonaBDEs <3 DecaBDEs 97-98	24500	44

Since its production in the 1970s increased, information on these flame retardants has indicated some disturbing data. PBDEs have been found to persist and accumulate in the environment. PBDE concentrations in North American people were found to be 17 times higher than that of European people (5). Numerous scientists made connections with exposure to these flame retardants with various health issues (6).

The exact pathway to which the environment is exposed is not completely known, however numerous pathways are proposed. Consumer products introduce the flame retardant either as a reactive or additive chemical. Reactive flame retardants are covalently bound to the polymer, and the additive flame retardants are essentially dissolved in the initial formulation (7). The additive flame retardant tends to be volatile and ultimately leaches off the product and enters its surrounding environment. This outcome can occur during the aging and wear of a consumer good, during production and formulation of the flame retardant and consumer good, and leeching can also occur during its waste disposal or recycling stage. Human exposure is primarily looked at through three different exposure pathways: dietary, direct, and air. Having a lipophilic property, PBDEs tend to accumulate in the fatty tissues of all wild life, resulting in exposure to humans through food consumption, fish being the primary source. Direct exposure stems from directly interacting with consumer goods that contain this flame retardant such as automobiles and furniture. Exposure through air and dust of indoor environments whether it be at home or a work place also contribute to the accumulation.

Since its initial production in 1970 increasing information on these flamer retardants has drawn a lot of attention from various organizations as well as environmental groups resulting in policymaking at international, national, state and local level . In Europe, production of pentaBDE and octaBDE mixtures has been phased out and in certain countries all PBDEs have been discontinued. In 2004, penta and octa mixtures were voluntarily eliminated out of production in United States, and EPA has imposed a Significant New User Rule (SNUR) (8). Chemtura Corporation, the only company to synthesize PBDEs still produces decaBDEs. Although lower brominated mixtures have been phased out, the bioaccumulation of these toxins in the environment have been dominated by three main congeners; BDE-47, BDE-99, BDE-100. Studies suggest that through photolytic and microbial degradation in the environment the decaBDE is able to break down to lower congeners that are more stable and have a longer half-life (9). Therefore, the use of only decaBDE does not remedy the situation entirely.

Numerous remediation technologies have been explored but fall short due to limited degradation, long remediation times, and slow debromination. A recent study by Robrock explored three different biological cultures (ANAS195, *D.hafniense, D.restrictus*) for debromination in which it showed similar pathways on all three and had strong similarities to degradation of PCBs (10). However, in a three month study, the relative degradation was fairly slow and exhibited smaller scale degradation in comparison to PCBs, micro molar conversion of PCBs to nano molar conversion of PBDEs. The study found the anaerobic treatment to be slower for higher brominated congeners and faster for lower brominated congeners, as a result of increasing hydrophobicity with higher congeners. A majority of the degradation was impartial resulting in formation of congeners that are higher in toxicity.

Photolytic degradation is another technique explored for PBDEs. Photolytic degradation plays a large role in the flame retardants ability to break down in the environment into lower congeners, more so than anaerobic degradation. Numerous studies have proven that the degradation rate is based on solvent system, matrix ratios, and bromination degree. The path goes through debromination, having a higher rate with higher brominated congeners over the lower ones, and being faster in organic solvents over slower rates in water (11). Whether in a solvent system or through solid phase micro extraction, UV exposure produces highly toxic dibenzofurans (12). Incomplete debromination is another drawback, as it produces congeners with higher toxicity than the parent toxins.

Recent interest in dehalogenation of waste by the use of zero valent metals (ZVMs) has brought much focus upon halogenated hydrocarbons. ZVM have been proven to be effective against trichloroethylene (TCE) and perchloroethylene (PCE). A 2005 study by Keum and Li also concluded promising data in debromination of PBDEs by the use of zero valent iron. In a 40 day study 90% BDE-209 was converted in lower substituted congeners dominated mainly by triBDEs and tetraBDEs (13). The use of zero valent metals to treat PBDEs is new and still requires numerous studies to explore its feasibility, however it has proven effective with PCBs (14) and ultimately is projected onto PBDEs due to their chemical and structural similarities. As a result of its corrosive nature, extensive pre-treatment and anaerobic storage, iron is not an ideal metal for this type of remediation. Magnesium much like iron can remediate but does not require such extensive treatment, it also exhibits a higher oxidation potential with 2.372V in comparison to Fe 0.44V therefore having a greater electropotential driving force.

$$Mg^{2+} + 2e^{-}\ Mg^{0} \qquad E^{0} = -2.37V$$
$$Fe^{2+} + 2e^{-}\ Fe^{0} \qquad E^{0} = -0.44V$$

Zero-valent magnesium and iron form hydrogen gas by reacting with a protic solvent, as seen below.

$$2\ Mg^{0} \rightarrow 2\ Mg^{2+} + 4e^{-}$$
$$2\ ROH \rightarrow 2H^{+} + 2R^{-}$$
$$2H^{+} + 2e^{-} \rightarrow H_2\,(g)$$

As seen previously with zero-valent iron treatment of halogenated hydrocarbons, the hydrogen gas goes on to replace the halogen of the toxin (14).

Experimental

Chemicals and Reagents

2,2',4,4'- tetrabromodiphenyl ether (BDE-47, CAS No:5436-43-1)) were purchased from Accustandard, Inc. Palladium coated magnesium, 0.08% based on weight, ball milled March 13th of 2007 were obtained from UCF industrial chemistry labs. Methanol and toluene solvents utilized in the experiment were all obtained from Fischer scientific, Inc, and were all optima grade.

Palladium Coated Magnesium

Magnesium (~ 4 μm) was obtained from Hart Metals, Inc (Tamaqua, PA). 1% palladium on graphite was obtained from Engelhard (Iselin, NJ), while 10% palladium on graphite was obtained from Acros Organics. A ~0.08 wt%

palladium-magnesium mixture was prepared by ball-milling 78g Mg with 7g of 1% palladium on graphite in a stainless steel canister (inner dimensions 5.5 cm by 17 cm) with 16 steel ball bearings (1.5 cm diameter, at a total mass of 261.15g). The material was milled for 30 minutes using a Red Devil 5400 series paint mixer. A 0.8% palladium-magnesium mixture was prepared in a similar fashion using 10% palladium on graphite.

Treatment of BDE-47 by Zerovalent Mg/Pd

Five mL of 20 ppm BDE-047 in methanol (20mg/L) were placed in a septum sealed (PTFE lined) vial containing 0.25 grams of 0.08% Mg/Pd. During the reaction period the vials were placed on a shaker table (Cole Parmer 51704 Series) until appropriate extraction time. Control vials were set up without Mg/Pd, and the experiment was set up in duplicates.

Extraction and Analysis

Extraction was performed at 0, 0.25, 0.50, 1, 1.5, 2, 2.5, 3, 3.5, 4, 5 hours of reaction with the palladized metal. The extraction process included placing 5 ml of Toluene (Optima) in the vials and shaking for approximately 2 minutes followed by extraction of 4 mL through a syringe filter (Puradisc 25mm with 0.45 μm pore size). To the extracted 4 ml of 50/50 methanol/toluene 2 ml of deionized water was applied and again shaken for a minute. Afterwards the vial was placed in a centrifuge for two minutes, at which point the top layer was pulled off using glass pipettes and placed in an autosampler vial for analysis. All samples were analyzed by Shimadzu GC-17A/MS-QP5000 equipped with DB-XLB column (30m length, 0.25mm I.D., 0.25μm Film) and an autosampler. Mobil phase consisted of helium with a column flow of 1.3 mL/min, injection port temperature was set at 260°C with a mass spectrometer interface of 225°C. Oven temperature ramp was held for 2 minutes at 80°C, followed by a 30°C/min ramp to 200°C, and finally a 5°C/min ramp to 320°C.

An extended study was set up by following the same parameters but extending reaction time to 24 hours. For kinetic analysis, a secondary study was set up with similar parameters as well but was slowed down by reducing the amount of 0.8% Mg/Pd to 0.10g.

Results

Several different bimetallic systems were explored in a preliminary study in order to determine the ideal system for further extensive study of 2,2',4,4'-tetrabromodiphenyl ether (BDE-047). This congener is one of the most persistent PBDE congeners in the environment,

2,2',4,4'- tetrabromodiphenyl ether
BDE-047

The bimetallic systems utilized were 1%, 10% and 30% Mg/Pd, Fe/Ni, Fe/Pd, Mg, and acid washed μFe. Three hour studies were set up for each system to determine degradation percent of the parent compound as shown in Figure 1.

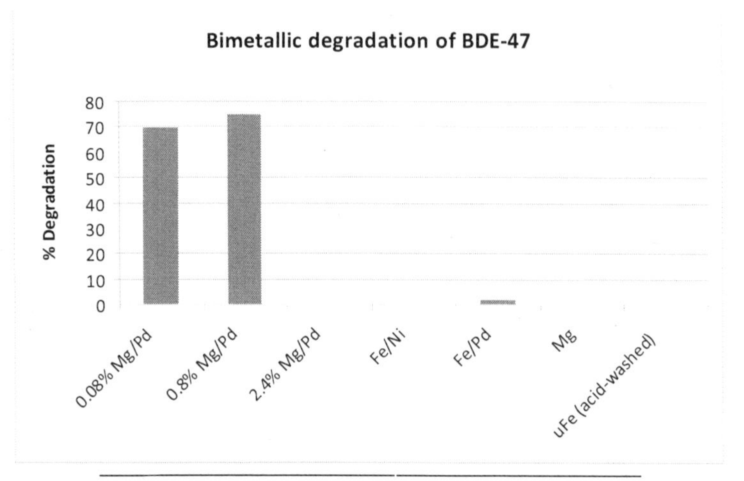

Bimetallic System	% Degradation
0.08% Mg/Pd	69.39
0.8% Mg/Pd	75.01
2.4% Mg/Pd	N/A
Fe/Ni	<1
Fe/Pd	1.94
Mg	<1
microFe (acid-washed)	<1

Figure 1: BDE-47 degradation by various systems in a three hour study.

In the initial studies, 0.8% Mg/Pd proved to be the most effective with 75.01% degradation of 20ppm BDE-047 in methanol in three hours. 2.4% Mg/Pd showed the highest potential in terms of degradation rates, however due to its reactive and vigorous nature it was difficult to work with due to high heats and large pressure build ups and was omitted.

A five hour study of 20ppm BDE-047 in methanol reacting with 0.10g of 0.8% Mg/Pd proved to be effective in degrading congener 47. The results showed 80% degradation within the first 30 minutes followed by 99% within the remaining 4.5 hours as shown in Figure 2.

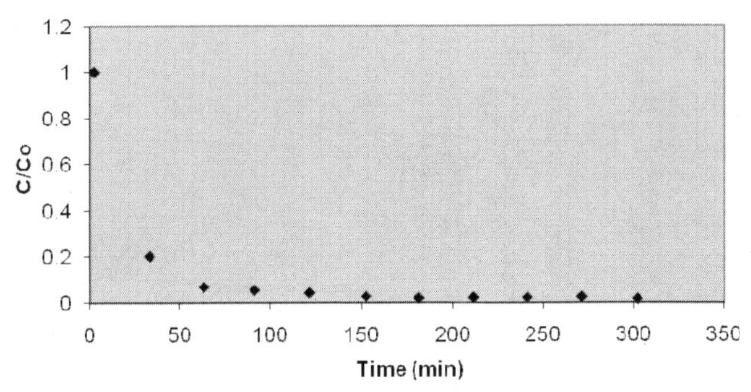

Figure 2: Degradation of BDE-47 with Mg/Pd in Methanol with 0.8% Pd

The reaction proved to be a step-wise debromination, the parent BDE-47 debrominates down into tri substituted congeners, then a di- and mono-, and finally into bare diphenyl ether.

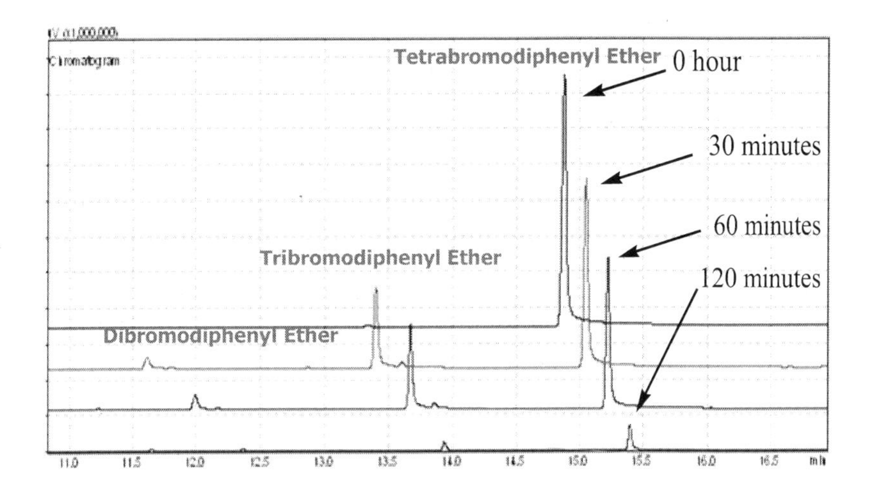

Figure 3: GC-MS chromatogram of BDE-47 degradation

The exact byproduct congeners were not identified, but have been limited down based on their parent congeners bromine orientation, 2,4,4'-tribromodiphenyl ether (BDE-028) or 2,2',4-tribromodiphenyl ethers (BDE-017) for the tri substituted congeners, 2,4-dibromodiphenyl ethers (BDE-007), 2,4'-dibromodiphenyl ether (BDE-008), 4,4'-dibromodiphenyl ether (BDE-015) for di substituted congeners and finally to either 2-monobromodiphenyl ether (BDE-001) or 4-monobromodiphenyl ether (BDE-003) for the mono substituted congeners as shown in Figures 3 and 4. Identification of congeners will be of focus in future studies in hopes of determining stereo selectivity of ortho, para, and meta oriented bromines, and determining major and minor byproducts.

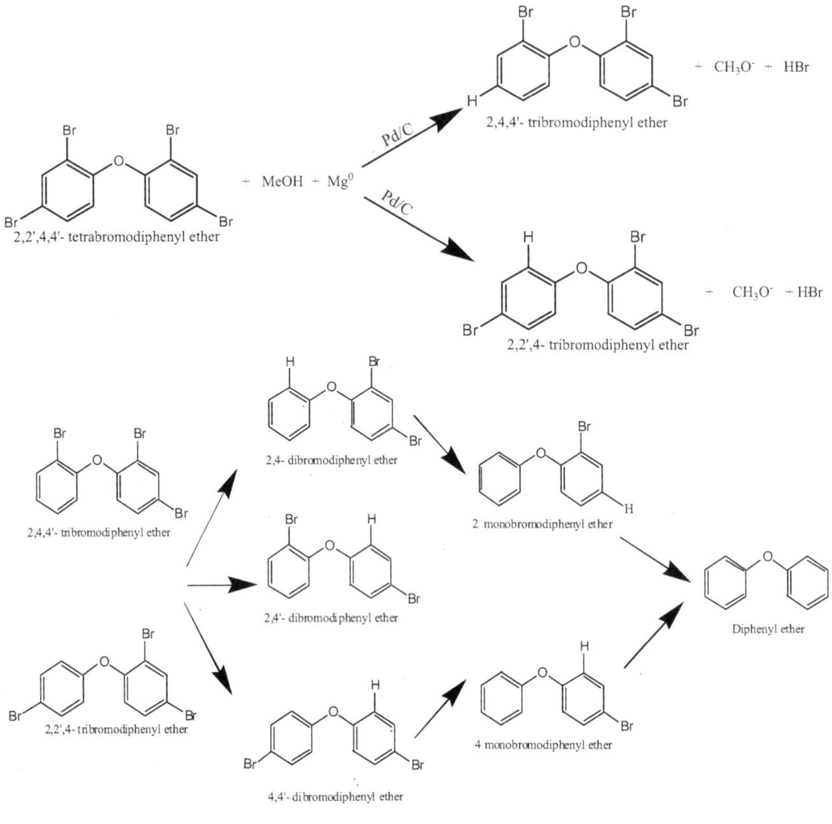

2,2',4,4'- tetrabromodiphenyl ether + MeOH + Mg^0 →(Pd/C) 2,4,4'- tribromodiphenyl ether + CH_3O^- + HBr

2,2',4,4'- tetrabromodiphenyl ether + MeOH + Mg^0 →(Pd/C) 2,2',4- tribromodiphenyl ether + CH_3O^- + HBr

2,4,4'- tribromodiphenyl ether
2,2',4- tribromodiphenyl ether
2,4- dibromodiphenyl ether
2,4'- dibromodiphenyl ether
4,4'- dibromodiphenyl ether
2 monobromodiphenyl ether
4 monobromodiphenyl ether
Diphenyl ether

Figure 4: Stepwise debromination of BDE-47.

The hydrogen gas produced from the zero-valent magnesium reacts with the toxin and substitutes the bromines through an unknown mechanism, this can be seen below (14).

$$M^0 + 2CH_3OH \rightarrow M^{2+} + H_2 + 2CH_3O^-$$

$$2RBr + H_2 \xrightarrow{Pd} 2RH + Br_2$$

$$H_2 + 2Br_2 \rightarrow 2HBr$$

After 45 minutes the degradation rate began to slowdown. This can be attributed to competition for active sites by two factors. As previously seen in degradation of PCBs with Mg/Pd systems (14), the parent toxin competes for active sites on the metal, once the active sites have been saturated the reaction is halted until the degraded parent toxins departs the active site. The newly formed byproducts then are also competing and are favored on the reaction site, the graph below supports this theory. The initial byproduct, either 2, 4, 4'-tribromodiphenyl ether (BDE-028) or 2, 2', 4-tribromodiphenyl ether (BDE-017), concentration maximizes around 45 minutes of reaction time and steadily starts degrading by the Mg/Pd system. This result directly correlates with the slowdown of the parent congener 2, 2', 4, 4'- tetrabromodiphenyl ether (BDE-047) around 45 minutes, indicating that lower substituted congeners may already be formed on the active sites of the metal and do not leave until diphenyl ether is formed..

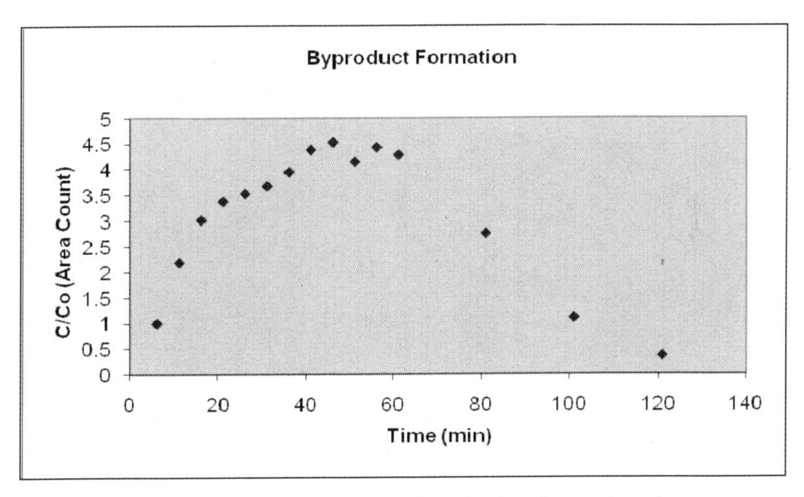

Figure 5: Plot of tribrominated diphenyl ether byproduct formation.

An additional experiment was set up with altered parameters to focus on the kinetics. The 0.8% Mg/Pd concentration was lowered from 0.10g to 0.05g and the first 60 minutes where the focus of the reaction. The first order, second order and zero order kinetic plots are shown in Figure 6.

The reaction follows 1st order degradation with an R^2 value of 0.9485 however it is followed by very close 2nd order degradation with an R^2 value of 0.9483. With such close results further studies are needed to confirm these finding.

Unlike previous studies published that indicate impartial debromination resulting in more hazardous byproducts, the use of Mg/Pd proved to undergo complete debromination to diphenyl ether within 30 minutes as shown in Figure 7. This is promising as previous remediation techniques like microbial, photolytic, and zero-valent Fe were not able to achieve complete debromination.

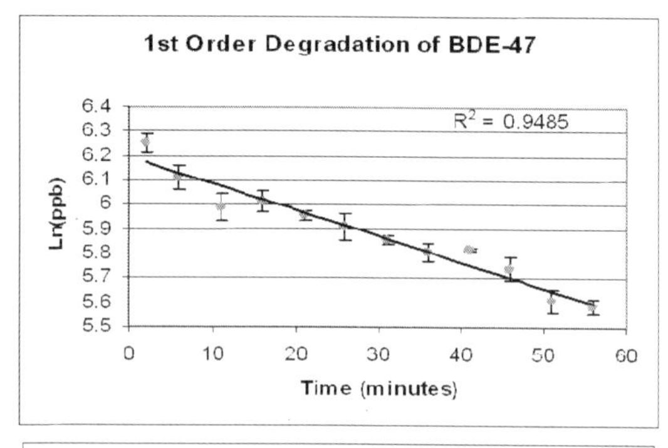

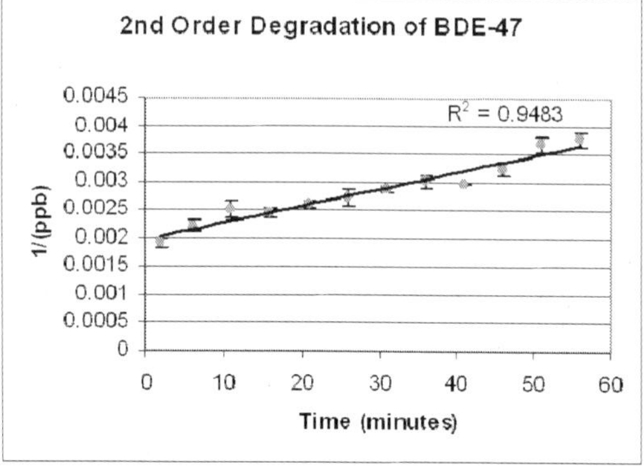

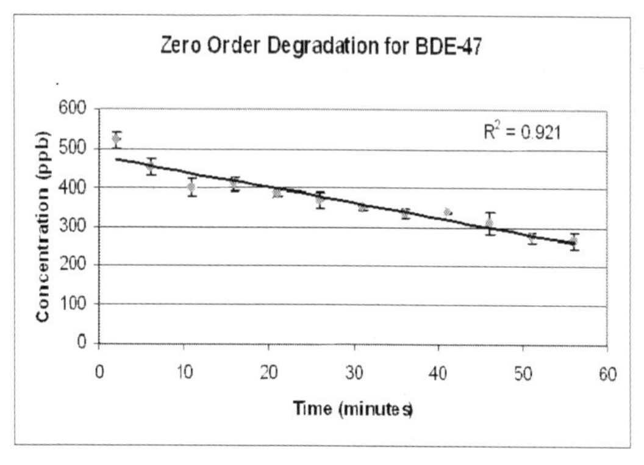

Figure 6: Degradation kinetic comparison of BDE-47.

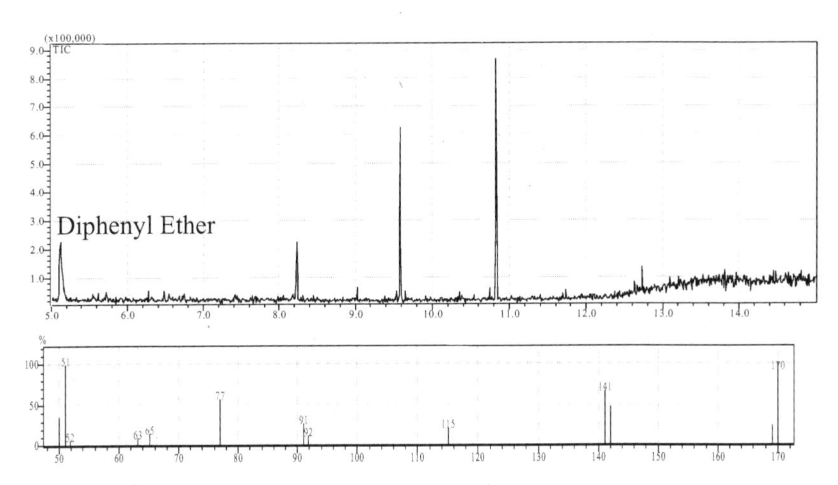

Figure 7: Chromatogram and MS fragmentation for Diphenyl Ether.

One final parameter examined was the production of dibenzofuran. Dibenzofuran is a volatile hazardous air pollutant that can be found as a byproduct of smoking. It is believed that this toxin is not necessarily a byproduct of BDE-47 but rather a product of thermal degradation on GC-MS. Due to the nature of the instrumentation, it is crucial to inject samples at high temperatures, however with such sensitive compounds exceeding threshold temperatures, a breakdown of the byproducts into dibenzofuran is inevitable. A significant amount of dibenzofuran was also detected as shown in Figure 8.

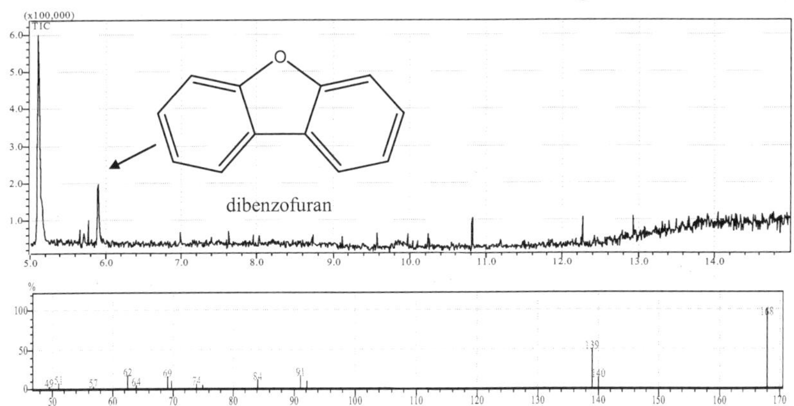

Figure 8: Chromatogram and MS fragmentation of Dibenzofuran

Extensive modification of the instruments methods minimized the formation of dibenzofuran but did not eliminate it, further analysis using cold-on-column injection will be needed to truly determine the source of dibenzofuran.

An additional extended study was set up to observe how dibenzofuran and diphenyl ether faired up against the Mg/Pd bimetallic system. The five hour study was extended to 24 hours, the concentration of both compounds reached maximum concentrations at 5 hours and was designated as initial to obtain percent degradation. Table II and Figure 9 show the ratio change resulted in 75.63% degradation for dibenzofuran and 55.60% degradation for diphenyl ether. This illustrated that Mg/Pd bimetallic systems are capable of degrading dibenzofuran effectively.

Table II: Percent degradation of Dibenzofuran and Diphenyl Ether

	C/C_0 *5 hours*	C/C_0 *24 hours*	*% Degradation*
Dibenzofuran	1.0	0.24	75.6
Diphenyl Ether	1.0	0.44	55.6

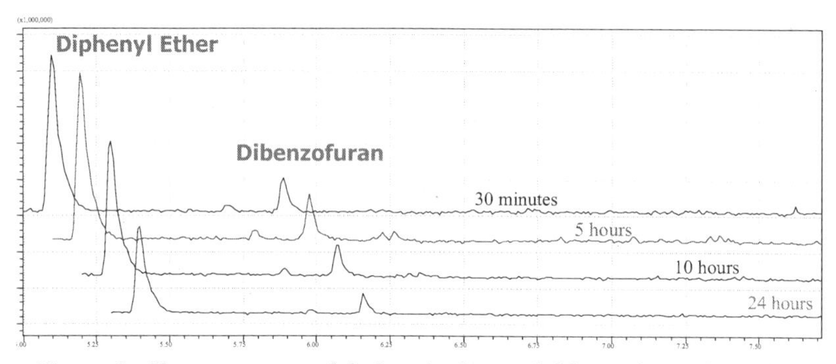

Figure 9: Chromatogram of diphenyl ether and dibenzofuran degradation.

Conclusion

Treatment of one of the most persistent BDE congeners, BDE-47, by Mg/Pd bimetallic systems proved to be an effective way of remediation. In a five hour study 99% of BDE-47 was debrominated, with 80% of remediation taking place within the first 30 minutes. The slowdown is attributed to competition for active sites on magnesium by the parent toxin (BDE-47) as well as newly formed byproducts. The remediation reaction followed a step-wise debromination path with pseudo 1st order kinetics. Unlike other studies utilizing zero-valent metals, Mg/Pd also proved to be capable of obtaining complete debromination to diphenyl ether. Dibenzofuran formation was detected however it is believed to be produced by thermal degradation on the GC-MS. Extended studies indicated that Mg/Pd bimetallic systems can remediate dibenzofuran if it were to be present.

In improving Mg/Pd remediation technologies, future research will focus on several factors that will help optimize conditions for enhanced liquid membrane remediation and possible future applications. The established congener library on the GC-MS will be utilized to determine major and minor byproducts. In addition, studies on BDE-1, BDE-2, and BDE-3 will be conducted to determine stereo selectivity of ortho, meta, and para positioned bromines. Finally, select congeners in different solvent systems will be explored to determine the exact path and mode of debromination.

References

1. Rahman F. *Sci. Total Environ.* **2001**, *275*, 1-17.
2. Birnbaum, S.L; Staskal, F.D. *Environ Health Perspect.* **2004**, *112*(1), 9-17.
3. Turnbull, Aidon. Environ. http://www.b2bweee.com (accessed September 1, 2008). Industry Use of Brominated Flame Retardants in Electronic Equipment.
4. Bromine Science and Environmental Forum. www.bsef.com (accessed September 1, 2008). Major Brominated Flame Retardants Volume Estimates: Total Market Demand By Region in 2001. 21 January **2003**.
5. Hites, R. A. *Environ Sci Technol.* **2004**, *38*, 945-956.
6. Alee, M; Wenning, R. J. *Chemosphere.* **2002**, *46*, 579-582.
7. Alaeea, M.; Ariasb, P.; Sjödin, A.; Bergman, A. Environ. Int. **2003**, *29*, 683-689.
8. Environmental Protection Agency. www.epa.gov (accessed October 1,2006). US EPA PBDE Project Plan, March 2006.
9. Eriksson, J.; Green, N.; Marsh, G. *Environ Sci Technol.* **2004**, *38*, 3119-3125.
10. Robrock, K.; Korytar, P.; Alvarez-Cohen, L. *Environ Sci Technol.* **2008**, *42*, 2845-2852.
11. Mas, S.; De Juan, A.; Lacorte, S.; Tauler, R. *Anal Chim Acta.* **2008**, *618*, 18-28.
12. Sanchez-Prado, L.; Llompart, M.; Lores, M.; Garcia-Jares, C.; Cela R. *J. Chromatogr.* **2005**, *1071*, 85-92.
13. Keum, Young-Soo; Li X. Qing. *Environ Sci Technol.* **2005**, *39*, 2280-2286.
14. Devor, R.; Carvalho-Knighton, K.; Aitken, B.; Mloney, P.; Holland, E.; Talalaj, L.; Elsheimer, S.; Clausen, A. C.; Geiger, L. C. *Chemosphere.* **2008**, *73*(6), 896-900.

Chapter 6

Rapid Dechlorination of Polychlorinated Dibenzo-p-dioxins by Nanosized and Bimetallic Zerovalent Iron

Effect of Palladization and Toxicity Change

Yoon-Seok Chang

[1]School of Environmental Science and Engineering, POSTECH, San 31, Hoyojadong, Namgu, Pohang, Korea 790-784
E-Mail: yschang@postech.ac.kr

Polychlorinated dibenzo-p-dioxins (PCDDs) and polychlorinated dibenzofurans (PCDFs) are the terms, collectively and commonly known as dioxins. Dioxins are highly toxic, and persistent compounds therefore, remain as significant environmental pollutants. Dioxins can be generated from different sources. Their detrimental health and environmental effects, low aqueous solubility, relatively high stability, and chlorinated nature contribute to their persistence and resistance to degradation. Hence, removal of dioxins is a global priority. Zero-valent metals are known to carry out catalytic reductive dehalogenation and transformations of many organic pollutants. Among the zero-valent metals, zero valent iron (ZVI) is a preferred choice for pollutant degradation as it is low in cost and environmentally benign in nature. Use of zero-valent iron in nano form for the treatment of such difficult-to-degrade compounds has proven to be very beneficial. Nanomaterials have unique properties which make them highly potent catalysts. In the present study, both microsized and nanosized ZVI were used for dechlorination of dioxins. ZVI (both micro- and nanosized nZVI, without palladization) was able to dechlorinate PCDD congeners with four chlorines in an aqueous system. Although dechlorination of dioxins is thermodynamically feasible, the focus of the work presented here is on the kinetics of dechlorination in the absence of any previous reports. It was observed that unamended nZVI could dechlorinate PCDD congeners, however the rate of reaction proceeded slowly and complete dechlorination was not achieved within practically acceptable time. In contrast, palladized nanosized zero-valent iron (Pd/nZVI) rapidly dechlorinates PCDDs, including the mono- to tetrachlorinated congeners. The rate of 1,2,3,4-tetrachlorodibenzo-*p*-dioxin (1,2,3,4-TeCDD) degradation using Pd/nFe was about 3 orders of magnitude faster than with only nZVI.The distribution of products obtained from dechlorination of 1,2,3,4-TeCDD suggests that palladization of nZVI particles shifts the pathways of contaminant degradation toward a greater role of hydrogen atom transfer than electron transfer. The decision between choosing nZVI or Pd/nZVI for treatment of PCDD/Fs (Polychlorinated dibenzodioxins or polychlorinated dibenzofurans) is still ambiguous because there are no reports for the toxic equivalent quantity (TEQ) changes during dechlorination. We have also observed that dechlorination of higher chlorinated congeners such as octachlorodibenzo-*p*-dioxin (OCDD) increased the overall TEQ by 3 fold with the formation of lower chlorinated compounds that had higher TEF. Predictions made on the basis of the modeling studies show that it might take more than 100 years for complete and efficient removal of PCDDs from the environment. Therefore, the choice of using nZVI or Pd/nZVI for dechlorination should not only be decided by the rate of dechlorination but also by the toxicity of the break-down products.

Introduction

Polychlorinated dibenzo-p-dioxins (PCDDs) and dibenzofurans (PCDFs) belong to the class of tricyclic, planar, aromatic compounds containing one to eight chlorine atoms (Figures 1 and 2). PCDDs contain seventy-five congeners and PCDFs contain one hundred and thrity five. Commonly they are known as dioxins. PCDD/Fs are introduced into the biosphere as a result of incineration processes *(1,2)*, as by-products of chemical reactions such as chlorine bleaching of paper and pulp, and during the manufacture of some chlorinated pesticides, herbicides, fungicides, wood preservatives, and textile dyes (*3,4*), and as a result of some natural processe, such as volcanic activity and forest fires (*5,6*).

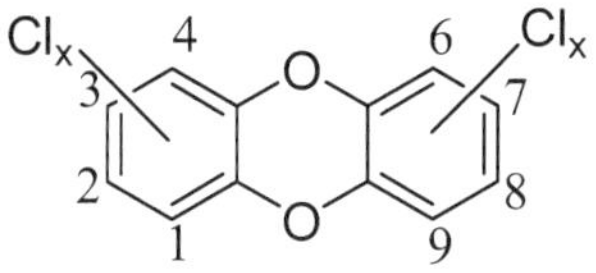

Figure 1. Polychlorinated dibenzo- p –dioxins (PCDDs).

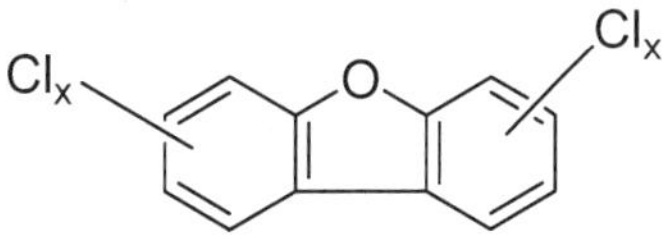

Figure 2. Polychlorinated dibenzofurans (PCDFs).

In the past decade, PCDD/Fs have created huge environmental problems because of their chemical inertness, recalcitrance and resistance to degradation, high lipid solubility, and toxicity (*7-9*). The primary human health problems associated with PCDD/Fs are immunotoxicity, developmental toxicity, neurotoxicity, carcinogenicity of skin and liver, and gastrointestinal toxicity (*10-12*). Specially, congeners substituted with lateral chlorine atoms, such as 2,3,7,8-tetrachlorodibenzo-p-dioxin (2,3,7,8-TeCDD) are highly toxic to mammals and other organisms (*13-15*). Large amounts of highly toxic 2,3,7,8-TeCDD and other dioxins have also entered into the environment as a result of wars and accidents (*16-18*).

Research following these disasters indicated that dioxins were not only extremely toxic, but could also be generated by incineration of organic compounds in the presence of chlorine. Use of agrochemicals has also caused widespread pollution of agricultural and terrestrial environments with dioxins (*19*). Additionally, the observed natural formation of PCDD/Fs in soil sediments is most likely due to polymerization of chlorophenols into dioxins in the presence of fungal peroxidases (*3,20*). Even if the concentration of PCDD/Fs in the hydrosphere is lower than in other matrices due to their extremely low water solubility (10-4 ~ 10-8 ng/ (*21-22*), some industrial wastewater or leachates from landfill sites contain higher concentrations of PCDD/Fs than estimated (*23-24*). This is because the hydrosphere contains natural organic matters (NOM) and particulate matters (PM). NOM, especially humic and fulvic acids, act as surfactants to enhance water solubility of organic pollutants (*25,26),* and PM suspended in wastewater are one of the well-known sorption sites for hydrophobic compounds (*27*).

Clean up of sites and wastewaters, which are contaminated with such pollutants, are difficult due to their toxicity and long half-lives. Since PCDD/Fs are resistant to thermal, chemical, and biological degradations, they accumulate in natural environments (*7,8*). Constant efforts have been made towards the removal and remediation of such recalcitrant and persistent compounds. Different approaches are formulated for the elimination of dioxins and its congeners. These include several physico-chemical methods, but their successful applications are still uneconomical.

Another approach is the biotransformation of dioxins using specific microorganisms to less harmful, non-hazardous substances, which are then integrated into natural biogeochemical cycles . The dechlorination of PCDD/Fs has been reported by pure (*28)* and mixed (*29)* cultures of anaerobic bacteria, anaerobic cultures with added electron transfer mediators (*30,31*), and in abiotic systems using ZVI in subcritical water (*32)*. However, successful bioremediation techniques for field-scale application are still in their infancy. Since the available degradation methods do not provide a complete viable solution, use of zerovalent nanomaterials could be a technology of choice.

Reductive dehalogenation using zerovalent iron (ZVI) has been extensively studied for the remediation of halogenated aliphatic contaminants including carbon tetrachloride, 1,1,1-trichoroethane, and trichloroethylene (*33-35*). In contrast, few studies have examined the dehalogenation of halogenated aromatics by ZVI, and they involved only a limited range of chlorinated phenols (*36-38*), polybrominated diphenylethers (*39*), and polychlorinated biphenyls (*40*). The only report published on the dechlorination of polychlorinated Dibenzo-*p*- dioxins (PCDDs) using ZVI is a preliminary study was performed with PCDD/Fs-contaminated fly ash by Chang *et.al.* (*41*).

The toxicity of PCDD/F congeners varies greatly, with 2,3,7,8-TeCDD being the most toxic, so incomplete dechlorination starting with the more highly chlorinated PCDD/Fs can increase the net toxicity of mixtures. To avoid this, effective remediation of PCDD/Fs requires not only fast dechlorination of the parent compounds but also rapid conversion of the parent compounds in nontoxic products. With a variety of environmentally relevant reductants, dechlorination by dissociative electron transfer is thermodynamically favorable for all chlorines on all PCDD/Fs congeners (*42,43*). However, the kinetics of PCDD dechlorination are expected to be slow, based on analogy to other chlorinated aromatics (*44,45*) and on available data on reductive dechlorination of PCDDs.

In the study described here, we sought to determine the kinetics and pathways of PCDD degradation using microsized iron (mZVI) and nanosized iron particles (nZVI), with (Pd/mZVI and Pd/nZVI) and without palladization and the environmental acceptability of the dechlorinated products. Rate kinetics and pathways obtained were compared, and mechanisms for PCDD dechlorination by ZVI and palladized ZVI were proposed.

Although PCDD/Fs can be dechlorinated via electron transfer by nZVI and atomic hydrogen substitution by Pd/nZVI, overall toxicity of PCDD/Fs can increase by incomplete dechlorination because 2,3,7,8-substituted congeners have greater toxicities, known as toxic equivalent factors (TEF). Our studies have revealed that electron transfer by unamended ZVI led to toxicologically feasible dechlorination pathways because the lateral position chlorines (2,3,7,8 positions) were preferentially removed rather than the chlorines at the peri-positions (1,4,6,9 positions). Although the atomic hydrogen transfer by bimetallic iron (Pd/nZVI) preferred the toxicologically unfeasible pathway (peri-position dechlorination), Pd/nZVI showed enhancement in the rate of degradation by almost 3-4 orders of magnitude. While developing a strategy for

PCDD/F dechlorination, their toxicity should be taken into account. The decision between using the unamended nZVI or Pd/nZVI for the efficient treatment of PCDD/Fs is still ambiguous because there are no reports for the toxic equivalent quantity (TEQ) change during the dechlorination of 2,3,7,8-substitued congeners that have very high TEF.

The TEQ changes were predicted using linear free energy relationsips (LFER) modeling methods and the pattern for dechlorination of PCDDs were estimated. Simulations done with this model predict that more than 100 years would be required for complete dechlorination of octachlorinated dibenzo-p-dioxin (OCDD) to the non-chlorinated compound (dibenzo-p-dioxin, DD) under conditions optimum for treatment with nZVI. The intermediates obtained during the dechlorination process will increase the net TEQ by 10 fold within 3~6 years, mainly due to the predominance of 2,3,7,8-substituted congeners.

Our studies on dioxin treatment using nZVIand Pd/nZVI, determination of their dechlorination profiles and TEQ changes both by modeling techniques and experimental data were noteworthy. This study will provide the basis for choosing the correct iron nanocatalyst (naked or modified) for treatment of PCDD/Fs contaminated sites and wastewaters in terms of overall toxicity.

Experimental Section

This section describes the synthesis and characterization of nanoscale zero-valent iron (nZVI) and bimetallic (Pd/nZVI) particles used in this study. It also explains the chemistry behind the catalytic properties of such nano materials. Dechlorination reactions using PCDDs as target compounds and the method of analysis of the dechlorinated products are also mentioned.

Iron Nanoparticles for Dechlorination of Chloro-organics

With a standard reduction potential (E_h^0) of -0.44 V, zero-valent iron (Fe^0) primarily acts as a reducing agent in the dechlorination reaction. The iron is oxidized (electron donor) (Equation 1) whilst alkyl halides (RX) are reduced (electron acceptor) (Equation 2). Because the estimated standard reduction potentials of the dehalogenation half-reaction of various alkyl halides range from +0.5 to +1.25 V at pH 7 (*46*), the net reaction is thermodynamically very favorable under most conditions (Equation 3).

$$Fe^0 \rightarrow Fe^{2+} + 2e- \quad (1)$$

$$RX + 2e^- + H^+ \rightarrow RH + X^- \quad (2)$$

$$Fe^0 + RX + H^+ \rightarrow Fe^{2+} + RH + X^- \quad (3)$$

The dechlorination of chlorinated organics by zero-valent iron is a surface mediated reaction, and therefore by increasing the surface area, the rate of reaction will increase, resulting in multiplicative reaction rates of dechlorination.

Nanoscale iron particles (size 1-100 nm) are characterized by high surface-area-to-volume ratios, high levels of stepped surface and high surface energies. The large fraction of atoms residing at the surfaces and at the grain boundaries of iron nanoparticles provides a huge surface reactivity, a manyfold increase from that available with bulk metals. Thus, they possess tens to hundreds times higher specific surface area than those of commercial-grade iron particles. This unique property makes these nanoparticles the best candidate for surface-mediated dechlorination.

However, zero-valent iron nanoparticles can form a hydroxide or oxide layer on the surface during the reaction or upon contact with air, which significantly reduces their reactivity and decreases the effective use of the nanoparticles. Additionally, the intermediate chlorinated byproducts produced during the reduction process can be more toxic than the parent compounds.

Iron-based bimetallic nanoparticles offer a better alternative to iron nanoparticles. These nanoparticles contain a thin layer of catalytic metal (e.g. Pd and Pt, which are not active by themselves) doped onto the surface of the active (reducing) metal Fe. As well as catalyzing the dechlorination reaction, such bimetallic particles have been found to reduce the amount of toxic chlorinated by-products, indicating that the dechlorination reactions continue further towards completion.

Incorporation of a second catalytic metal like Pd, Zn, Ni or Pt onto the iron surface solves the problem associated with particle agglomeration. Iron nanoparticles undergo surface oxidation within a few hours (black to reddish-brown) whereas palladium modified iron particles have no visible color change in air – suggesting stability and longevity (*47*). Correspondingly, the nZVI particles lose reactivity within a few days while Pd/nZVI particles remain active for longer duration. Physically mixing the two metals does not increase the rate of reaction; the palladium must be doped onto the surface. Doping palladium on the surface sets up a galvanic couple, which increases the rate of corrosion of the iron and increases the rate of dechlorination (*47*).

Synthesis of ZVI and Bimetallic ZVI Particles

Nanoscale iron (nZVI) was synthesized from $FeCl_3.6H_2O$ (0.15 M) by reduction with $NaBH_4$ (0.24 M) as shown in Equation 4, following established procedures (*48, 49*). The particles thus prepared were used freshly without drying. For each set of experiments, fresh nZVI particles were synthesized in sealed vials. The iron dosage was controlled by the amount of $FeCl_3.6H_2O$ solution used (replicates showed that the variability in dry weight of nZVI resulting from this procedure was 1%). After the reaction was completed, the remaining $NaBH_4$ solution was decanted and the particles were washed thoroughly with degassed water to remove excess $NaBH_4$ and other impurities.

$$4\text{-}Fe^{3+} + 3BH_4^- + 9H_2O \rightarrow 4Fe^0 + 3H_2BO_3^- + 12H^+ + 6H_2 \quad (4)$$

Excessive borohydride was the key factor for rapid and uniform growth of iron nanocrysrtals (*49*).

Palladization was performed electrochemically, following established procedures (*50, 51*). An aliquot (10-3000 μL) of deoxygenated, 5mM Pd (II) - acetate solution (in acetone) was added to the bottles containing mZVI or nZVI, and the mixture was then allowed to react for 30 minutes on an orbital shaker. The palladized iron was then rinsed with acetone to remove the unreacted Pd (II)-acetate. This caused the reduction and subsequent deposition of Pd on the Fe surface (*48*) as shown in Equation 5:

$$Fe^0 + Pd^{2+} \rightarrow Fe^{2+} + Pd^0 \quad (5)$$

To control the Pd concentration on the iron particles, a calibration was obtained by adding various amounts of Pd to the iron particles and determining the Pd/Fe ratio (w/w) with inductively coupled plasma emission spectroscopy (ICP). The dechlorination experiments were carried out mainly with particles having 0.5% palladium loading unless otherwise stated.

Characterization of ZVI and Bimetallic ZVI Particles

Brunauer-Emmett-Teller (BET) surface area analysis of the synthesized nanoparticles was performed using nitrogen adsorption method with surface analyzer system as shown in Table I.

The elemental compositions of the four types of iron particles (mZVI, Pd/mZVI, nZVIand Pd/nZVI) were determined by X-ray photoelectron spectroscopy (Table I), using Mg Kα line (1253.6 eV) as an excitation source and using inductively coupled plasma-atomic emission spectrometry.

The composition and structure of the nZVI and Pd/nZVI were characterized by transmission electron microscopy and electron dispersive X-rays (EDX) operated at 400 kV. For TEM analysis (Figures 3a and 3b), 300-mesh Cu TEM grids with a carbon film were dipped into acetone that contained the dispersed nZVI or Pd/nZVI.

Palladization did not have any effect on the surface area of the nanoparticles. The physico-chemical properties of the iron nanoparticles synthesized in this study are similar to those reported in literature on comparable materials (*40,48,52*).

The Pd 3d region (data not shown) of the XPS spectra showed peaks at Pd 3d3/2 (340.3 eV) and Pd 3d5/2 (335.0 eV), reported by others (*51*). Pd^{2+} was reduced to Pd^0 on the iron surface. The TEM images of nZVI (Figure 3a) revealed that the fresh, unpalladized particles consisted of roughly spherical cores covered with a shell that is uniformly 2-3 nm thick (*53*). The core/shell structure resulting from borohydride reduction has been shown (for similar materials) to consist predominantly of α-Fe^0 and a mixture of iron oxides and borates (*54*). In contrast, the shell of the palladized nZVI (Figure 3b) had a very different morphology; its oxide shell was thicker (10-30 nm) and less uniform, which was consistent with the effect of bimetal plating on other types of nZVI (*55*). Figure 3b includes EDX obtained at two places in the image domain, and these show that the Pd is localized in a rough, outer layer, which is similar to morphologies that have been reported previously (*52, 56*).

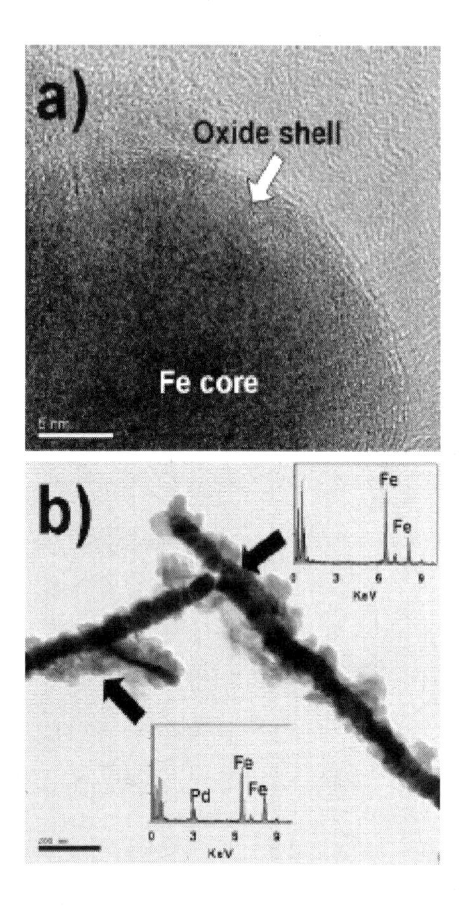

Figure 3a. TEM image of core-shell morphology of the nZVI particles and the formation of oxide film (x 1,000,000).
Figure 3b. TEM image of Pd/nZVI and EDX spectra of the points indicated by arrows in the image (x25,000)

Table I. Physical and Chemical Properties of Iron Particles

	Type	BET (surface area m^2/g)[a]	Elemental composition (wt%)[c]
mZVI	Commercially available fisher iron	0.10±0.01	Fe>99[b]
Pd/mZVI	Palladized mZVI	0.09±0.01	Fe 87 O 12 Pd 0.6[d]
nZVI	Nanosized iron prepared by reduction of ferric chloride by borohydride	33.2±1.2	Fe 44 O 50 B 6
Pd/nZVI	Palladized nZVI	33.6±1.3	Fe 39 O 55 B 5 Pd 0.5[d]

[a] From duplicate measurements. [b] From Fisher product information sheet. [c] Fe,O,B and Pd were measured with XPS using the scofield method. [d] The amount of Pd that was analyzed using ICP.

(Reproduced with permission from reference *51*, copy right, 2008 American Chemical Society)

Dechlorination of PCDDs and Analysis of Degradation Products

For dechlorination experiments, 30-200 nmol of PCDDs (1,2-DiCDD, 1,2,3-TriCDD, or 1,2,3,4-TeCDD) in acetone was injected into the reaction vials containing the iron particles. The acetone was evaporated under a stream of argon. The vials were then filled with 20 mL of degassed deionized water, leaving ~1 mL of headspace unfilled to control pressure in the vials (due to hydrogen produced during corrosion of the Fe^0). The vials were sealed with Teflon-lined septa inserted screw caps and placed on a roll mixer (15 rpm) in a dark room at room temperature until they were extracted for analysis. Experiments were conducted in triplicates to ensure reproducibility of the data. After the stipulated time intervals (0-2 years for mZVI and nZVI, 0-2 days for Pd/mZVI and Pd/nZVI), the reaction vials were sacrificed for analysis. The aqueous phase in each reaction vial was decanted into 30 mL glass tubes with the iron particles held back via a strong magnet, and both phases (aqueous and particle) were spiked with hexachlorobenzene(HCB) to serve as an internal standard. The aqueous phase was extracted with toluene. The particles were extracted by sonicating for 5 min in acetone two times, then once in a mixture of toluene/acetone (1:2 v/v). Before analysis, 1,2,3,4-TeCN was added to the extracts to serve as the recovery standard. Separate experiments were conducted to evaluate the efficiency of the extraction procedure, recovery percentage was found to be more than 85%. The identification and the quantification of the initial congeners and their dechlorinated products were performed by gas chromatography with ion-trap mass spectrometry. However, the concentrations of the congeners formed by dechlorination of PCDDs with unpalladized iron (mZVI and nZVI) were too low to be detected using IT-MS, so these samples were analyzed by high resolution GC with high-resolution MS. Three-point calibrations were used to obtain the response factors for all PCDD congeners included in this study (1,2,3,4-TeCDD; 1,2,3- and 1,2,4-TriCDD; 1,2-,1,3-,1,4-, and 2,3- DiCDD; 1- and 2-MCDD; and DD). The approximate detection limits were 10 picomoles (pmol) with IT-MS and 10 attomoles (amol) with HR-MS.

Results and Discussion

Kinetics of PCDD Dechlorination

At the end of 2 months (reaction time), mZVI and nZVI did not produce any dechlorinated products from 1,2-DiCDD and 1,2,3-TriCDD. However, dechlorination of 1,2,3,4-TeCDD gave quantifiable amounts of dechlorinated intermediates (MCDDs to TriCDDs and DD) as shown in Figure 4A. The sum of dechlorinated products from 1,2,3,4-TeCDD, however, was only ~2 mol% of the initial 1,2,3,4-TeCDD (after 50 days of reaction time). No further change in the distribution of the dechlorinated products was observed after 2 years of PCDD dechlorination using mZVI, but the recoverable quantity of 1,2,3,4-TeCDD continued to decline, eventually resulting in a total mass balance of all PCDDs congeners that was ~40% (after 2 years). Previous reports (*40*) have also stated the total decline in mass balance in experiments conducted to study the degradation of PCBs with nZVI, where it was attributed to the irreversible adsorption of the initial congener to the walls of glass vials and the Teflon lined septa. In our studies, control experiments performed without iron particles, total mass balance obtained was about 80-85% for TeCDD (over 60 days). Hence, we concluded that most of the unrecoverable TeCDD was irreversibly adsorbed onto the unreactive areas of the ZVI particles. Our finding that tetra- (or higher) chlorinated congeners reduce more rapidly rather than tri- and di- chlorinated congeners is consistent with structure-activity relationships showing that dechlorination rates generally correlate with the degree of chlorination of halogenated organics (*57*), and that the degree of chlorination generally correlates with the reduction potential of closely related halogenated organics (*58*).

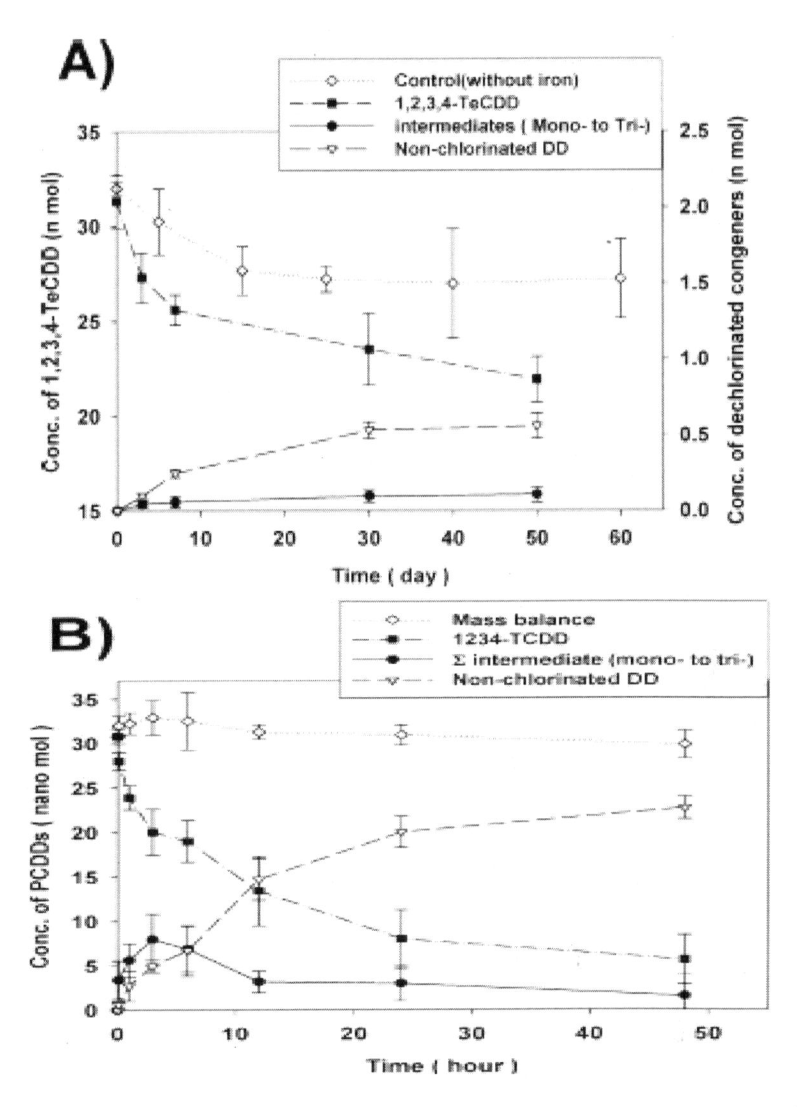

Figure 4a. Degradation of 1,2,3,4-TeCDD (initial concentration 31.1nmol) by mZVI (0.4g)
Figure 4b. Degradation of 1,2,3,4-TeCDD (initial concentration 31.1nmol) byPd/nZVI (0.1g nano sized iron , 0.5% Pd loading)

Palladization of the iron particles resulted in enhanced rates of dechlorination for all PCDD congeners. Pd/mZVI and Pd/nZVI dechlorinated 90% 1,2,3,4-TeCDD to DD within 48 hours, with pseudo first order constants (k_{obs}) equal to 3.3 (+/- 0.8) × 10^{-2} h^{-1} and 5.1 (+/- 1.3) × 10^{-2} h^{-1}, respectively (Tables IIA and IIB).

The dechlorination efficiency with palladized iron was almost 3 orders of magnitude higher when compared with unpalladized iron. Figure 4B shows the time profile for reaction of 1,2,3,4-TeCDD with Pd/nZVI. Highest concentration of intermediates (tri- to mono-) was obtained after 3 h of reaction, which decreased gradually. At the end of two days, di-chloro -dioxin (DD) was the only congener, which accumulated and accounted for most of the products. The data (not shown for each intermediate congener separately) implies stepwise dechlorination. Dechlorination rates were faster for the less-chlorinated congeners: 1.5-5 times faster for 1,2,3- TriCDD compared with 1,2,3,4-TeCDD (Figure 5), and 2-4 times faster for 1,2-DiCDD compared with 1,2,3-TriCDD (Table II). The data (including data notshown for each intermediate congener, separately) implies stepwise dechlorination. Dechlorination rates were faster for the less-chlorinated congeners: 1.5-5 times faster for 1,2,3- TriCDD compared with 1,2,3,4-TeCDD , and 2-4 times faster for 1,2-DiCDD compared with 1,2,3-TriCDD (Table IIA and B). Such differences in dechlorination rates have also been observed for PCBs and palladized granular iron (*41*) and chlorinated ethenes reacting with nanoscale (*59*) and microscale iron (*60*).

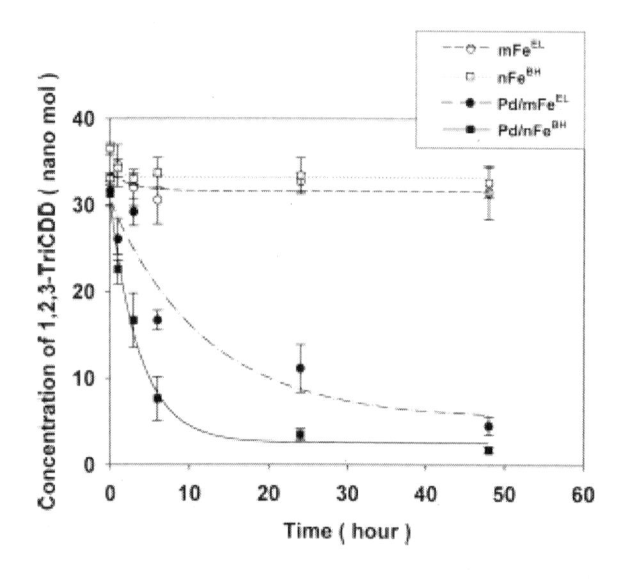

Figure 5. Degradation of 1,2,3-TriCDD (initial concentration 34.78 nmol) by mZVI (0.4g), nZVI (0.1g), Pd/mZVI (0.4g,0.5% Pd loading) and Pd/nZVI (0.1g nano sized iron , 0.5% Pd loading)

It is difficult to assess quantitatively the advantages of palladization of the iron particles, as some key factors, such as the proportion of particle surface area that Pd occupies are not known (*61*). Nevertheless, the enhanced rates of dechlorination in the present study have not been observed in studies reported earlier for comparable systems. For example, palladization of nZVI has been reported to increase dechlorination rates by 1.5-2 orders of magnitude for PCBs and chlorinated benzenes (*41*). In studies concerning halogenated aliphatics, it was found that the increase in reduction rates due to palladization is significantly affected by the pattern of chlorination even within a set of congeners (*62*). The data in our studies is in accordance with the above-mentioned fact that palladization significantly enhances the rate of PCDD dechlorination in all the cases.

However, the use of nanosized iron instead of micro sized iron for PCDD dechlorination is also questionable. Based on k_{obs}, Pd/nZVI yielded degradation rates for PCDDs congeners that were 2-5 times greater than for Pd/mZVI. However, the surface area normalized rate constants (*k*sa, calculated using the BET-specific surface areas) given in Table I, suggest that Pd-mZVI is about 20-60 times more reactive than Pd/nZVI.

Being in the nano size does not necessarily impart higher reactivity when comparisons are made on a (total) surface area basis (*63*). However, in this case, the lower *k*sa obtained with Pd/nZVI compared to Pd/mZVI are likely due to differences in how the Pd is distributed on the surface of the iron particles, which cannot be quantified. Despite the relatively high rates of PCDD dechlorination by Pd/nZVI, our data suggest that a small fraction of PCDDs is recalcitrant to degradation over considerable exposure times.

After 2 days, 10% of initial 1,2,3,4-TeCDD, 5% of 1,2,3-TriCDD, and 2% of 1,2- DiCDD remained unaccounted for (Figures 4B and 5). Since we silanized (all reaction vials in this research were silanized with 1,1,1,3,3,3-hexamethyldisilazane [HMDS] the glass wares prior to use , the chances of adsorption of PCDDs onto the glass vials was negligible. The irreversible adsorption of these residues onto the unreactive sites on the iron surface caused by heterogeneous plating of Pd is highly possible (as shown in the TEM figures). Hence, we favor the hypothesis that PCDD residues were sequestered onto nonreactive surface areas of Pd/nZVI. This hypothesis is consistent with the observation that the amount of sequestered PCDD increases with the degree of chlorination which is accompanied by decreased solubility (*63*).

Table IIA. The first-order PCDD dechlorination rate constants (k_{obs}) and surface area- normalized PCDD dechlorination rate constants (k_{sa}) of mZVI, nZVI

	mZVI		nZVI	
	k_{obs} (h^{-1})[a]	k_{sa} ($Lh^{-1}m^{-2}$)[b]	k_{obs} (h^{-1})[a]	k_{sa} ($Lh^{-1}m^{-2}$)[b]
1,2,3,4-TeCDD	$3.3 \pm 0.8 \times 10^{-2}$	$1.9 \pm 0.4 \times 10^{-2}$	$5.1 \pm 1.3 \times 10^{-2}$	$3.0 \pm 0.9 \times 10^{-4}$
1,2,3-TriCDD	$5.2 \pm 1.5 \times 10^{-2}$	$2.9 \pm 0.9 \times 10^{-2}$	$2.6 \pm 0.4 \times 10^{-1}$	$1.5 \pm 0.3 \times 10^{-3}$
1,2-DiCDD	$2.3 \pm 0.4 \times 10^{-1}$	$1.3 \pm 0.2 \times 10^{-1}$	$4.5 \pm 0.6 \times 10^{-1}$	$2.7 \pm 0.5 \times 10^{-3}$

[a] Reported uncertainties are 95% confidence intervals for pseudo first –order fit of product data. [b] Reported uncertainties are calculated from 95% confidence limits using error-propagation methods. [c] Not detected. (Reproduced with permission from reference *51*, copy right 2008, American Chemical Society)

Table IIB. The first-order PCDD dechlorination rate constants (k_{obs}) and surface area normalized PCDD dechlorination rate constants (k_{sa}) of Pd/mZVI and Pd/nZVI

	Pd/mZVI		Pd/nZVI	
	k_{obs} (h^{-1})[a]	k_{sa} ($Lh^{-1}m^{-2}$)[b]	k_{obs} (h^{-1})[a]	k_{sa} ($Lh^{-1}m^{-2}$)[b]
1,2,3,4-TeCDD	$1.7 \pm 0.4 \times 10^{-5}$	$8.7 \pm 2.0 \times 10^{-5}$	$3.7 \pm 1.3 \times 10^{-5}$	$2.2 \pm 0.8 \times 10^{-7}$
1,2,3-TriCDD	ND[c]	ND	ND	ND
1,2-DiCDD	ND	ND	ND	ND

[a] Reported uncertainties are 95% confidence intervals for pseudo first –order fit of product data. [b] Reported uncertainties are calculated from 95% confidence limits using error-propagation methods. [c] Not detected.

Dechlorination Pathways of PCDDs

The main dechlorination pathway of 1,2,3,4-TeCDD by unpalladized iron (mZVI and nZVI) was 1,2,3,4-TeCDD to 1,2,4-TriCDD to 1,3-DCDD to 2-MCDD (Figure 6 and Table III). This pathway is consistent with that proposed by Huang et al. (43), based on the calculated difference between the free energies for each possible reductive dechlorination step. This agreement suggests that dissociative electron transfer probably is the dominant pathway for dechlorination of 1,2,3,4-TeCDD by unpalladized iron. Another rationale for the observed pattern of dechlorination comes from the calculated properties of each C-Cl bond. Calculations done with density functional theory showed that the lateral chlorines (2-, 3- positions) on PCDDs have greater intramolecular C-Cl repulsion energies than the peri-chlorines (1-, 4- positions) (65). Therefore, steric repulsion favors the dissociation of C-Cl bonds in the lateral positions, which is consistent with the pattern of sequential dechlorination steps that we observed with unpalladized iron.

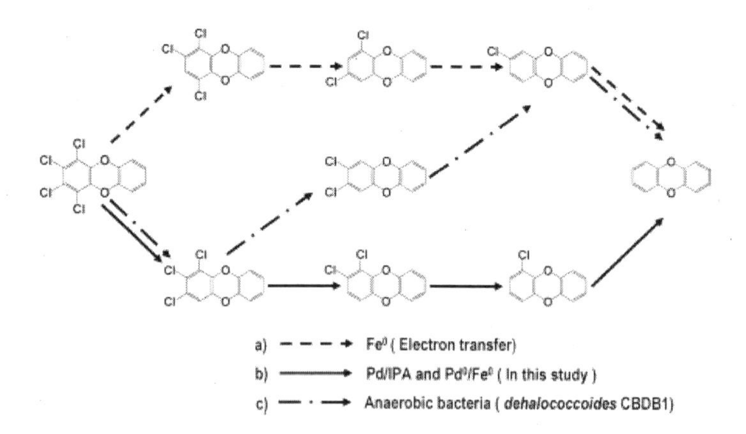

Figure 6. Comparision of dechlorination pathways for 1,2,3,4-TeCDD

Table III. Distribution of Dechlorinated Products

Products		1,2,3,4-TeCDD Unpalladized	Palladized
Tri-	123	37 ± 4	**77 ± 8**
	124	**63 ± 4**	23 ± 8
Di-	12	14 ± 2	**75 ± 7**
	13	**44 ± 3**	4 ± 2
	14	28 ± 3	5 ± 2
	23	14 ± 2	16 ± 4
Mono-	1	38 ± 4	**69 ± 11**
	2	**62 ± 5**	31 ± 11
Products		**1,2,3-TriCDD Palladized**	**1,2-DiCDD Palladized**
Tri-	123	-	-
	124	-	-
Di-	12	**51 ± 9**	-
	13	6 ± 3	-
	14	-	-
	23	43 ± 10	-
Mono-	1	**61 ± 2**	**54 ± 15**
	2	39 ± 12	46 ± 15

NOTE: Each number is the percentage of dechlorinated products for each homologue; standard deviation was estimated from triplicates. Bold indicates the major congener for each homologue. (Reproduced with permission from reference *51*, copy right 2008, American Chemical Society)

This pattern of PCDD dechlorination has implications for remediation. Dechlorination of PCDDs by unpalladized iron apparently favors less toxic products because the initial step removes lateral-chlorines preferentially, as expected for reduction by dissociative electron transfer. Removal of peri-chlorines during the dechlorination of higher chlorinated dioxins causes the formation of 2,3,7,8-substituted congeners, which increases the toxicity of the mixture as observed in the dechlorination pattern of PCDDs by one particular bacterial strain *dehalococcoides* sp. strain CBDB1as shown in Figure 6 (*66*).

The pathway of 1,2,3,4-TeCDD by palladized iron (Pd/mZVI and Pd/nZVI) was 1,2,3,4-TeCDD to 1,2,3-TriCDD to 1,2-DiCDD to 1-MCDD (Figure 6). This pathway is not the thermodynamically favored route for dissociative electron transfer, but it is consistent with the pathway reported for PCDD dechlorination in alkaline isopropyl alcohol (IPA) with Pd catalysts (*67*). Dechlorination by the Pd/IPA system involves the transfer of hydrogens donated by the R-hydrogen in IPA, catalyzed on the Pd surface, and involving H (atomic or "nascent" hydrogen) or possibly H^- (hydride) and not electron transfer (*67*). It is possible that a similar mechanism might be involved in the reduction of 1,2,3,4-TeCDD by palladized iron, as reported here. The relative significance of electron transfer versus hydrogen atom transfer has long been part of the debate on why palladized iron (and other bimetallic combinations of ZVI with noble, catalytic metals) often reduces contaminants more rapidly and to more desirable products than unpalladized ZVI (*62, 68*).

The galvanic coupling between the iron and noble metal increases the rate at which H_2O/H^+ is reduced to H_2 (the anoxic/aqueous corrosion reaction), which occurs via absorbed atomic hydrogen (H_{ads}), which, in turn, can contribute to increased rates of contaminant reduction (*62, 68*).

The data presented in this chapter indicates that the pathways of PCDD reduction supports the latter interpretation of the bimetallic effect: that galvanic coupling favors formation of activated H^-species, and reduction by these species is responsible for accelerated rates of contaminant reduction. Another argument in favor of H-atom transfer as the dominant pathway in PCDD dechlorination by palladized iron can be made on steric considerations. The distance between the atomic nuclei of neighboring chlorines in a dioxin molecule is about 3-4 Å (*69*) and the size of an H atom is about 1 Å, so steric hindrance could inhibit the formation of any precursor complex between H atom and PCDDs. Steric effects do not play a major role in dechlorination by electron transfer, because the rate-limiting step (electron attachment) precedes any atom transfer step.

Modeling the Reductive Dechlorination of Polychlorinated Dibenzo-*p*-Dioxins: Kinetics, Products, and Equivalent Toxicity

As mentioned above, the dechlorination of PCDD congeners by dissociative electron transfer is thermodynamically favorable. Gibbs free energies for reductive dechlorination have been calculated for 76 PCDDs congeners (*43*). It would be desirable to use these data to explain, and perhaps predict, the pathways and kinetics of dechlorination of PCDDs. However, the previously reported thermodynamic data on PCDDs exhibit significant discrepancies: for example, reported values of ΔH_f differ depending on the calculation method by as much as 30 kcal/mol, (*43,70*). Moreover, the available data may contain significant deviations because of the various assumptions included in the calculation process. Previous studies had assumed that PCDD congeners with same the number of chlorines had same value for vapor pressure and aqueous solubility (*43*).

Linear free energy relationships (LFERs) describing reductive dechlorination have been reported recently for chemical reductants like ZVI (*71, 72*). To assess the overall effect of PCDD contamination, we developed a method which included a complete and quantitative solution to this problem. This method allows the calculation of the toxic equivalent quantity (TEQ) of the mixture of PCDD congeners resulting from dechlorination pathways predicted by LFER that relates rate constants of dechlorination to reduction potentials calculated from molecular structure theory. The reduction potentials were derived from improved Gibbs free energies, calculated using density functional theory (DFT), vapor pressures and solubility's for individual congeners. The LFER was calibrated based on the rate constants we reported for PCDD dechlorination by nZVI. LFER describes the relative reaction rates that agree

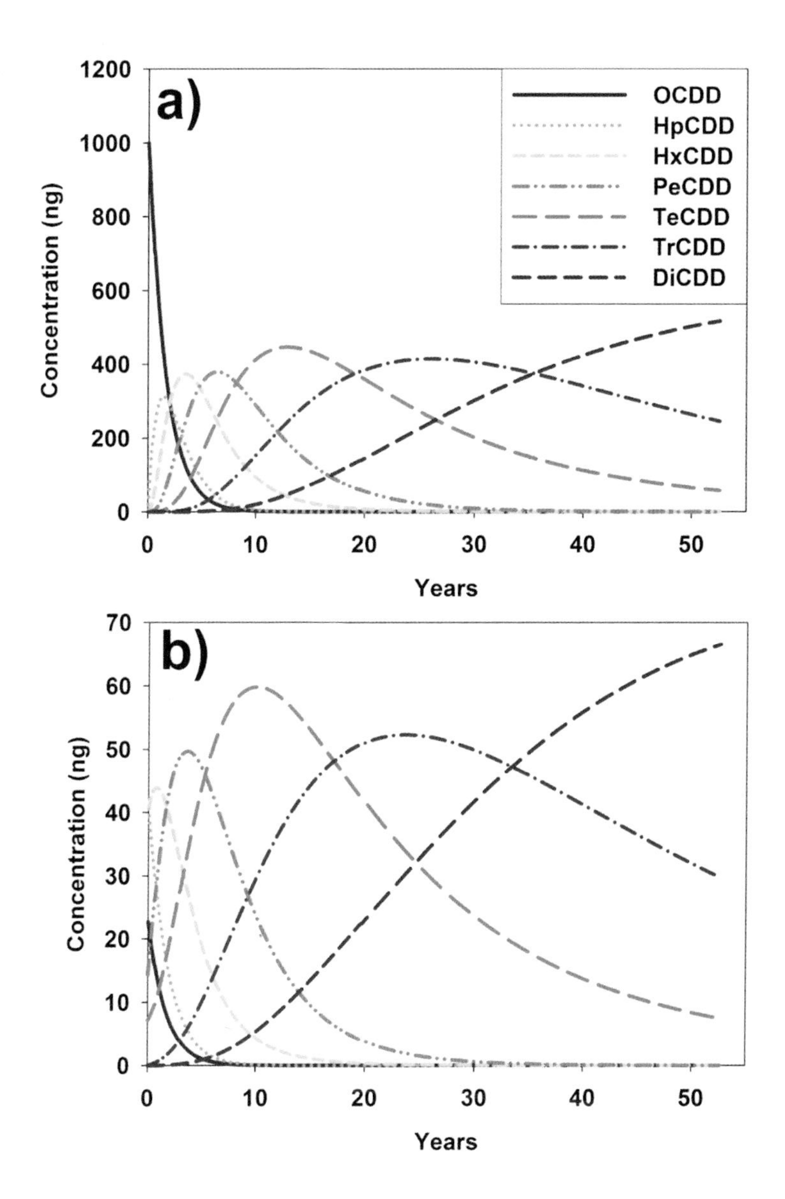

Figure 6.7. Predicted homologue profiles of PCDDs dechlorination (A) OCDD (initial conc. = 1000 ng/g, (B) PCDDs mixture from typical fly ash. The concentrations of MCDD and DD were less than 10 ng/g, so they are not shown in the figure.

well with available data on microbiological dechlorination of PCDDs. Simulation, based on the above model, predicts that more than 100 years would be required for complete dechlorination of higher chlorinated congeners like octachlorinated dibenzo-p-dioxin (OCDD) to non-chlorinated compound (dibenzo-p-dioxin, DD) under conditions optimum for treatment with nZVI. It also states that the TEQ of the mixture of intermediates obtained during dechlorination process increases 10-folds in 3~6 years mainly due to the higher toxicity of the 2,3,7,8-substituted congeners.

Measurement of PCDD dechlorination kinetics and calibration of LFERs

If a compound [S] is dechlorinated to the lesser chlorinated compounds (P_1, P_2), with first order rate constant k_0, Equation 7 can be obtained from Equation 6 by considering the rate constants of k_1 and k_2 for generation of P_1 and P_2.

$$-\frac{dS}{dt} = k_0 \cdot S = (k_1 + k_2) \cdot S \qquad (6)$$

$$k_1 + k_2 = k_0 \qquad (7)$$

The rates of dechlorination of P_1 and P_2 are affected by the concentration of S (Equations 8 and 9), but the ratio k_1:k_2 is proportional to the ratio P_1:P_2 at time t (Equation 10). For a particular PCDD congener, experimental data for the observed dechlorination rate constant (k_0) and of the observed product ratio for P_1 and P_2 (α, β), were used to calculate k_1 and k_2 according to equation 7 and equation 10.

$$-\frac{dP_1}{dt} = k_1 \cdot S \quad , \quad -\frac{dP_2}{dt} = k_2 \cdot S \qquad (8, 9)$$

$$P_{1,t} : P_{2.,t} = \frac{k_1}{k_0} S_0 (1 - e^{-k_0 t}) : \frac{k_2}{k_0} S_0 (1 - e^{-k_0 t}) = k_1 : k_2 = \alpha : \beta \qquad (10)$$

The only congener for which k_0, α, and β have been reported previously is for 1,2,3,4-TeCDD reduction by ZVI (*53*). Same methods were used (*53)* to measure k_0, α, and β for 2,3,7,8-TeCDD and for OCDD as shown in Table IV.

Table IV. **Experimental Rate Constants for the PCDDs: Dechlorination and Product Ratio**

Substrate	k_{obs} [h^{-1}] (Experimental)	Products	Product ratio (Experimental)	k_1, k_2 [h^{-1}]	Ref
1,2,3,4-TCDD	$3.7 \pm 1.3 \times 10^{-5}$	1,2,3-TriCDD 1,2,4-TrCDD	37 63	1.4×10^{-5} 2.4×10^{-5}	[10]
2,3,7,8-TCDD	$8.0 \pm 0.9 \times 10^{-6}$	2,3,7-TrCDD	100	8.0×10^{-6}	This Study
OCDD	$8.5 \pm 1.5 \times 10^{-5}$	1,2,3,4,6,7,8-HpCDD 1,2,3,4,6,7,9-HpCDD	71 29	2.5×10^{-5} 6.1×10^{-5}	This Study

(Reproduced with permission from Kim *et. al.*, unpublished data)

Extra thermodynamic correlations are often successful because the free energy change from reactants to the transition state is generally proportional to the free energy change from reactants to products (*73*). Over a limited range, this relationship may be approximately linear, as calculated by Equation 11. In this study, the LFER was calibrated using the experiment rate constants in Table IV (including 5 dechlorination reactions for the 3 PCDDs) for the response variable and $E'_{0,n}$ calculated for each congener as the descriptor variable. Once the LFER was established, values of $E'_{0,n}$ for each dechlorination step (combination of parent congener and dechlorination product) were used to calculate values of k_n (n = 1–256).

$$\log k_n = \alpha \cdot E'_{0,n} + \beta \tag{11}$$

Simulation of PCDD dechlorination

The concentration of congener i (i = 1–76) at time t ($C_{i,t}$) can be expressed as the sum of concentrations formed from dechlorination of more-highly chlorinated PCDDs minus the concentrations lost by dechlorination to less-highly chlorinate PCDDs. For each congener, $C_{i,t}$ was calculated from the dechlorination rate constants (k_n) obtained by equation 11. For example, the concentration of 2,3,7,8-TeCDD at time t was expressed according to Equation 12 as:

$$C_{49,t} = C_{49,t-1} + C_{55,t-1} \cdot e^{k_{99}} - C_{49,t-1} \cdot e^{k_{127}} \quad (12)$$

assuming 1-hour time steps, C_{55} = concentration of 1,2,3,7,8-PeCDD [g·L^{-1}] at time t, k_{99} = the rate constant for dechlorination of 1,2,3,7,8-PeCDD to 2,3,7,8-TeCDD [h^{-1}], and k_{129} = rate constant for dechlorination of 2,3,7,8-TeCDD to 2,3,7-TriCDD [h^{-1}].

Reduction potential of PCDD Dechlorination in the Aqueous Phase

We calculated standard-state Gibbs free energy for all the 76 PCDD congeners in aqueous phase ($\Delta_f G^0_{sol,i}$, i = 1–76) and the reduction potentials for the 256 possible dechlorination reactions ($E'_{0,n}$, n = 1–256, mV) using the values of $\Delta_f G^0_{sol,}$ along with the corresponding values of $E'_{0,n}$ (standard reduction potentials)reported previously by Huang *et al.*(*43*). A direct comparison between the two data sets for $E'_{0,n}$ shows that values obtained by our study are 100 mV more positive than those from Huang *et al.*

However the absolute values for reduction potential are not significant as they will offset each other in LFER process. Relative differences of reduction potential in the dechlorination pathway are more important as they directly affect the toxicity change.

Qualitative inspection of the data for trends within these sets of congeners suggests that congeners with more fully chlorinated rings (e.g., 1,2,3,4,6,9-HxCDD, 1,2,3,4,7-PeCDD and 1,2,3,4-TeCDD) have relatively higher $E'_{0,n}$ than the congeners with chlorines distributed over both aromatic rings. Dissociative electron transfer of the former group is presumably favored by the greater localization of electron density caused by the one-sided chlorination pattern. The difference of reduction potential between OCDD to 1,2,3,4,6,7,9-HpCDD(473.2mV) and to 1,2,3,4,6,7,8-HpCDD(410.5mV) in previous studies was 62mV . In contrast, according to our calculations the difference in reduction potential was only 20mV. Quantitatively, the predicted toxicity would be higher in our study as the dechlorination of chlorines from the peri position (1, 4, 6, and 9) was predominant. The reduction potentials of PCDD dechlorination in our case was much more linear (R^2=0.4437) in comparison to previous study (R^2=0.1815) (*43*).

The approach taken in this study for assessing the results of PCDD dechlorination (including the calibration of the LFER, prediction of rate constants, simulation of product distribution, and calculation of mixture toxicity) could be applied to other families of contaminants with analogous patterns of congeners, including the polychlorinated biphenyls (PCBs), benzenes, and phenols, and polybrominated diphenylethers (PBDEs). As with the PCDDs, patterns of relative reactivity among congeners are likely to be fairly general, even though absolute rates can only be estimated for systems that match the specific conditions used in the model calibration. The fact that the data are not available to fully validate any specific application of this modeling approach is a concern, but it also illustrates its importance, because it is the only practical way to address key questions regarding the overall effect of treatments for materials contaminated with complex mixtures of congeners.

Prediction of PCDDs Dechlorination: Congeners, Homologue and Toxic Equivalent Quantity Profiles

Figure 7a shows the decrease in the concentration of OCDD (to 1% of the initial OCDD in 8.5 years), but then OCDD gives a series of appearance/disappearance profiles with increasingly long time periods between the peaks (because the reaction becomes less favorable with each dechlorination). After 50 years of simulated reaction, the concentration of Di-CDD congeners was still roughly 50% of the original OCDD concentration, and 90% conversion to the fully-dechlorinated product (DD) required more than 200 years (data not shown). Figure 7b shows the simulation of PCDDs in the PCDD mixture of typical fly ash showing similar results: the mixture is mostly converted to DiCDD congeners in 50 years and 90% conversion to DD takes over 150 years. However, within the time period of 5-15 years, the combined effect of the relative rates of the dechlorination steps for each congener causes a peak in concentration of TCDD congeners.

Considering the highly variable toxicities of the individual PCDD congeners, the simulation results suggest that the overall risk from PCDD contamination may not decrease as a result of dechlorination. This was investigated by calculating TEQ for two cases. For OCDD at an initial concentration of 1000 ng (Figure 8a), the TEQ increased from 1 ng/g to 12.7 ng/g in 5.3 years, however vice versa scenario (decrease in concentration to 1 ng/g) will take about 42.3 years. In the case of PCDD mixture in fly ash (Figure 8b), the TEQ increased from 0.79 to 1.49 ng/g in 3.4 years. Decrease in this trend will take almost 11.3 years.

According to the above predictions, removal of PCDDs from contaminated sites with nZVI (the catalyst for which the model was calibrated) will require more contact time than most treatment methods are designed to provide. As noted above, microbiological reductive dechlorination of PCDDs will, in general, follow pathways similar to those computed here but at slower rates. To overcome such challenges, rapid and efficient reductive dechlorination is required. Hence, newer catalysts should be explored, which are more effective than nZVI. There are not many candidates for this, except Pd/nZVI and zero-valent zinc which have been shown to dechlorinate 1,2,3,4-TeCDD very rapidly (*53, 71*). For this reason, the use of hybrid technology (anaerobic dechlorination followed by aerobic oxidation by suitable microorganisms) would be the best solution for effective PCDD removal from various compartments in the environment. As reductive dechlorination of PCDDs makes successive intermediates that are less labile to further reduction, they are more prone to oxidation.

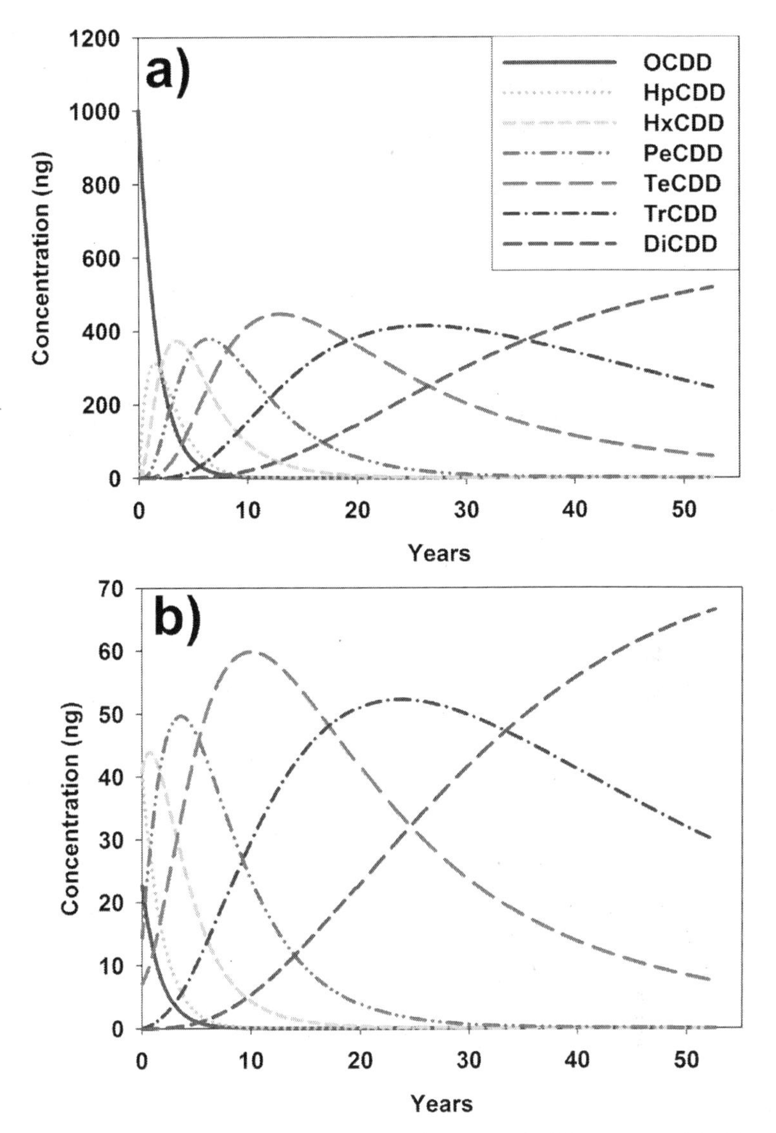

Figure 7. Predicted homologue profiles of PCDDs dechlorination (A) OCDD (initial conc. = 1000 ng/g, (B) PCDDs mixture from typical fly ash. The concentrations of MCDD and DD were less than 10 ng/g, so they are not shown in the figure.
(See page 1 of color insert.)

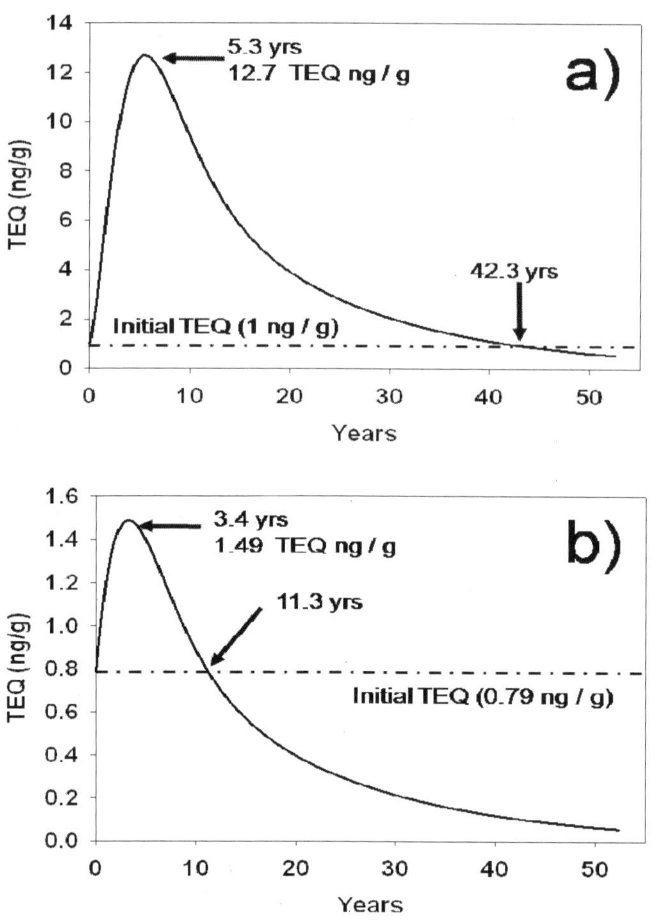

Figure 8. Predicted TEQ changes of PCDD dechlorination (A) OCDD (initial conc. = 1000 ng/g, 1 ng TEQ, (B) PCDDs mixture (initial 0.79 ng/g TEQ) from typical fly ash.

Conclusion

From our studies, we conclude that PCDD dechlorination by Pd/nZVI favors the H atom transfer mechanism. These mechanistic considerations are advantageous for PCDD degradation by unpalladized iron, because preferential removal of the lateral chlorines lowers the overall toxicity of the PCDD mixtures. However with palladised iron PCDD dechlorination is much faster. Choice of the catalyst for efficient and effective removal of PCDDs should be considered also on the basis of overall toxicity pattern of the dechlorinated intermediates formed. Predictions made on the basis of LFER calculations point out that it might take more than 100 years for complete removal of PCDDs from the environment, hence hybrid technology could prove to be a holistic solution.

This model can also be applied to other species of contaminants with analogous patterns of congeners, including the polychlorinated biphenyls (PCBs), benzenes, and phenols, and polybrominated diphenylethers (PBDEs). And, overall assessment of dehalogination process for various matrices such as soil, sediment, ground water etc can be made.

References

1. Evans C.S.; Dellinger B. Environ Sci Technol. **2005**; 39, 2128–2134.
2. Lustenhouwer, J. W. A.; Olie, K.; Hutzinger, O. Chlorinated dibenzo-p-dioxins and related compounds in incinerator effluents: A review of measurements and mechanisms of formation. *Chemosphere* **1980,** *9*, 501–522.
3. Laine, M. M.; Ahtiainen, J.; Wagman, N.; Oberg, L. G.; Jørgensen, K. S. Fate and toxicity of chlorophenols, polychlorinated dibenzo-p-dioxins, and dibenzofurans during composting of contaminated sawmill soil. *Environ. Sci. Technol.* **1997**, *31*, 3244–3250.
4. Krizanec, B.; Le Marechal, A. M.; Voncina, E.; Brodnjak-Voncina, D. Presence of dioxins in textile dyes and their fate during the dyeing processes. *Acta Chim. Slov.* **2005,** *52*, 111–118.
5. Gribble, G. W. The diversity of naturally produced organohalogens. *Chemosphere* **2003,** *52*, 289–297
6. Kim, E. J.; Oh, J. E.; Chang, Y. S. Effects of forest fire on the level and distribution of PCDD/Fs and PAHs in soil. *Sci. Tot. Environ.* **2003,** *311*, 177–189.
7. Alcock, R. E.; Jones, K. C. Dioxins in the environment: A review of trend data. *Environ. Sci. Technol.* **1996,** *30*, 3133–3143.
8. Hutzinger, O.; Blumich, M. J.; Berg v.d, M.; Olie, K. Sources and fate of PCDDs and PCDFs: an overview. *Chemosphere* **1985,** *14*, 581–600.
9. Mandal, P. K. Dioxin: A review of its environmental effects and its aryl hydrocarbon receptor biology. *J. Comp. Physiol. B* **2005,** *175*, 221–230.
10. Kogevinas, M. Carcinogenicity of dioxins. *Lancet* **1999,** *354*, 429–430.
11. Lorenzen, A.; Okey, A. B. Detection and characterization of Ah receptor in tissue and cells from human tonsils. *Toxicol. Appl. Pharm.* **1991,** *107*, 203–214.
12. Peper, M.; Klett, M.; Frentzel-Beyme, R.; Heller, W. D. Neuropsychological effects of chronic exposure to environmental dioxins and furans. *Environ. Res.* **1993,** *60*, 124–135.
13. Boening, D. W. Toxicity of 2,3,7,8-tetrachlorodibenzo-p-dioxin to several ecological receptor groups: A short review. *Ecotox. Environ. Safe.* **1998,** *39*, 155–163.
14. Landers, J. P.; Bunce, N. J. The Ah receptor and the mechanism of dioxin toxicity. Biochem. J. 1991, 276, 273–287.
15. Pohjanvirta, R.; Tuomisto, J. Short-term toxicity of 2,3,7,8-tetrachlorodibenzo-p-dioxin in laboratory animals: Effects, mechanisms, and animal models. *Pharmacol. Rev.* **1994,** *46*, 483–549.
16. Hay, A. Vietnam's dioxin problem. *Nature* **1978**, *271*, 597–598.

17. Hay, A. Seveso: dioxin damage. *Nature* **1977**, *266*, 7–8.
18. Bertazzi, P.A.; Bernucci, I.; Brambilla, G.; Consonni, D.; Pesatori, A. C. The Seveso studies on early and long-term effects of dioxin exposure: a review. *Environ. Health Persp.* **1998**, *106*, 625–633.
19. Masunaga, S.; Takasuga, T.; Nakanishi, J. Dioxin and dioxin-like PCB impurities in some Japanese agrochemical formulations. *Chemosphere* **2001**, *44*, 873–885.
20. Oberg, L. G.; Glas, B.; Swanson, S. E.; Rappe, C.; Paul, K. G. Peroxidase-catalyzed oxidation of chlorophenols to polychlorinated dibenzo- *p* –dioxins and dibenzofurans. *Arch. Environ. Contam. Toxicol.* **1990**, *19*, 930–938.
21. Govers, H. A. J.; Krop, H. B. Partition constants of chlorinated dibenzofurans and dibenzo-p-dioxins. *Chemosphere* **1998,** *37*, 2139–2152.
22. Shiu, W. Y.; Doucette, W.; Gobas, F. A. P. C.; Andren, A.; Mackay, D. Physical-chemical properties of chlorinated dibenzo-p-dioxins. *Environ. Sci. Technol.* **1988,** *22*, 651–658.
23. Assmuth, T.; Vartiainen, T. Concentrations of 2,3,7,8-chlorinated dibenzo-p-dioxins and dibenzofurans at landfills and disposal sites for chlorophenolic wood preservative wastes. *Chemosphere* **1994,** *28*, 971–979.
24. Zhang, Q. H.; Xu, Y.; Wu, W. Z.; Xiao, R. M.; Feng, L.; Schramm, K. W.; Kettrup, A. PCDDs and PCDFs in the wastewater from Chinese pulp and paper industry. *Bull. Environ. Contam. Toxicol.* **2000,** *64*, 368–371.
25. Quagliotto, P.; Montoneri, E.; Tambone, F.; Adani, F.; Gobetto, R.; Viscardi, G. Chemicals from wastes: Compost-derived humic acid-like matter as surfactant. *Environ. Sci. Technol.* **2006,** *40*, 1686–1692.
26. Tanaka, S.; Oba, K.; Fukushima, M.; Nakayasu, K.; Hasebe, K. Water solubility enhancement of pyrene in the presence of humic substances. *Anal. Chim. Acta* **1997,** *337*, 351–357.
27. Harner, T.; Green, N. J. L.; Jones, K. C. Measurements of octanol - Air partition coefficients for PCDD/Fs: A tool in assessing air - Soil equilibrium status. *Environ. Sci. Technol.* **2000,** *34*, 3109–3114.
28. Bunge, M.; Adrian, L.; Kraus, A.; Opel, M.; Lorenz, W. G.; Andreesen, J. R.; Görisch, H.; Lechner, U. Reductive dehalogenation of chlorinated dioxins by an anaerobic bacterium. *Nature* **2003**, *421*, 357–360.
29. Adriaens, P.; Fu, Q.; Grbic-Galic, D. Bioavailability and transformation of highly chlorinated dibenzo-*p*-dioxins and dibenzofurans in anaerobic soils and sediments. *Environ. Sci. Technol.* **1995**, *29*, 2252–2260.
30. Shiang Fu, Q.; Barkovskii, A. L.; Adriaens, P. Reductive transformation of dioxins: An assessment of the contribution of dissolved organic matter to dechlorination reactions. *Environ. Sci. Technol.* **1999**, *33*, 3837–3842.
31. Adriaens, P.; Chang, P. R.; Barkovskii, A. L. Dechlorination of PCDD/F by organic and inorganic electron transfer molecules in reduced environments. *Chemosphere* **1996**, *32*, 433–441.
32. Kluyev, N.; Cheleptchikov, A.; Brodsky, E.; Soyfer, V.; Zhilnikov, V. Reductive dechlorination of polychlorinated dibenzo-*p*-dioxins by zerovalent iron in subcritical water. *Chemosphere* **2002**, *46*, 1293–1296.
33. Gillham, R. W.; O'Hannesin, S. F. Enhanced degradation of halogenated aliphatics by zero-valent iron. *Ground Water* **1994**, *32*, 958–967.

34. Johnson, T. L.; Scherer, M. M.; Tratnyek, P. G. Kinetics of halogenated organic compound degradation by iron metal. *Environ. Sci. Technol.* **1996**, *30*, 2634–2640.
35. Miehr, R.; Tratnyek, P. G.; Bandstra, J. Z.; Scherer, M. M.; Alowitz, M. J.; Bylaska, E. J. Diversity of contaminant reduction reactions by zerovalent iron: Role of the reductate. *Environ. Sci. Technol.* **2004**, *38*, 139–147.
36. Morales, J.; Hutcheson, R.; Cheng, I. F. Dechlorination of chlorinated phenols by catalyzed and uncatalyzed Fe(0) and Mg(0) particles. *J. Hazard. Mater.* **2002**, *90*, 97–108.
37. Ravary, C.; Lipczynska-Kochany, E. Abiotic aspects of Zerovalent iron induced degradation of aqueous pentachlorophenol In *Preprint Extended Abstracts, Division of Environmental Chemistry, 209th National Meeting*; American Chemical Society: Anaheim, CA,1995; Vol *35*, pp738-740.
38. Kim, Y. H.; Carraway, E. R. Dechlorination of pentachlorophenol by zero valent iron and modified zero valent irons. *Environ. Sci. Technol.* **2000**, *34*, 2014–2017.
39. Keum, Y. S.; Li, Q. X. Reductive debromination of polybrominated diphenyl ethers by zerovalent iron. *Environ. Sci. Technol.* **2005**, *39*, 2280–2286
40. Lowry, G. V.; Johnson, K. M. Congener-specific dechlorination of dissolved PCBs by microscale and nanoscale zerovalent iron in a water/methanol solution. *Environ. Sci. Technol.* **2004**, *38*, 5208–5216.
41. Kim, J. H.; Lim, Y. K.; Chang, Y. S. Reductive dechlorination of PCDD/Fs on fly ash in aqueous solution using zero-valent iron metal, In *Preprint Extended Abstracts, Division of Environmental Chemistry, 228th National Meeting*; American Chemical Society: Philadelphia, Pennsylvania, 2004; Vol *44*, pp 235-238.
42. Lynam, M. M.; Kuty, M.; Damborsky, J.; Koca, J.; Adriaens, P. Molecular orbital calculations to describe microbial reductive dechlorination of polychlorinated dioxins. *Environ. Toxicol. Chem.* **1998**, *17*, 988–997.
43. Huang, C. L. I.; Keith Harrison, B.; Madura, J.; Dolfing, J. Gibbs free energies of formation of PCDDs: Evaluation of estimation methods and application for predicting dehalogenation pathways. *Environ. Toxicol. Chem.* **1996**, *15*, 824–836.
44. Dolfing, J.; Harrison, B. K. Gibbs free energy of formation of halogenated aromatic compounds and their potential role as electron acceptors in anaerobic environments. *Environ. Sci. Technol.* **1992**, *26*, 2213–2218.
45. Peijnenburg, W. J. G. M.; Hart, M. J.; Den Hollander, H. A.; Van De Meent, D.; Verboom, H. H.; Wolfe, N. L. QSARs for predicting reductive transformation rate constants of halogenated aromatic hydrocarbons in anoxic sediment systems. *Environ. Toxicol. Chem.* **1992**, *11*, 301–314.
46. Ghauch, A.; Charef, A.; Rima, J.; Martin-Bouyer, M. Reductive degradation of carbaryl in water by zero valent iron. *Chemopshere.* **2001**, *42*, 419-424.
47. Zhang, W.X.; Wang, C.B.; Lien, H.L. Treatment of chlorinated organic contaminants with nanoscale bimetallic paticles. *Catalysis today.* **1998**, *40*, 387-395.
48. Wang, C. B.; Zhang, W.-X. Synthesizing nanoscale iron particles for rapid and complete dechlorination of TCE and PCBs. *Environ. Sci. Technol.* **1997**, *31*, 2154–2156.

49. Glavee, G. N.; Klabunde, K. J.; Sorensen, C. M.; Hadjipanayis, G. C. Chemistry of borohydride reduction of iron(II) and iron(III) ions in aqueous and non-aqueous media. Formation of nanoscale Fe, FeB, and Fe_2B powders. *Inorg. Chem.* **1995**, *34*, 28–35.
50. Grittini, C.; Malcomson, M.; Fernando, Q.; Korte, N. Rapid dechlorination of polychlorinated biphenyls on the surface of a Pd/Fe bimetallic system. Environ. Sci. Technol. 1995, 29, 2898– 2900.
51. Muftikian, R.; Nebesny, K.; Fernando, Q.; Korte, N. X-ray photoelectron spectra of the palladium-iron bimetallic surface used for the rapid dechlorination of chlorinated organic environmental contaminants. *Environ. Sci. Technol.* **1996**, *30*, 3593–3596.
52. Zhou, H. Y.; Xu, X. H.; Wang, D. H. Catalytic dechlorination of chlorobenzene in water by Pd/Fe bimetallic system. *J. Environ. Sci.* **2003**, *15*, 647–651.
53. Kim, J.H.; Tratnyek, P.G.; Chang, Y.S. Rapid Dechlorination of Polychlorinated dibenzo-p-dioxins by Bimetallic and Nanosized Zerovalent Iron. Environ. Sci. Technol.2008, 42, 4106-4112.
54. Nurmi, J. T.; Tratnyek, P. G.; Sarathy, V.; Baer, D. R.; Amonette, J. E.; Pecher, K.; Wang, C.; Linehan, J. C.; Matson, D. W.; Penn, R. L.; Driessen, M. D. Characterization and properties of metallic iron nanoparticles: Spectroscopy, electrochemistry, and kinetics. *Environ. Sci. Technol.* **2005**, *39*, 1221–1230.
55. Chun, C. L.; Baer, D. R.; Matson, D. W.; Amonette, J. E.; Penn, R. L. Characterizations and reactivity of metal-doped iron and magnetite nanoparticles. In *Preprint Extended Abstracts, Division of Environmental Chemistry, 233rd National Meeting*; American Chemical Society: Chicago, Illinois, 2007, Vol *47*, pp 408-412.
56. Kim, Y. H.; Carraway, E. R. Dechlorination of chlorinated ethenes and acetylenes by palladized iron. *Environ. Tech.* **2003**, *24*, 809– 819.
57. Scherer, M. M.; Balko, B. A.; Gallagher, D. A.; Tratnyek, P. G. Correlation analysis of rate constants for dechlorination by Zerovalent iron. *Environ. Sci. Technol.* **1998**, *32*, 3026–3033.
58. Tratnyek, P. G.; Weber, E. J.; Schwarzenbach, R. P. Quantitative structure-activity relationships for chemical reductions of organic contaminants. *Environ. Toxicol. Chem.* **2003**, *22*, 1733– 1742.
59. Song, H.; Carraway., E. R. Catalytic hydrodechlorination of chlorinated ethenes by nanoscale zero-valent iron. *App. Catal., B.* **2008**, *78*, 53–60.
60. Arnold,W.A.; Roberts., A. L. Pathways and kinetics of chlorinated ethylene and chlorinated acetylene reaction with $Fe^{(0)}$ particles. *Environ. Sci. Technol.* **2000**, *34*, 1794–1805.
61. Cwiertny, D. M.; Bransfield, S. J.; Livi, K. J. T.; Fairbrother, D. H.; Roberts, A. L. Exploring the influence of granular iron additives on 1,1,1-trichloroethane reduction. *Environ. Sci. Technol.* **2006**, *40*, 6837–6843.
62. Cwiertny, D. M.; Bransfield, S. J.; Roberts, A. L. Influence of the oxidizing species on the reactivity of iron-based bimetallic reductants. Environ. Sci. Technol. **2007**, 41, 3734–3740.
63. Tratnyek, P. G.; Johnson, R. L. Nanotechnologies for environmental cleanup. *Nano Today* **2006**, *1*, 44–48.

64. Shiu, W. Y.; Doucette, W.; Gobas, F. A. P. C.; Andren, A.; Mackay, D. Physical-chemical properties of chlorinated dibenzo-*p*dioxins. *Environ. Sci. Technol.* **1988**, *22*, 651–658.
65. Lee, J. E.; Choi, W.; Mhin, B. J. DFT calculation on the thermodynamic properties of polychlorinated dibenzo-*p*dioxins: Intramolecular Cl-Cl repulsion effects and their thermochemical implications. *J. Phys. Chem. A* **2003**, *107*, 2693– 2699.
66. Bunge, M.; Adrian, L.; Kraus, A.; Opel, M.; Lorenz, W. G.; Andreesen, J. R.; Go¨risch, H.; Lechner, U. Reductive dehalogenation of chlorinated dioxins by an anaerobic bacterium. *Nature* **2003**, *421*, 357–360.
67. Ukisu, Y.; Miyadera, T. Hydrogen-transfer hydrodechlorination of polychlorinated dibenzo-*p*-dioxins and dibenzofurans catalyzed by supported palladium catalysts. *App. Catal., B.* **2003**, *40*, 141–149.
68. Schrick, B.; Blough, J. L.; Jones, A. D.; Mallouk, T. E. Hydrodechlorination of trichloroethylene to hydrocarbons using bimetallic nickel-iron nanoparticles. *Chem. Mater.* **2002**, *14*, 5140–5147.
69. Hirokawa, S.; Imasaka, T.; Urakami, Y. Ab initio MO study on the S1rS0 transitions of polychlorinated dibenzo-*p*-dioxins. *J. Mol. Struct.: Theochem* **2003**, *622*, 229–237.
70. Perlinger, J. A.; Venkatapathy, R.; Harrison, J. F. Linear Free Energy Relationships for Polyhalogenated Alkane Transformation by Electron-transfer Mediators in Model Aqueous Systems. *J. Phys. Chem. A* **2000,** *104*, 2752-2763.
71. Tratnyek, P. G.; Weber, E. J.; Schwarzenbach, R. P. Quantitative structure-activity relationships for chemical reductions of organic contaminants. *Environ. Toxicol. Chem.* **2003,** *22*, 1733-1742.
72. Scherer, M. M.; Balko, B. A.; Gallagher, D. A.; Tratnyek, P. G. Correlation analysis of rate constants for dechlorination by zero- valent iron. *Environ. Sci. Technol.* **1998**, *32*, 3026-3033.
73. Wang, Z.; Huang, W.; Fennell, D. E.; Peng, P. Kinetics of reductive dechlorination of 1,2,3,4-TCDD in the presence of zero-valent zinc. *Chemosphere* **2008,** *71*, 360-368.

Chapter 7

Degradation of TNT, RDX, and TATP using Microscale Mechanically Alloyed Bimetals

Rebecca Fidler[1], Tamra Legron[1], Kathleen Carvalho-Knighton[2], Cherie L. Geiger[1], Michael E. Sigman[1] and Christian A. Clausen[1]

[1] Department of Chemistry, University of Central Florida, Orlando, FL, 32816
[2] Department of Environmental Science, Policy and Geography, University of South Florida-St. Petersburg, St. Petersburg, FL, 33701

Microscale mechanically alloyed bimetals, particularly Mg/Pd, are being explored as alternative remediation methods for the catalytic reduction of energetic materials triacetone triperoxide (TATP), a peroxide explosive, and environmental contaminants: trinitrotoluene (TNT), and cyclo-trimethylenetrinitramine (RDX). TNT and RDX have been found to contaminate soil and water near industrial production sites, therefore a method for rapid and cost effective remediation is needed. Mg/Pd, as well as other bimetals, Fe/Ni, and Fe/Pd, were shown to reduce TNT and RDX contamination in water samples with varying reactivities. For the degradation of TNT using the microscale mechanically alloyed bimetals, the normalized rate constants obtained were 5.6×10^{-4}, 5.0×10^{-3}, and 2.7×10^{-4} L g^{-1} min^{-1} for Mg/Pd, Fe/Pd, and Fe/Ni, respectively. The normalized rate constants for the degradation of RDX obtained from the vial studies were as follows 1.5×10^{-4}, 4.4×10^{-5}, and 3.2×10^{-5} L g^{-1} min^{-1} for Mg/Pd, Fe/Pd, and Fe/Ni, respectively. Another explosive that has become more prevalent is triacetone triperoxide (TATP). TATP although not environmentally recalcitrant has proved difficult to treat due to its sensitivity to heat and friction. Mechanically alloyed Mg/Pd was shown to degrade TATP in a methanol and water solution with a normalized rate constant of 1.2×10^{-3} L g^{-1} min^{-1}, and acetone was observed as the major product.

Introduction

RDX and TNT Contamination

The incomplete discharge of explosives in munitions and waste streams in industrial production areas has brought about the distribution of TNT and RDX in soil, groundwater, and explosive residue on structures (*1-3*). The toxicity and mutagenicity of RDX and TNT (*4,5*) has resulted in EPA regulations to develop remediation technologies and safeguard the environment from present and future contamination of these energetic compounds (*6,7*). Increased attention to the future environmental and health concerns of the explosives 2,4,6-trinitrotoluene (TNT, Figure 1) and cyclo-1,3,5-trimethylene-2,4,6-trinitramine (RDX, Figure 2) has augmented development of technologies that remediate contaminated groundwater, soil, and structures.

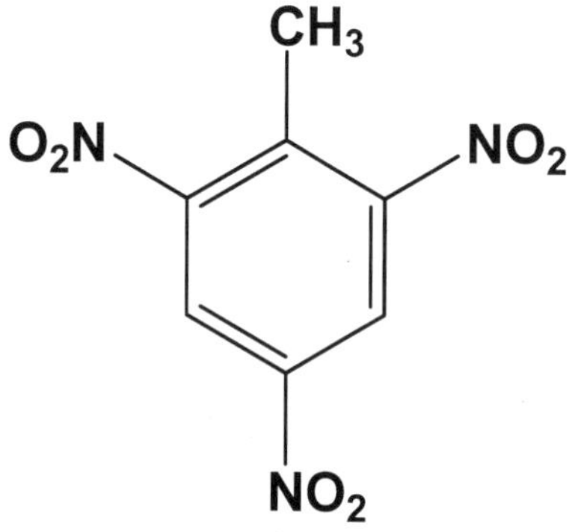

CH_3

O_2N NO_2

NO_2

Figure 1. TNT

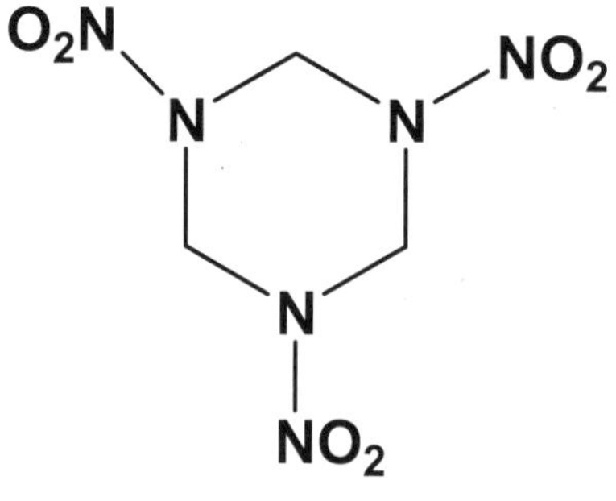

O_2N NO_2

N N

N

NO_2

Figure 2. RDX

Remediation technologies have focused on removal of explosives from contaminated soil (*8*) and groundwater contamination (*9,10*) while other work has concentrated on the decontamination of explosives from ordnance and scrap metal (*11*). Research on explosive remediation has utilized microbial (*12*) and phytoremediation techniques (*13*). Although these biological remediation technologies have economical advantages and low energy usage, they have limitations that include slower kinetic rates (*12*) and the suitability for remediating only low-level contamination (*14*).

Other remediation techniques have included use of complexed (*15*) and zero-valent iron (ZVI) (*10,16,17*), Fenton reactions (*18*), transition metals including TiO_2 palladium photocatalysis (*19,20*), and nickel catalysts (*21*). Remediation techniques using ZVI are typically limited to abiotic conditions, controlled pH (*22*), and limited by surface corrosion, as well as, the production of 2,4,6-triaminotoluene (TAT) (*16*). Exploring commonly known hydrogenation catalysts, such as Pd and Ni, allows for the successful remediation of these explosives under ambient temperatures, pressures, and pH conditions. These metals are also, overall, less susceptible to surface corrosion. However, these remediation techniques may require hydrogen sources and reaction vessels purged with hydrogen gas (*21*). The reaction criteria for these remediation techniques would make them less suitable for *in situ* treatment of explosives contamination.

Combining transition metal catalysts with ZVI has proved successful in reducing chlorinated compounds (*23*). The effect on the degradation mechanics of TNT and RDX by using mechanically alloying Pd and Ni to ZVI was explored in this paper. A zero-valent metal that has been used as a substitute for ZVI is magnesium, which has its advantages over the widely used ZVI as a reductive metal since Mg has a greater reduction potential as that compared to ZVI:

$$Mg^{2+} + 2e^- \rightarrow Mg^0 \qquad E^0 = -2.37V$$

$$Fe^{2+} + 2e- \rightarrow Fe^0 \qquad E^0 = -0.44V$$

Magnesium also possesses a self-limiting oxide layer unlike the easily corroded ZVI. The thermodynamically favored magnesium combined with the hydrogenation catalyst palladium has been explored as a reductive catalytic system and has been shown to reduce chlorinated aromatics (*24,25*) thus the use of Mg/Pd as a reductive catalytic system for nitroaromatics and other nitro explosives appears promising.

Bimetals, Mg/Pd, Fe/Ni, and Fe/Pd, in combination with other technologies including emulsified zero valent metal (EZVM) (*26*) and bimetallic treatment systems (BTS) (*27*) will provide a *in situ* method for treating TNT and RDX contamination in groundwater, soil, and structures. The future field deployment of these technologies to remediate explosive contamination is the ultimate objective in exploring these metals.

Degradation of TATP

Another energetic material that has become more commonly used is triacetone triperoxide (TATP, Figure 3) even though not considered environmentally hazardous. TATP is a cyclic peroxide explosive that is readily synthesized using hydrogen peroxide, acetone, and an acid catalyst (*28*). TATP has become more widely used in terrorist acts (*29*), as well as used by clandestine chemists (*30*). Typically, TATP is found in its dry crystal form or wet crystals obtained from reaction mixtures. TATP has a high vapor pressure and is especially sensitive to decomposition from heat or friction, making it unsuitable for industrial production. Although TATP is not environmentally recalcitrant, TATP contamination is still found in areas of underground production as well as contaminates targets of terrorist attacks, which could pose a threat to both the public as well as the law enforcement personnel. The instability of TATP makes the clean up of a contaminated area a challenging problem resulting in the need for a safe and rapid *in situ* destruction and clean-up method. Currently, TATP destruction is limited to thermal decomposition (*31*), refluxing in toluene with $SnCl_2$ (*32*), and exposing TATP to copper at a low pH (*33*). Although the methods are successful in cleaving TATP to produce non-explosive byproducts, these methods are not *in situ* techniques and thus results in disturbing the TATP contamination.

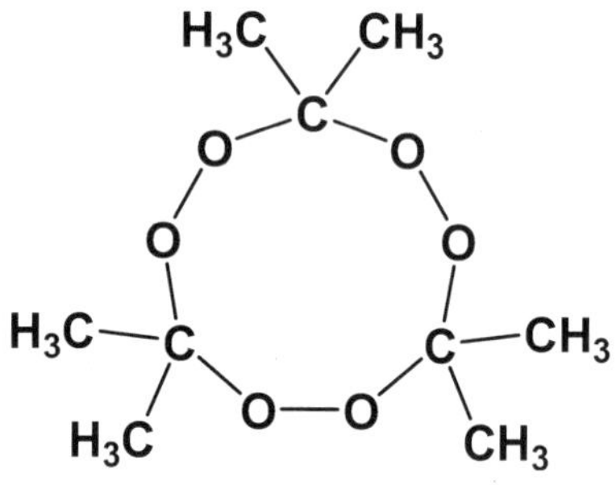

Figure 3. TATP

Mg/Pd has also been explored in this paper as a reductive catalytic system that cleaves the TATP ring to produce non-explosive byproducts. A technology that can aid in the *in situ* treatment of TATP contaminated areas is the use of Mg/Pd in accordance with the EZVM technology. The outer oil membrane of the EZVM has been observed to absorb TATP crystals, which can then be degraded by Mg/Pd that is contained within the inner aqueous layer. This EZVM technology allows for the treatment of both wet and dry TATP contamination.

Experimental Procedure

Metal Preparation

The metals used included microscale iron (1-3 μm diameter); microscale magnesium (2-4 μm diameter); 1% palladium on carbon; and 75% Nickel on graphite (<75 μm). Mg/Pd, Fe/Ni, and Fe/Pd were prepared by ball-milling 78 g of Mg or Fe and 7 g of the respective bimetal. The optimal ball-milling process used was developed in-house (*34*). The catalyst loading for the prepared bimetals are found in Table I. Initial characterization of the Mg/Pd was performed using scanning electron microscopy (SEM) and energy X-ray dispersion spectroscopy (EDS).

Table I. Catalyst Loading for the Bimetals

Bimetal	ZVM (%)	Catalyst (%)	Carbon (%)
Mg/Pd	91.8	0.08	8.20
Fe/Pd	91.8	0.08	8.20
Fe/Ni	91.8	6.20	2.00

RDX and TNT Neat Metal Studies and Analysis

Standard solutions (5000 μg/mL RDX or TNT in acetonitrile) were diluted in deionized water to concentrations of 10 ppm RDX and 5 ppm TNT. All vial metal studies were performed in duplicate, under ambient conditions (temperature and pressure) and normal aerobic conditions. Five mL samples of the appropriate aqueous explosive sample were exposed to 0.25 g of the appropriate metal in 12 mL glass amber vials that were continuously shaken using a reciprocating shaker. Samples were extracted using 5 mL toluene and shaken by hand for 2 minutes followed by an ultrasonic bath for 5 minutes. The toluene layer was filtered and removed from metal and water using Puradisc™ 25 mm diameter 0.45 μm syringe filter attached to a glass syringe.

GC analysis conditions were modified from methods reported in literature (*35-39*). Samples were analyzed on a Perkin Elmer AutoSystem gas chromatograph equipped with an electron capture detector (GC-ECD). An Rtx-5 column (30 m, 0.25 mm i.d., 0.25 μm df) was used. The ECD makeup gas was ultra high purity nitrogen at 30 mL/min, and helium was used as carrier gas with a constant pressure of 20 psi and 40 psi for the analysis of TNT and RDX, respectively. The injector temperature was 180°C and the detector temperature was 325°C. For TNT analysis, the oven was programmed at an initial temperature of 50°C, held 1 minute, ramped at 8°C/minute until reaching 250°C and was held 3 minutes. For RDX analysis, the oven was programmed at an initial temperature of 50°C held for 1 minute, ramped at 20°C/min. until reaching a final temperature of 300°C where it is held for 3 minutes. A Thermo Finnagin

Trace gas chromatograph DSQ mass spectrometer (GC-MS) was also used. The column, oven temperature program, carrier gas, and injector port were the equivalent to the GC-ECD conditions. The GC-MS transfer line and the source were both at 200°C.

TATP Neat Metal Studies and Analysis

TATP was prepared in the laboratory using acetone, hydrogen peroxide, and sulfuric acid according to previously published procedures (*28,40*), and the prepared TATP crystals were dissolved in methanol and deionized water. Neat metal studies were performed using 0.25 g Mg/Pd in 20 mL vials and 5 mL of a 4:1 water: methanol solution containing TATP. Samples were syringe extracted using 5 mL of toluene, hand-shaken for 2 minute and centrifuged. The top organic layer was removed and analyzed using a modified method (*39*). A Thermo Trace Gas Chromatograph (GC) equipped with a DSQ Mass Spectrometer (MS) was used for the analysis. The GC was outfitted with a 30m Rtx-5® column (0.25 mm ID, 0.25 μm df). Helium was used as the carrier gas at a flow rate of 1.2 mL/min. The sample was injected into a 110°C injector port. The initial GC oven temperature was held at 50°C for 1 minute and then ramped at 10°C/min until reaching a final temperature of 140°C (held for 0.5 minutes). The transfer line was held at 150°C and the source temperature was set at 150°C. The MS was programmed for selective ion monitoring (SIM) of m/z 43, 59, and 75. Acetone was analyzed using a modified literature method using a Perkin Elmer Clarius GC with flame ionization detection (FID) (*40*). The GC/FID was equipped with a Stabilwax® capillary column (30 m, 0.53 mm ID, 1 μm df) using helium (5.2 mL/min flow rate) as the carrier gas. The sample was introduced into an injector held at 170°C with a 2:1 split ratio. The initial oven temperature was 50°C (held for 1 min), and then ramped at 10°C/min until reaching a final temperature of 150°C (held for 1 min).

Experimental Results

Characterization of Metal

Initial testing for the use of bimetals for the degradation of explosives was utilizing mechanically alloyed Mg/Pd. Characterization of the metal was conducted using SEM (Figures 4 and 5) and EDS (Figure 6). The bulk of the metal particle sizes produced from the mechanical alloying process fell within 2-20 μm as seen in Figure 5. The EDS determined the presence of Mg, Pd, C, and O, and the calculated abundances are for in Figure 6.

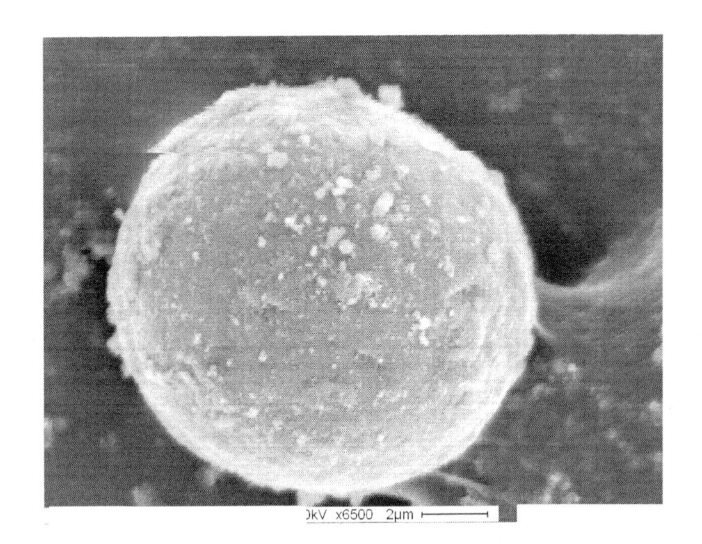

Figure 4: SEM image of individual Mg/Pd particle

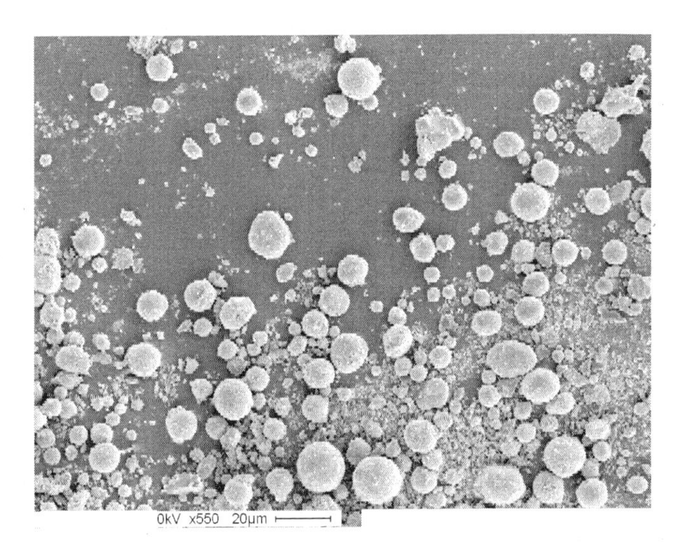

Figure 5: SEM image of the bulk Mg/Pd particles

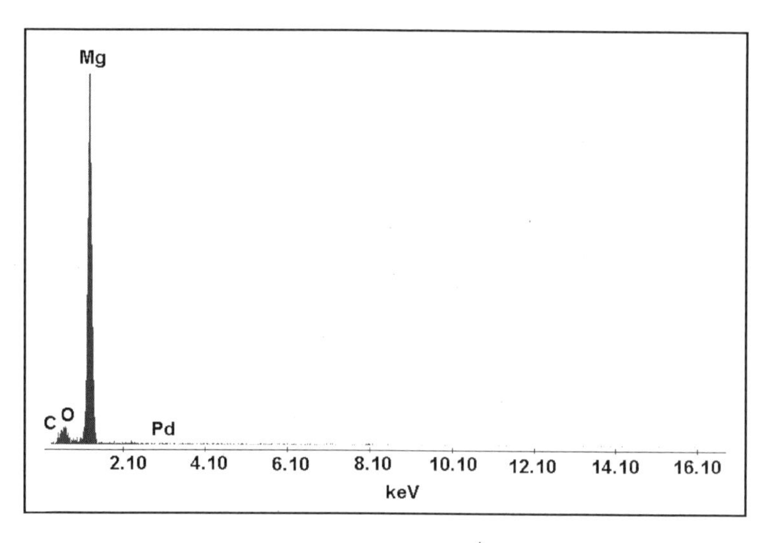

Element (α-Band)	Weight %
C (K)	1.46
O (K)	12.28
Mg (K)	86.09
Pd (L)	0.17

Figure 6. EDS of the mechanically alloyed Mg/Pd particles

TNT Neat Metal Kinetic Studies

The kinetic data of the TNT degradation with Mg/Pd particles (Figure 7) appear to demonstrate a pseudo-first-order rate law, which agrees with experimental data using comparable metals with explosives (*15,16,21*). The pseudo-first-order rate is:

$$\frac{d[TNT]}{dt} = -k_{TNT}[TNT]$$

The rate constant (k_{TNT}) was determined using the following first-order kinetic equation:

$$\frac{[TNT]}{[TNT]_0} = e^{-k_{TNT}t}$$

The rate constants calculated from the pseudo-first order equation are normalized with the metal concentration (ρ_m=50 g L^{-1}). The normalized rate constants for the degradation of TNT are given in Table II.

As observed in Figure 7, TNT in water was degraded by Mg/Pd. Initial work was completed to identify the byproducts of TNT degradation with Mg/Pd. This work was completed in a 50% (v/v) methanol/water solution in order to increase the solubility of the TNT in the aqueous layer. GC-MS was used to identify 2, 4-dinitrotoluene, 4-amino-2, 6-dinitrotoluene and 2-amino-4, 6-dinitrotoluene. These byproducts were observed to increase then diminish with Mg/Pd expected to have degraded these byproducts in addition to TNT (Figure 8). The final byproducts of all these reactions are being explored as well.

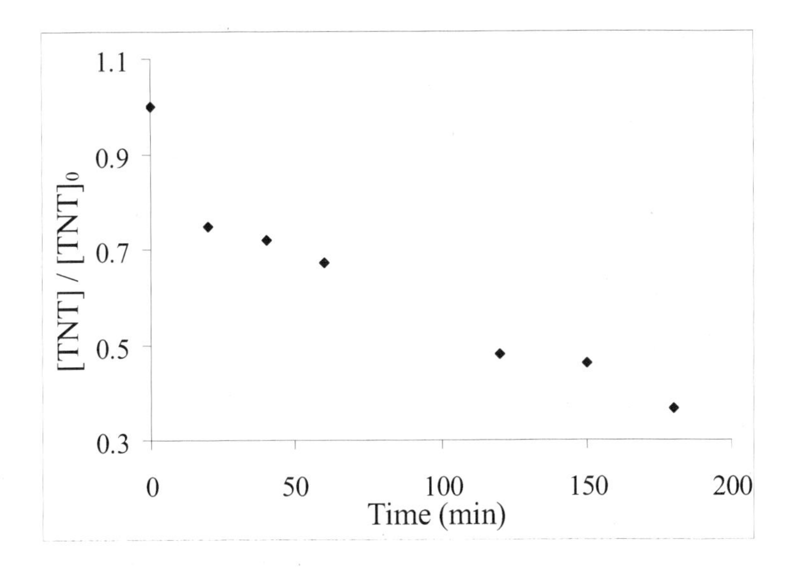

Figure 7. Pseudo-first order kinetic degradation plots of TNT in water

Additional work was completed to test the ability of microscale iron and iron based bimetals to degrade TNT. As seen if Figure 9, TNT was not degraded in the presence of the micro-scale iron at ambient conditions, yet the addition of the hydrogenation catalysts activated the iron to degrade TNT in water. The degradation of TNT with Fe/Pd occurred rapidly; consequently, the rate constant of the degradation of TNT with Fe/Pd was only calculated during the initial drop (first 20 minutes) of TNT concentration.

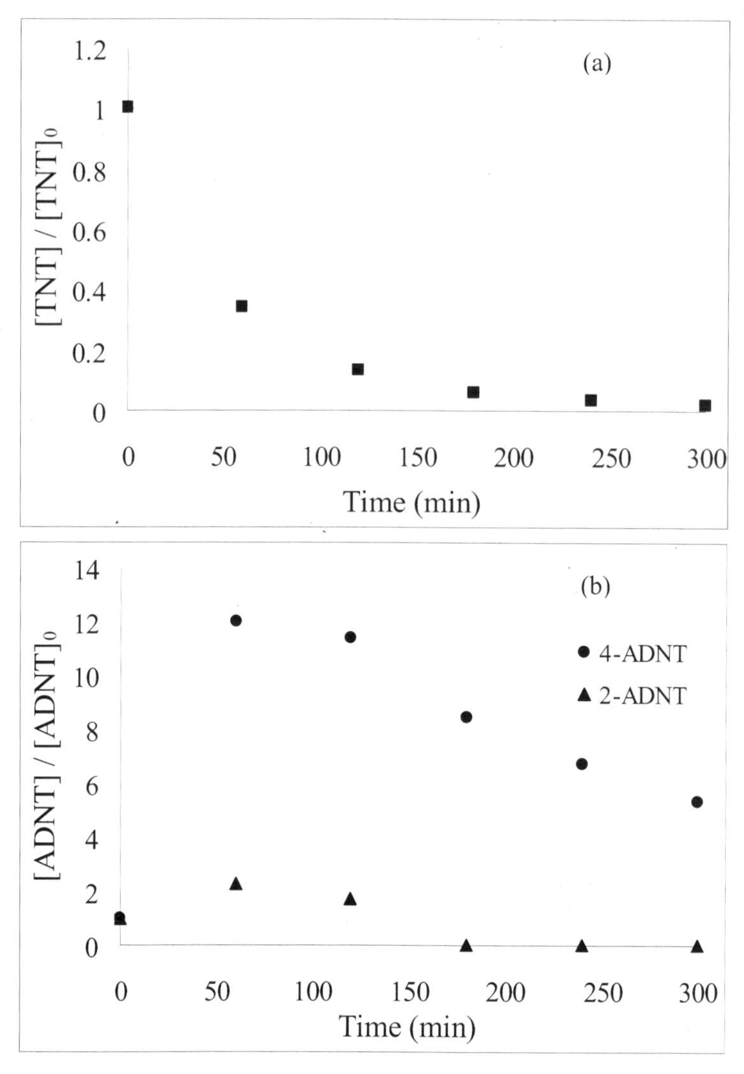

Figure 8. Pseudo-first order kinetic degradation of TNT using Mg/Pd particles in

50% (v/v) water/methanol (a) and byproduct production (b)

Table II. Normalized Pseudo-First-Order Rate Constants (k_{TNT}) of TNT Degradation in Water.

Metal	Normalized Rate Constant (k_{TNT}) (L g^{-1} min^{-1})	Coefficient of Determination (R^2)
Fe	NDO*	NA
Mg/Pd	5.6×10^{-4}	0.96
Fe/Pd	5.0×10^{-3}	0.99
Fe/Ni	2.7×10^{-4}	0.84

NOTE: NDO = No degradation observed

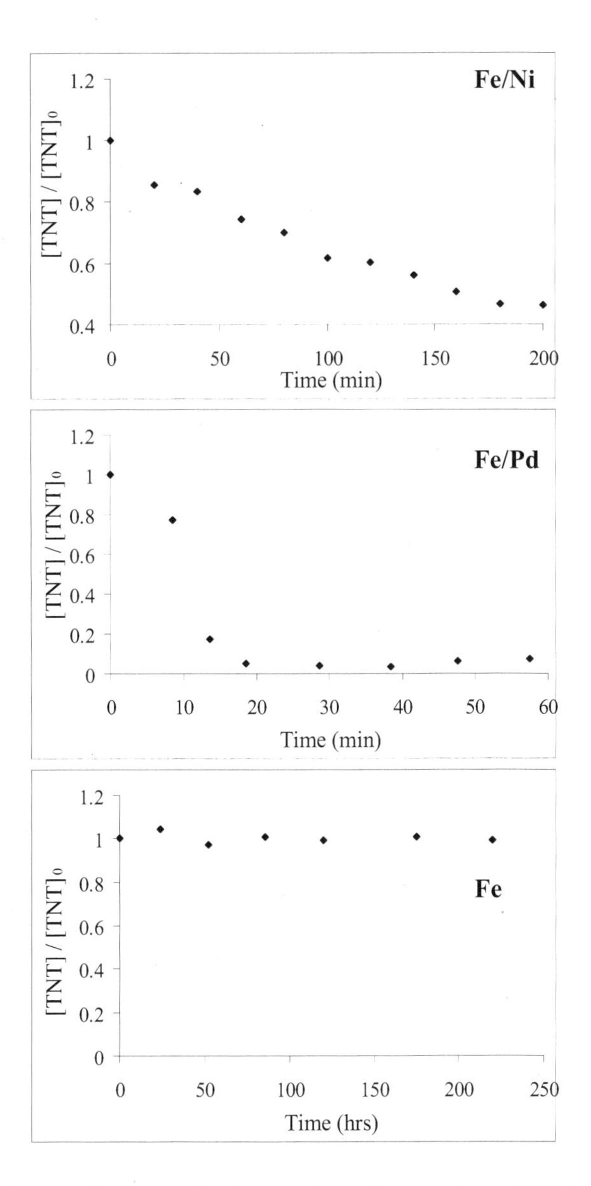

Figure 9. Pseudo-first order kinetic degradation plots of TNT in water

RDX Neat Metal Kinetic Studies

Similar to TNT, RDX degradation also exhibited pseudo-first-order kinetic degradation, thus the pseudo first order rate law of RDX degradation was expressed as

$$\frac{d[RDX]}{dt} = -k_{RDX}[RDX]$$

A lag-period was observed for the first 100 minutes after exposing the RDX to Fe/Pd and within the initial 45 minutes in the Mg/Pd studies with RDX as seen in Figure 10a and 10b. This lag-period may be due to adsorption of the hydrogen to the catalyst's surface. This adsorption of hydrogen is required for the reduction reaction to occur. These lag-periods were ignored when calculating the pseudo-first order rate constants for the Mg/Pd and Fe/Pd studies. The rate constants (k_{RDX}) for all the metals are determined using the following first-order kinetic equation:

$$\frac{[RDX]}{[RDX]_0} = e^{-k_{RDX}t}$$

The rate constants for RDX degradation are found in Table III. The data from each metal study with RDX was fit to the previous first-order rate constant and the exponential fit had coefficients of determination at 0.90 or above. The rate constants were normalized to the metal concentration (ρ_m=50 g L^{-1}) that was used in the vial studies. The zero-order kinetic plots of RDX in water, Figures 10-13, demonstrate RDX degraded by Mg/Pd, Fe/Pd, and Fe/Ni as compared that of micro-scale Fe. The micro-scale Fe showed no degradation after 9 days in ambient conditions. No byproducts were identified in analyzing for RDX with the GC-ECD. Further work and additional analytical technique needs to be implemented to identify the final byproducts of the RDX degradation.

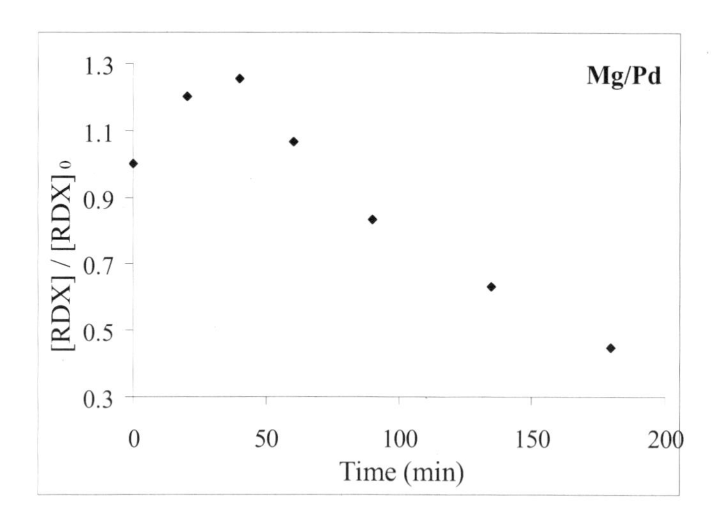

Figure 10. Pseudo-first order kinetic degradation of RDX in water using Mg/Pd

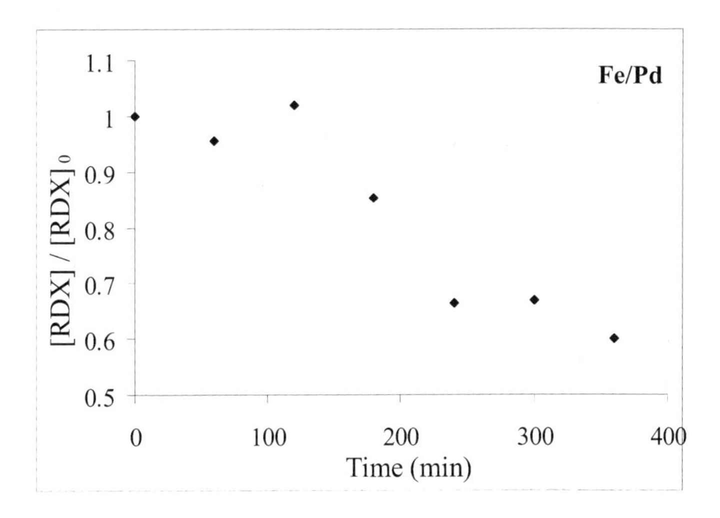

Figure 11. Pseudo-first order kinetic degradation of RDX in water using Fe/Pd

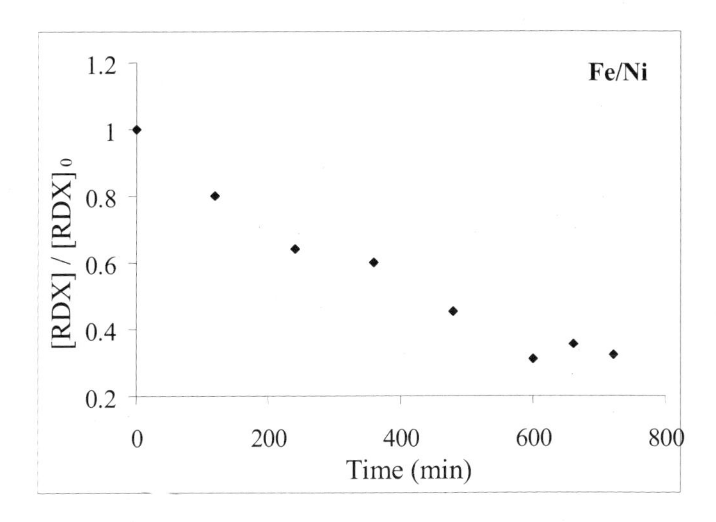

Figure 12. Pseudo-first order kinetic degradation of RDX in water using Fe/Ni

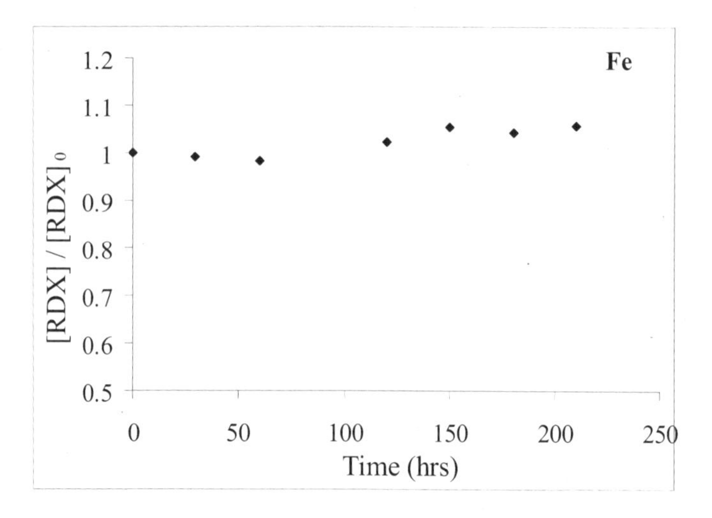

Figure 13. Pseudo-first order kinetic degradation of RDX in water using Fe

Table III. Normalized pseudo-first-order rate constants (k_{RDX}) of RDX degradation in water.

Metal	Normalized Rate Constant (k_{RDX}) (L g^{-1} min^{-1})	Coefficient of Determination (R^2)
Fe	NDO*	NA
Mg/Pd	$1.5x10^{-4}$	0.99
Fe/Pd	$4.4x10^{-5}$	0.97
Fe/Ni	$3.2x10^{-5}$	0.90

NOTE: NDO = No degradation observed

TATP Neat Metal Kinetic Studies

TATP was degraded by mechanically alloyed Mg/Pd particles as seen by the pseudo-zero order kinetic data (Figure 14). The kinetics of the TATP disappearance has exhibited a pseudo-first order rate law:

$$\frac{d[TATP]}{dt} = -k_{TATP}[TATP]$$

The rate constant (k_{TATP}) is determined using the following first-order kinetic equation:

$$\frac{[TATP]}{[TATP]_0} = e^{-k_{TATP}t}$$

The rate constant listed in Table IV was calculated by fitting the pseudo first-order kinetic equation to the degradation data, and the rate constant was normalized to the metal concentration (ρ_m=50 g L^{-1}). There was an initial drop in TATP concentration observed followed by a slower rate of TATP degradation. This first initial point was not used in calculating the rate constant of TATP degradation.

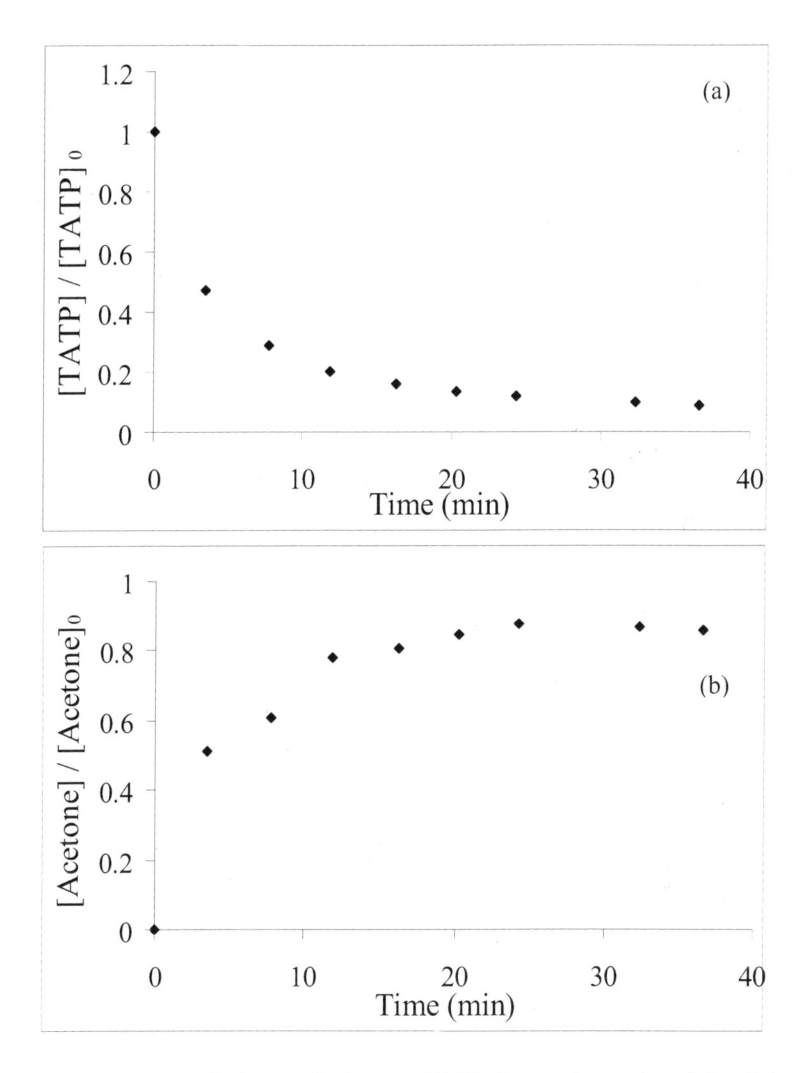

Figure 14. Pseudo first-order kinetic TATP degradation (a) with Mg/Pd particles and acetone production (b) in 4:1 water:methanol solution

Table IV. Normalized pseudo-first-order rate constant (k_{TATP}) of TATP degradation in 4:1 water:methanol solution

Metal	Normalized Rate Constant (k_{TATP}) (L g^{-1} min^{-1})	Coefficient of Determination (R^2)
Mg/Pd	$1.2x10^{-3}$	0.86

Acetone was observed as the major byproduct of the degradation of TATP with Mg/Pd and the production of acetone is shown in Figure 14.

Conclusions

All three bimetals, Fe/Pd, Fe/Ni, and Mg/Pd, were shown to degrade the explosives RDX and TNT in contrast to the observation that microscale ZVI showed no degradation at ambient temperatures and aerobic conditions. These metals show a varying degree of activity between the two explosives. For RDX degradation, the activity of the metal is observed as Mg/Pd > Fe/Ni > Fe/Pd whereas the metal activity with TNT is observed as Fe/Pd > Mg/Pd > Fe/Ni. This difference in activity may be due to an alternative degradation pathway, which requires future investigation to document if this is the reason. The activities of the metals with the explosives are compared in Figure 15.

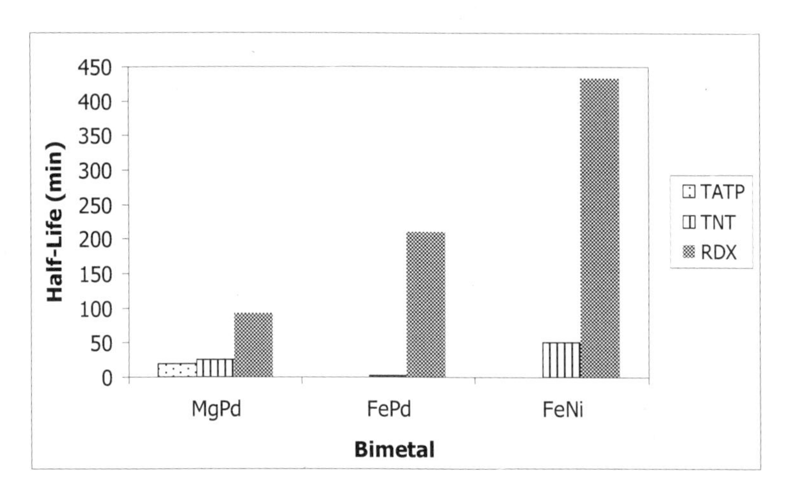

Figure 15. Comparison of the activity of different bimetals with different explosives

The final byproducts of the degradation of these explosives are very important in the bimetal's capacity for field type applications. Products, such as TAT, have been recognized as the major product of the reaction with TNT and ZVI under inert environments (*16*). TAT, however, is more hazardous than the TNT contamination aiming to be degraded. The nature of the tested bimetals may lead them to proceed in a different reductive pathway than that of ZVI, which could produce more reduced byproducts than that of ZVI. Other work with Ni catalysts has produced byproducts that did not include TAT (*21*), which appears promising for future use of these bimetals for field application utilizing the EZVM and BTS delivery technologies. Further work must be accomplished to determine the byproducts of these reactions, which would verify the suitability of field application.

TATP was also revealed to be degraded using the bimetal, Mg/Pd in which the major byproduct of that reaction was acetone. The Mg/Pd EZVM also was successful in absorbing and degrading the TATP from the aqueous solution. It also has been observed to absorb and degrade dry TATP crystals (not presented in this paper). Incorporating the Mg/Pd with the EZVM delivery technology provides a feasible *in situ* method for treating wet and dry crystals of TATP.

References

1. Park J., Comfort S. D., Shea P. J., Machacek T. A., *J. Environ. Qual.* (2004), **33**(4), 1305-13.
2. Pennington, J.C. and Brannon, J.M. (2002) *Thermochimica Acta* **384**, 163-72.
3. Pennington, J.C. et al. (2005). ERDC TR-05-2. April 2005.
4. Lachance, B., Robidoux, P.Y., Hawari, J., Ampleman, G., Thiboutot, S., and Sunahara, G. J. (1999) *Mutation Research* **444**, 25-39.
5. Peters, G.T., Burton, D.T., Paulson, R.L., Turley, S.D. (1991) *Environ. Toxicol. Chem.* **10**, 1073-1081.
6. USEPA (1988). Office of Drinking Water, Washington, D.C. CASRN 118-96-7.
7. USEPA (1997). Solid Waste and Emergency Response (5306W) EPA530-F-97-045.
8. Ahmad, F., Schnitker, S.P., Schnitker, S.P., Newell, C.J. (2007) *J. of Contaminant Hydrology* **90**, 1-20.
9. Best, E.P.H., Sprecher, S.L., Larson, S.L., Fredrickson, H.L., and Bader, D.F. (1999) *Chemosphere* **38**, 3383-96.
10. Oh, S.Y., Chiu, P.C., and Kim, B.J. (2006) *Water Sci. & Technol.* **54**, 47-53.
11. Jung, C.M., Newcombe, D.A., Crawford, D.L., and Crawford,R.L. (2004) *Biodegradation* **15**, 41-8.
12. Kulkarni, M. and Chaudhari, A. (2007) *J. of Environ. Management* **85**, 492-512.
13. Hannink, N.K., Rosser, S.J., and Bruce, N.C. (2002) *Critical Reviews in Plant Sciences* **21**, 511-38.
14. Snellinx, Z., Nepovim, A., Taghavi, S., Vangronsveld, J., Vanek, T., and van der Lelie, D. (2002) *Environ. Sci. Pollut. Res.* **9**, 48-61.
15. Kim, D. and Strathmann, T.J., (2007) *Environ. Sci. Technol.* **41**, 1257-64.
16. Bandstra, J.Z., Miehr, R., Johnson, R.L., and Tratnyek, P.G. (2005) *Environ. Sci. Technol.* **39**, 230-8.

17. Welch, R., Riefler, R. G., Environ. Eng. Sci. (2008), **25** (9), 1255-62.
18. Liou, M.-J., Lu, M.-C., Chen, J.-N. (2003) *Water Research* **37,** 3172-79.
19. Son, H.-S., Lee, S.-J., Cho, I., and Zoh, K.-D., (2004) *Chemosphere* **57**, 309-17.
20. Dillert, R., Brandt, M., Fornetfett, I., Siebers, U., and Bahnemann, D. *Chemosphere* **30**, 2333-41.
21. Fuller, M.E., Schaefer, C.E., and Lowey, J.M. (2007) *Chemosphere* **67**, 419-27.
22. Mu, Y., Yu, H.-Q., Zheng, J.C., Zhang, S.J., and Sheng, G.-P. (2004) *Chemosphere* **54**, 789-94.
23. Cwiertny, D.M., Bransfield, S.J., Livi, K.J.T., Fairbrother, D.H., Roberts, A.L. (2006) *Environ. Sci. Technol.* **40** 6837-43.
24. Hadnagy, E., Rauch, L.M., and Gardner, K.H. (2007) *J. of Environ. Sci and Health Part A* **42**, 685-95.
25. Patel, U.D. and Suresh, S. (2007) *J. of Hazardous Mat.* **147**, 431-8.
26. Quinn, J., Geiger, C., Clausen, C., Brooks, K., Coon, C., O'Hara, S., Krug, T., Major, D., Yoon, W.-S., Gavaskar, A., and Holdsworth, T. (2005) *Environ. Sci. and Tech.* **39**, 1309-18.
27. Brooks, K.B., Quinn, J.W., Clausen, C.A., Geiger, C.L.; Aitken, B.S., Captain, J., and Devor, R.W. (2006) Remediation of Chlorinated and Recalcitrant Compounds--2006, Proceedings of the International Conference on Remediation of Chlorinated and Recalcitrant Compounds, 5th, Monterey, CA, United States, May 22-25.
28. Milas N.A. and Golubovic A. (1959) *J. Am. Chem. Soc.* **82**, 3361-64.
29. Block, L. (2006) *Terrorism Monitor* **4,** 1-2.
30. Cannon, J.G. (2006) *The Oklahoman.* Mar 1.
31. Oxley, J.C., Smith J.L., Chen, H. (2002) *Propellants, Explosives, Pyrotechnics* **27**, 209-16.
32. Bellamy, J.A. (1999) *J. Forensic Sci.* **44**, 603-608.
33. Costantini, Michel. (1989) United States Patent 5003109.
34. Aitken, B., Geiger, C., Clausen, C., and Quinn, J. (2006) Remediation of Chlorinated and Recalcitrant Compounds--2006, Proceedings of the International Conference on Remediation of Chlorinated and Recalcitrant Compounds, 5th, Monterey, CA, United States, May 22-25.
35. Sigman, Michael E. and Clark, C. Douglas. (2005) *Rapid Comm. in Mass Spec.*, **19** (24), 3731-36.
36. USEPA (1996) Nitroaromatics and Cyclic Ketones by Gas Chromatography. EPA Method 8091.
37. USEPA (2007) Explosives by Gas Chromatography. EPA Method 8095.
38. Calderara, S., Gardebas, D., Martinez, F. (2003) *Forensic Sci. Int.*, **137**(1), 6-12.
39. Sigman, M.E., Clark, C.D., Fidler, R., Geiger, C.L., Clausen, C.A. (2006) *Rapid Comm. in Mass Spec.* **20** (19), 2851-57.
40. Zilly, M., Langmann, P., Lenker, U., Satzinger, V., Schirmer, D., Klinker, H. (2003) *J. of Chromatogr. B.* **798**, 179-86.

Chapter 8

Arsenic Removal by Nano-scale Zero Valent Iron and how it is Affected by Natural Organic Matter

Hosik Park[1], Sushil Raj Kanel[2], and Heechul Choi[1,*]

[1]Department of Environmental Science and Engineering, Gwangju Institute of Science and Technology (GIST), 261 Cheomdan-gwagiro, Buk-gu, Gwangju 500-712, Korea
[2]School of Civil and Environmental Engineering, Georgia Institute of Technology, 311 Ferst Drive Northwest, Atlanta, Georgia 30332-0512, USA

Nano scale zero-valent iron (NZVI) is attracting a great deal of attention for use in treatment of contaminated soil and wastewater. NZVI has been used as an excellent adsorbent or reductant for organic and inorganic contaminants treatment due to its small size, high surface area, and very high reactivity. Specifically, NZVI has been used as an adsorbent to remove heavy metals and metalloids from aqueous phase depending on the experimental conditions. In a natural water treatment system, the presence of natural organic matter (NOM) may influence the targeted contaminant removal efficiency. Therefore, removal of NOM and the effects of NOM on contaminants removal have been widely studied to efficiently remove the target compounds. In this chapter, we introduce arsenic and NOM removal using NZVI as well as the interaction of NOM on the arsenic removal by NZVI. In addition, applications of NZVI for treatment of contaminated water and arsenic in the environment are also reviewed.

Introduction

Arsenic contamination in natural water and groundwater has been receiving significant attention recently, and its removal has become a considerable challenge for environmental scientists and engineers. Arsenic, a common constituent of the earth's crust, is a well-known carcinogen, and it is naturally present in water in different oxidation states and acid-base species depending on the redox and pH conditions (*1*). Typically, arsenic has been introduced into the environment through a combination of natural processes, and this natural occurrence of arsenic in groundwater is of great concern due to the toxicity of arsenic and the potential for chronic arsenic exposure (*2*).

Currently, potable groundwater supplies in many countries around the world contain dissolved arsenic in excess of 10 $\mu g\ L^{-1}$. To address this problem, the World Health Organization (WHO) has set a maximum guideline concentration of 10 $\mu g\ L^{-1}$ for arsenic in drinking water (*3,4*).

Natural organic matter (NOM) is ubiquitous in all natural water sources and soil systems. However, in water treatment facilities, NOM can potentially cause unexpected problems in the pollutant treatment process, and is a great threat in terms of its ability to create carcinogenic problems. Therefore, NOM removal is a primary focus of environmental treatment facilities. Even if NOM is not specifically targeted for removal, macromolecular dissolved organic matter has been shown to compete with low molecular weight synthetic organic chemicals, reducing their adsorption rates and equilibrium capacities (*5*). Therefore, the adsorption of NOM has been widely investigated in order to optimize its removal from solutions and to minimize its impact on the adsorption of other compounds targeted for removal (*6-8*). Given the many ways in which NOM interacts with the environment through biotic and abiotic processes, it is important to know the concentrations NOM present in various environmental compartments and also its physical and chemical characteristics, including how these factors impact environmental conditions.

Consequently, efficient contaminated water treatment before release into the environment is required, using contaminant treatment techniques such as physical, chemical, or biological processes. However, there are some disadvantages to these techniques—most notably their inherent high operation and maintenance costs (*9,10*). To compensate for these costs, use of nanomaterials in pollutant removal is an emerging new technology that can provide cost-effective solutions to some of the most challenging environmental clean-up problems. Among the nanomaterials being used for environmental pollutant removal, nano-scale zero valent iron (NZVI) is getting the most attention from environmental scientists and engineers; several researchers have already reported on the effects of this nanoparticle, as it has shown an amazing capacity to remove some of harmful organic and inorganic pollutants. In particular, NZVI has been very effective for the transformation and detoxification of a variety of common environmental contaminants, such as TCE (*11-15*), PCB (*16*), heavy metals (e.g., chromium, lead) (*17*), metalloids (*18,19*), and nitrates (*20-22*), among others.

In this chapter, NZVI application to arsenic and NOM removal and the effects of NOM on arsenic adsorption by NZVI will be introduced.

Arsenic in the Environment

Arsenic is a metalloid that is naturally present in the environment in a variety of forms (organic and inorganic), oxidation states, and valances depending on both natural and anthropogenic sources (*1*). Arsenic in soil and water can naturally occur from the weathering of soil, volcanic activity, or forest fires as well as from anthropogenic sources such as arsenical pesticides, disposal of fly ash, mine drainage, and geothermal discharge (*23*). In terms of anthropogenic sources, mining activities, ore smelting, sulfuric acid manufacturing processes, and pesticides are the most predominant sources (*2*). Dissolved arsenic in groundwater, in which the concentration is in excess of 10 μg L^{-1}, has been detected in many countries (including Bangladesh, India, Taiwan, Mongolia, Vietnam, Argentina, Chile, Mexico, Ghana and the United States). The groundwater arsenic found in these countries includes natural sources of enrichment as well as mining-related sources; recent reconnaissance surveys of groundwater quality in Nepal, Myanmar, and Cambodia have also revealed concentrations of arsenic exceeding 5 mg L^{-1} (*3*).

The trivalent and pentavalent ions of arsenic are known to be acutely toxic; for this reason, the WHO has set a maximum guideline concentration of 10 μg L^{-1} for arsenic in drinking water (*4*). Since it is extremely toxic to humans, long-term exposure can cause health disorders such as cancer, hyperpigmentation, circulatory disorders, and neurological damage at aqueous concentrations as low as 0.1 mg L^{-1} (*24-26*).

Genesis of Arsenic

Arsenic is widely distributed in nature, with origins that include geological formation from the soil, industrial waste discharge, and the agricultural use of arsenical pesticides. Arsenic is ranked as the 20th most abundant element in the Earth's crust (4.01×10^{16} kg) (*27*) and the natural content of arsenic in soil is 5–6 mg kg^{-1}; when potential anthropogenic arsenic input is included, the arsenic content in soil was recorded as 2.18 mg kg^{-1} in 2000 (*27,28*). Arsenic can occur as a major constituent in more than 200 minerals, including elemental As, arsenides, sulphides, oxides, arsenate, and arsenites, with the most abundant arsenic ore mineral being arseno pyrite (FeAsS). The concentrations of arsenic in varieties of rock such as igneous, metamorphic, and sedimentary rocks are 1.5 mg kg^{-1}, 5 mg kg^{-1} and 5–10 mg kg^{-1}, respectively. However, arsenic concentration in the atmosphere is generally low, ranging from 10^{-5} to 10^{-3} μg m^{-3} in unpolluted areas, 0.003 to 0.18 μg m^{-3} in urban areas, and greater than 1 μg m^{-3} close to industrial plants (*3*).

Geochemistry of Arsenic

Arsenic occurs in the Earth's crust at levels ranging from 0.1 mg L^{-1} to several hundred mg L^{-1} as arsenopyrite (FeAsS), realgar (AsS) and orpiment (AsS_3). Inorganic forms of arsenic are present as arsenate (As(V)), arsenite

(As(III)), arsenic metal (As^0), and arsine gas (AsH_3). In organic forms, arsenic exists as compounds such as tetramethylarsonium salts, arsenocholine, arsenobetaine, dimethyl (rivosyl) arsine oxide, and arsenic containing lipids (*29*).

Under normal conditions, arsenic exists in two oxidation stages, arsenite (As(III)) and arsenate (As(V)), where these trivalent and pentavalent ions are known to be acutely toxic. Arsenic exists as different species depending on the pH and redox conditions. At natural pH values, arsenite exists in solution as H_3AsO_3 and $H_2AsO_3^{-1}$, whereas arsenate exists in solution as H_3AsO_4, $H_2AsO_4^{-1}$, $HAsO_4^{-2}$, and AsO_4^{-3}. However, since the redox reactions are relatively slow, both As(III) and As(V) are often found in soil and subsurface environments regardless of the environmental conditions (*2*).

Within a pH range of 6–9, As(V) exists as an anion ($H_2AsO_4^{-1}$, $HAsO_4^{2-}$) whereas As(III) exists as an uncharged molecule (H_3AsO_3) (*1*). The speciation of arsenic according to pH and redox condition is shown in Figure 1.

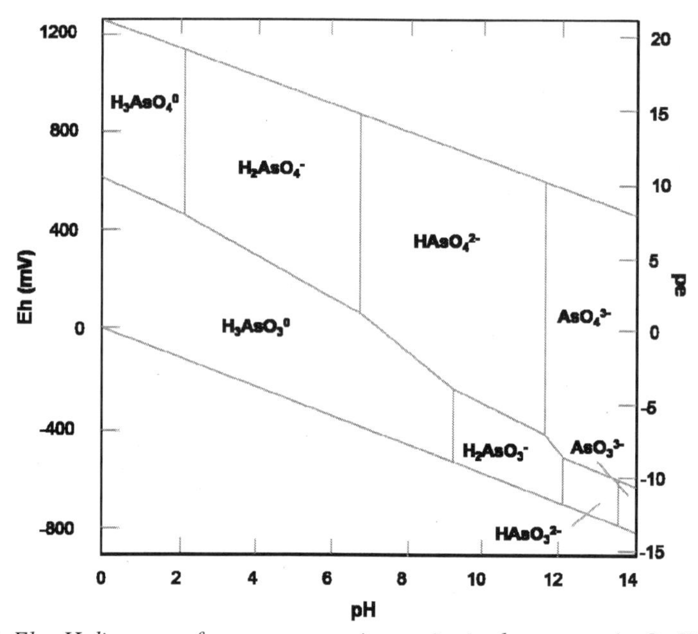

Figure 1. Eh-pH diagram of aqueous arsenic species in the system As-O_2-H_2O at 25 °C and 1 bar total pressure.

As(V) is dominant in oxygenated water, whereas in anoxic conditions As(III) is more stable. Possible change of arsenic from one form to another depends on thermodynamics, though accurately predicting the rate of change is quite difficult. For instance, even though the oxidation of As(III) to As(V) in oxygenated water is thermodynamically favored; the rate of transformation may take days, months, or even years depending on specific conditions. Strongly acidic or alkaline solutions, the presence of copper salts, carbon, unknown catalysts, and higher temperatures can increase the oxidation rate (*1*). Note that although both organic and inorganic forms of arsenic have been detected, organic species are rarely present at concentrations > 1 μg L^{-1} and are generally considered of little significance compared with inorganic arsenic species in drinking water treatment.

Occurrences of Arsenic

A large percentage of the natural water in various parts of the world has been identified as having arsenic problems ranging from less than 0.5 to more than 5000 μg L^{-1}. The most noteworthy occurrences are in parts of Argentina, Bangladesh, Chile, China, Hungary, India (West Bengal), Mexico, Romania, Taiwan, Vietnam, and many parts of the USA, particularly in the southwest (*3*).

In terms of the global population exposed to high arsenic concentrations, perhaps 50 million people in Bangladesh alone drink arsenic contaminated groundwater; correspondingly, thousands of new cases of severe arseniasis annually occur (*30,31*). In Bangladesh, sources of contamination include natural sources of enrichment as well as mining-related sources. Arsenic associated with geothermal water has also been reported in several areas, including hot springs in parts of Argentina, Japan, New Zealand, Chile, Kamchatka, Iceland, France, Dominica, and the USA. And reports of arsenic problems in localized groundwater are emerging in many countries, with many new cases likely yet to be discovered.

For these reasons, there have been numerous revisions to drinking-water regulations and guidelines for arsenic, which in turn has prompted a reassessment of the situation in many countries. Most notably, the recent discovery of the large-scale arsenic enrichment in Bangladesh has highlighted the need for an urgent worldwide assessment of alluvial aquifers. In mining areas, arsenic problems can be severe with concentrations in affected waters sometimes being in the range of mg per liter; mining-related arsenic contamination of local/regional water has also been identified in Ghana, Greece, Thailand, and the USA (*3*).

Technologies for Arsenic Removal

Several techniques for removing arsenic have been practiced, including precipitation, adsorption, oxidation, and membrane techniques, among others (*32-39*). Typically, these technologies for arsenic removal have been based on a few basic chemical processes, including oxidation/reduction, precipitation, adsorption and ion exchange, solid/liquid separation, physical exclusion and biological removal process, which are summarized below (Table I).

However, more efficient technologies, which have low cost, are effective, and can be easily implementable at local levels, need to be investigated. Thus, the versatility of nanoscale iron/iron oxide materials is currently being studied for potential use in arsenic removal due to their extremely large surface-area-to-volume ratio and high reactivity (*18,19,40,41*).

Table I. Arsenic Removal Technologies

Technologies	Advantage	Disadvantage
1. Oxidation		
a) Air oxidation	Simple, low cost	Slow, low efficiency
b) Chemical oxidation	Relatively simple and rapid, oxidizes other impurities	Kills microorganisms, costly
2. Coagulation-precipitation		
a) Alum	Relatively low cost	Toxic sludge, low efficiency in As(III)
b) Iron	Simple, chemical commonly available High efficiency	Toxic sludge, low efficiency in As(III)
c) Lime		pH dependent, produce sludge
3. Sorption		
a) Activated carbon	Economic	Large quantity of carbon required Produces toxic solid waste
b) Activated alumina	Well-known, commercially available	Regeneration required Hi-tech. operation and maintenance
c) Zero valent iron and iron oxide	Scope of development	Relatively high cost
d) Industrial by-products	Cheap	May produce other pollutants, further study of leaching required prior to use
4. Ion exchange	Fast reaction	Produces sludge
5. Stabilization	High efficiency High reduction rate	Difficult to predict stability Relatively costly
6. Cementation	High efficiency	Stability of adsorbed arsenic needs to be confirmed
7. Membrane	Well defined, high efficiency No toxic solid waste Capable of removing other contaminants	High capital and running costs Hi-tech operation and maintenance Toxic waste water produced

Nanoscale Zero Valent Iron

Recently, nanoscale zero-valent iron (NZVI) has drawn a great deal of attention due to its larger surface-area-to-volume ratio than bulk materials for a given amount, along with its high reactivity due to high surface area. For these reasons, NZVI has been applied in a number of research and industrial fields, including medical drug targeting materials in biomedicine, catalysts, and the adsorbent of environmental contaminants in soil and aqueous solutions. NZVI, an environmentally benign and reactive material, is present in most areas of water treatment, including drinking water treatments and groundwater remediation (*42*). For instance, NZVI has been used as a sorbent to successfully remove heavy metals and metalloids such as As(III), As(V), Cr(VI), Pb(II) and Ba(II) (*17-19,43*). In addition, NZVI has been used as a reductant to remove inorganic contaminants such as nitrate and perchlorate (*20,22,44-47*), and was also successfully used to remove carbothioate, herbicide, molinate, and humic acid (*48,49*). In recent research, NZVI was widely used to remove TCE and other halogenated compounds from groundwater contaminated with a dense-nonaqueous-phase-liquid (DNAPL) (*12-14,50*). In addition, due to its smaller size and higher surface-area-to-volume ratio than bulk zero-valent iron (ZVI) and its different surface structure, NZVI can be directly used in water systems and be effectively transported by water flow after its modification (*17,20,45,47*).

In this section, NZVI utilization for contaminated water treatment is further addressed.

Reductive Removal of Organic Compounds and Inorganic Ions

Dechlorination of organic compounds

NZVI has been utilized to rapidly dechlorinate various organic contaminants such as trichloroethylene (TCE) and perchloroethylene (PCE). The degradation principle of chlorinated organic compounds by NZVI is based on the corrosion of NZVI (*51,52*), as with a standard potential (Eh^0) of -0.44 V, NZVI primarily acts as a reducing agent; NZVI is effectively oxidized as an electron donor.

$$Fe^0 \rightarrow Fe^{2+} + 2e^- \quad (1)$$

On the other hand, chlorinated organic compounds are reduced.

$$RCl + H^+ + 2e^- \rightarrow RH + Cl^- \quad (2)$$

Thus, the following net reaction is thermodynamically very favorable under most conditions.

$$Fe^0 + RCl + H^+ \rightarrow Fe^{2+} + RH + Cl^- \quad (3)$$

For example, water contaminated by TCE and PCE has been treated using NZVI. In this case, the chlorinated organic compounds are reduced and the chloride ions are released as a result of the oxidation of metallic iron to ferrous ion (Fe^{2+}). However, this reaction undergoes several steps before reducing chlorinated organic compounds, via intermediates, to nontoxic compounds such as ethylene, ethane, and acetylene (*12*).

Hence, even though NZVI has demonstrated excellent efficiency, there are still a number of limitations to NZVI use. In attempts to overcome these limitations, research has focused on the synthesis of bimetallic NZVI using Pd and Ni, the best catalysts for decomposing chlorinated organic compounds (*12,53*). And bimetallic NZVI increases the reactivity and reaction with TCE without the production of by-products (*54*). However, existing methods for synthesizing bimetallic NZVI using catalysts requires expensive Pd and a high ratio of Ni—from 5 to 20 % (*55,56*). As such, challenges for efficiently synthesizing highly reactive NZVI for remediation of contaminated water and groundwater remain.

Reduction of inorganic anions

ZVI can be used to reduce relatively stable inorganic compounds such as nitrate, perchlorate and bromate (*57-60*). In addition, according to several studies, ZVI has drawn attention in nitrate reduction due to its effective reduction capacity. Nitrate reduction studies using ZVI were conducted in acidic conditions and were reported to produce by-products of nitrite, ammonia, or nitrogen gas. It was found that the acidic reaction conditions accelerated iron corrosion, but also resulted in ferrous ions being released during the reaction. This release then required a secondary treatment to remove ferrous ions and to adjust the pH. Meanwhile, Shon et al. reported that nitrate was reduced by NZVI under alkaline conditions (pH 9–10) (*61*). In their study, the denitrification reaction with NZVI follows pseudo-first-order kinetics, with the observed rate constant (k_{obs}) being 10.12 h^{-1} for 1 g of NZVI. During the reaction, NZVI was oxidized to crystalline magnetite (Fe_3O_4).

$$NO_3^- + 4Fe^0 + 10H^+ \rightarrow NH_4^+ + 4Fe^{2+} + 3H_2O \qquad (4)$$

Choe et al. studied the reduction of nitrate using NZVI with no pH control and observed that complete denitrification could be achieved in a few minutes (*20*). In their study, the concentration of reactants, nitrate, and NZVI as well as the mixing intensity were all dominant factors for nitrate reduction.

Bromate is an oxyhalide disinfection by-product (DBP) that is formed during the ozonation of water containing bromide (*62*). Bromate is regulated in the United States by the United States Environmental Protection Agency (USEPA) and the WHO, with an enforceable maximum contaminant level (MCL) of 10 μg L^{-1} in treated drinking water. However, despite these stated regulations and guidelines, the USEPA calculated the 1 in 1,000,000 cancer risk from bromate to be 0.05 μg L^{-1} (*63,64*). ZVI has also been applied to bromate reduction; the significant bromate reduction efficiency of ZVI was investigated by Li et al. (*65,66*). However, it was found that a large amount of ZVI must be used in the drinking water treatment process to effectively reduce bromate. Recently, Wang et al. (*67*) examined bromate reduction using NZVI. Their results revealed that a much lower dosage of NZVI (0.1 g L^{-1}) compared to ZVI

($25\ g\ L^{-1}$–$100\ g\ L^{-1}$) could effectively reduce bromate from 1000 ppb to less than 10 ppb in 20 min. The reduction reaction follows the pseudo-second-order kinetic model and the highest observed second order rate (K_{obs}) was 2.19×10^{-3} $\mu g^{-1}\ min^{-1}\ L$. In addition, they revealed that humic acid was the greatest factor influencing the decrease in NZVI reactivity during the bromate reduction.

Removal of metallic ions

NZVI has been studied for its potential use in the removal of metallic ions such as Cr(VI), Cu(II), Ag(II), Hg(II), Zn(II), Cd(II), Ni(II), and Pb(II) *(17,68,69)*. Using NZVI, the degradation mechanisms are based on transformation from toxic to non-toxic forms (reduction) or sorption/surface complex formation depending on the types of heavy metals. For example, heavy metals (Zn(II), and Cd(II)) in which the standard potential (Eh^0) is close to or more negative than that of iron (-0.41 V) are removed by sorption/surface complex formation. Conversely, the removal mechanism for metals (Cr(VI), Cu(II), Ag(I), and Hg(II)), which have a more positive standard potential than iron, is based on reduction (*69*).

The removal of Cr(VI) using NZVI is based on the reduction from toxic to non-toxic forms. Specifically, Cr(VI) is reduced to Cr(III) using NZVI and this reaction produces ferric ions (Fe(III)) and chromium ions (Cr(III)) (*17,68*).

$$Cr^{6+} + Fe^0 \rightarrow Cr^{3+} + Fe^{3+} \quad (5)$$

During this reduction, Cr(III) may be removed through precipitation/co-precipitation as $(Cr_xFe_{(1-x)})(OH)_3(s)$ or $Fe_{(1-x)}Cr_xOOH(s)$.

$$(1-x)Fe^{3+} + (x)Cr^{3+} + 3H_2O \rightarrow (Cr_xFe_{(1-x)})(OH)_3(s) + 3H^+ \quad (6)$$

$$(1-x)Fe^{3+} + (x)Cr^{3+} + 2H_2O \rightarrow Fe_{(1-x)}Cr_xOOH(s) + 3H^+ \quad (7)$$

Ponder et al. reported that 65–110 mg Cr(VI) could be reduced and precipitated uisng 1 g of NZVI (*17*). This result shows at least a 20 times higher removal efficiency than that of microscale ZVI. In addition, the reaction rate of NZVI for Cr(VI) removal was 25–30 times faster than that of ZVI. Another study of heavy metal removal using NZVI was conducted by Li et al. (*69*). They tested a mixture of 7 metal compounds (Ni, Cd, Zn, Ag, Cr, Pb, and Cu) with 5 $g\ L^{-1}$ of NZVI (each metal compound: $100\ mg\ L^{-1}$). The resulting removal efficiency ranged from 99 % to 37 % with Cd and Ni showing relatively lower removal efficiency due to the competition among the metal cations.

Thus, through the above studies, it can be confirmed that use of NZVI is an effective means of removing heavy metals in water, due primarily to its high removal efficiency and fast reaction.

Arsenic Removal from Groundwater using NZVI

The use of ZVI as a method for groundwater remediation started in the early 1990s when granular ZVI was first employed in permeable reactive barrier (PRB) systems; ZVI was tested for use in long term groundwater remediation as a PRB material (*70*). As ZVI can be used for shallow groundwater plume treatment—it is not suitable for source zone treatment and less effective to remove pollutants—NZVI was discovered. Because of its small size (< 100 nm), high surface area, and very high reactivity, NZVI has been increasingly studied for treating contaminated soil, groundwater, and wastewater. NZVI has been used to reduce, oxidize, and adsorb metalloids such as arsenic from the aqueous phase depending on experimental conditions.

In this section, we focus on arsenic (As(III) and As(V)) sorption using NZVI. In addition, experimental parameters such as kinetics and influence of pH during the adsorption of As(III) and As(V) using NZVI are discussed, as well as a microscopic and spectroscopic study of NZVI-arsenic products.

Influence of pH on As(III) and As(V) Adsorption

The effect of pH on As(III) and As(V) adsorption on NZVI is presented in Figure 2. For As(III), the extent of removal was 88.6–99.5 % in the pH range 4–10 and decreased sharply for pH values below 4 and above 10. For As(V), 100 % As_T (total arsenic) was found to be sorbed at pH 3–7 and the sorption decreased to 84.7 % at pH 9; a further pH increase to 11 led to a sorption decrease to 37.9 %. This pH-dependent behavior can be explained based on the ionization of both the adsorbate and adsorbent causing repulsion at the surface, which thereby decreases the net arsenic adsorption (*19*). Below pH 9.2, H_3AsO_3 is the predominant species and presumably the major species being adsorbed. When the pH is above 9.2, $H_2AsO_3^-$ becomes the predominant As(III) species, and NZVI corrosion product surfaces are predominantly negative (Fe(III)-O^-) causing electrostatic repulsion. A similar pH dependence trend in As(III) adsorption of amorphous iron oxide, goethite, and magnetite was observed by Dixit et al. (*35*) and on synthetic goethite (*71*), potentially attributed to the ionization of both adsorbates and adsorbents.

It should be noted that As(V) has pK_1, pK_2 and pK_3 values equal to 2.2, 7.08 and 11.5, respectively. In the pH range 2–7, $H_2AsO_4^-$ and $HAsO_4^{2-}$ are the predominant arsenic species (*36*), and presumably the major species being adsorbed. As the point of zero charge (PZC) of NZVI is pH 7.7 (*19*), the NZVI surface exhibits a net positive charge at a pH lower than pH 7.7, and adsorption of trace anionic As(V) species was enhanced by columbic interaction. However, it was found that, as 0.1 and 1 mg L^{-1} concentrations of As(V) was added to 0.1 g L^{-1} NZVI, the PZC of NZVI decreased from 7.7 to 7.5 and 7.2, respectively. This decrease indicates that As(V) might have attached to the NZVI corrosion product surface and formed inner sphere complexes (*2,19,72*).

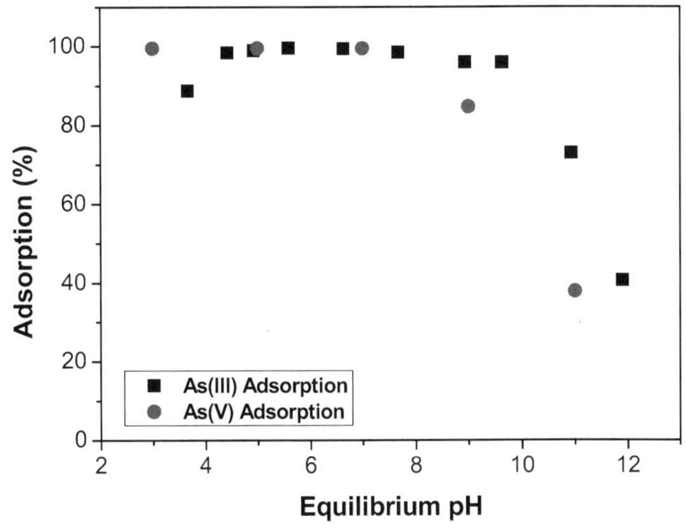

Figure 2. Adsorption of As(III) and As(V) on NZVI as a function of pH (envelope plot) (18, 19).

Kinetics of As(III) and As(V) adsorption

Kinetics of As(III) and As(V) adsorption using NZVI

The influence of NZVI concentration on the adsorption rate of As(III) was investigated using 1 mg L^{-1} of As(III) at pH 7 (Figure 3 and inset). For all treatments, except for the 0.5 g L^{-1} treatment, more than 80 % of the total arsenic (As_T) was adsorbed within 7 min, and ~99.9 % within 60 min. The As(III) adsorption kinetics data were then examined using pseudo-first-order reaction kinetics. This result shows that an initial faster rate of As(III) disappearance from an aqueous solution (~80 % As_T) takes place within 7 min followed by a slower uptake reaction. This sequence of reactivity is consistent with the results of the pseudo-first-order rate of As(V) adsorption using ZVI (73, 74). In our experiment, the K_{sa} for As(III) was 0.0042–0.0115 L m^{-2} min^{-1} at 0.5–10.0 g L^{-1} NZVI, respectively. Hence, the K_{sa} for NZVI was found to be much higher than that for microscale ZVI.

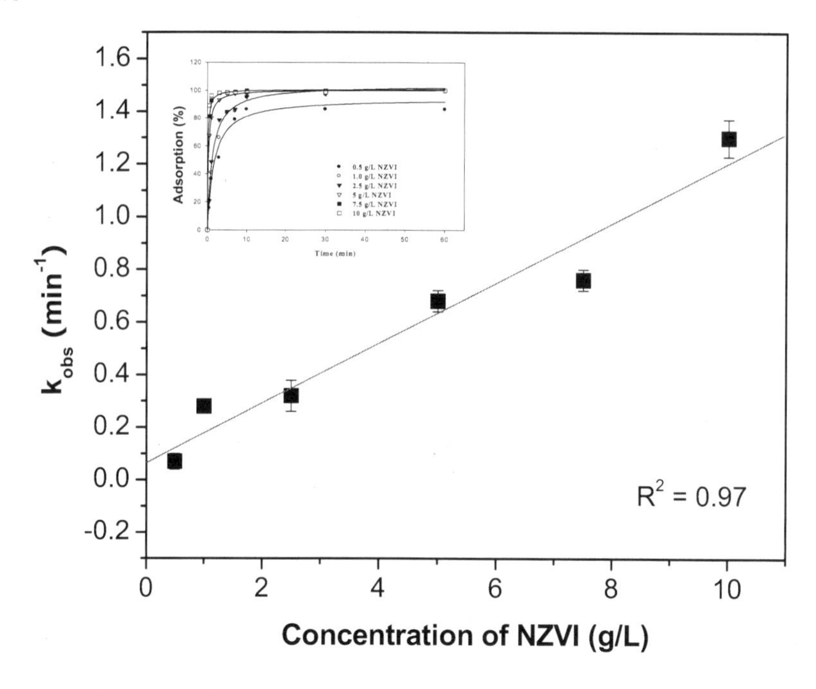

Figure 3. Kinetics of As(III) adsorption. The reaction is pseudo-first-order with respect to the total NZVI concentration (19).

Next, the As(V) adsorption kinetics were examined using NZVI (Figure 4), where the removal of As(V) could be fitted to pseudo-first-order reaction kinetics. The influence of NZVI concentrations (0.05–1.0 g L^{-1}) on the rate of adsorption of As(V) was investigated using 1 mg L^{-1} of As(V) at pH 7. The arsenic adsorption was observed to be spontaneous, as ~90 % As_T was found to be sorbed within 7 min (Figure 4 inset) and ~99.9 % within 60 min. Interestingly, As(V) adsorption increased as the adsorbent concentration was increased from 0.05 to 1.0 g L^{-1}; the dose of adsorbant increases the number of active sites, which then leads to an enhancement of As(V) adsorption. In this study, the k_{sa} for As(V) was 0.015–0.029 L m^{-2} min^{-1} at 0.05–1.0 g L^{-1} NZVI, respectively. This corresponds to 1 to 3 orders of magnitude higher k_{sa} for NZVI than that of ZVI (75, 76). This outstanding reactivity of NZVI is due to its large surface area and the nanosize effect.

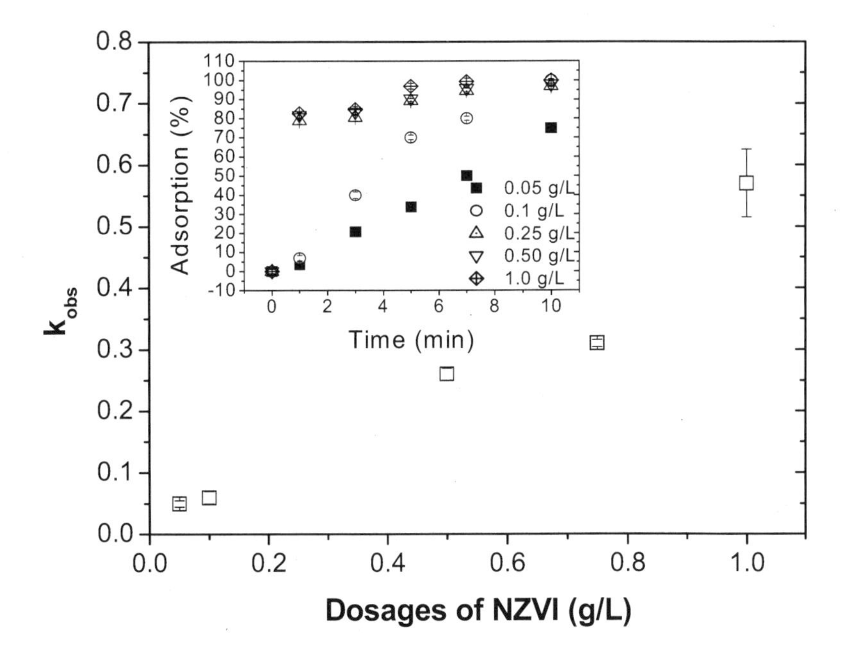

Figure 4. Effects of NZVI concentration on As(V) adsorption. The reaction is pseudo-first-order with respect to the total NZVI concentration (18).

As(III) and As(V)-NZVI XRD and SEM studies

The X-ray diffraction (XRD) and scanning electron microscope (SEM) results of As(III) and As(V) treated NZVI revealed that the NZVI reacting with As(III) gradually converted to magnetite/maghemite corrosion products mixed with lepidocrocite over a 60-day period. The collected samples of As(III) and As(V) treated NZVI after 7–90 days showed different surface textures and different pore sizes with respect to time of adsorption and precipitation of arsenic onto NZVI; aggregation of particles increases with reaction time due to Fe(III) oxide/hydroxide precipitation (Figure 5). The SEM images of As(III)-treated NZVI clearly show a growth of a fine needle-like crystallite, which may lead to an apparent amorphous phase. The very thin crystallites are expected to be energetically unstable and should eventually disappear and be replaced by more stable phases according to the Gay-Lussac-Oswald ripening rule (*19*).

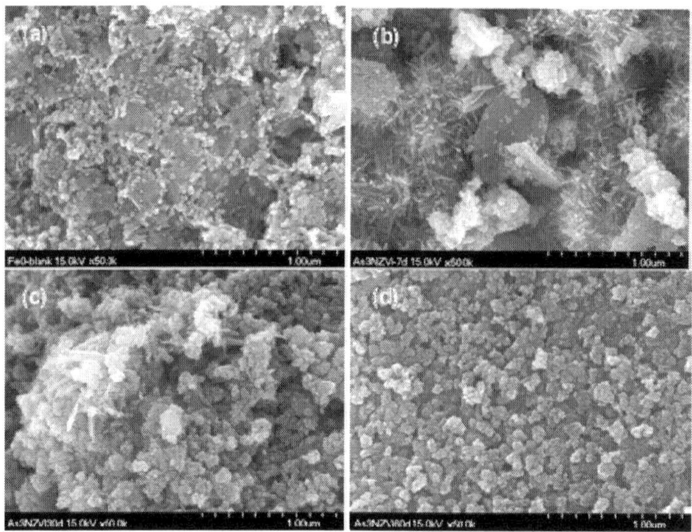

Figure 5. SEM images of (a)pristine NZVI, and As(III) sorbed on NZVI for (b) 7 days, (c) 30 days, and (d) 60 days (19).

The zero valence state and crystalline structure of NZVI were then confirmed by XRD analysis. The XRD data demonstrate that NZVI corrosion products are a mixture of amorphous Fe(III) oxide/hydroxide, magnetite (Fe_3O_4), and/or maghemite (γ-Fe_2O_3), and lepidocrocite (γ-FeOOH).

However, in the case of As(V) treated NZVI, the amorphous region almost disappeared and sharp crystalline peaks of lepidocrocite and magnetite/maghemite are seen (Figure 6) at 30 days. After 60 days, the main corrosion products of As(V) treated NZVI were magnetite/maghemite, but lepidocrocite completely disappeared. This same trend can be seen in the 90-day product, though interestingly the amorphous phase at 15–35 (2-theta) notably increased, as shown Figure 6. Also, a new peak appeared at 82 (2-theta degrees) in the 60- to 90-day XRD samples, which may be attributed to arsenic adsorbed Fe^0. Similarly, from the SEM-EDX quantitative peak analysis, we confirmed that 2 %, 3 % and 4 % of As(V) was adsorbed onto NZVI at 7, 60, and 90 days respectively; thus confirming the dynamic behavior of NZVI, which remained active, even after more than 3 months.

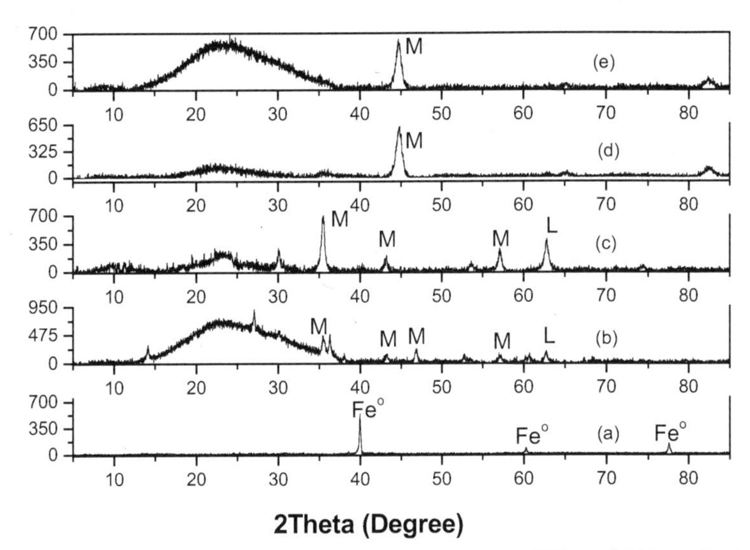

Figure 6. X-ray diffraction analysis of (a) pristine NZVI and 100 mg/L As(V) sorbed on 50 g/L NZVI in 0.01 M NaCl at pH 7 and 25°C for (b) 7 days, (c) 30 days, (d) 60 days, and (e) 90 days. Measured peaks are due to magnetite-maghemite (Fe_3O_4 / γ-Fe_2O_3) and lepidocrocite (γ-FeOOH). Peaks are referred to as magnetite/maghetite (M) (Fe_3O_4/γ-Fe_2O_3), lepidocrocite (γ-FeOOH) (L) and NZVI (Fe^o), respectively (18).

Natural Organic Matter Removal by NZVI and its Interaction during Arsenic Adsorption

The environmental fate and transport of arsenic species has received great deal of attention due to their environmental and human health effects. In natural water systems, the presence of NOM, anions, and cations may compete with arsenic species for adsorption sites on adsorbent surfaces, and thus strongly affect arsenic mobility.

To date, very few studies have reported on the effects of NOM on arsenic adsorption on an adsorbent (hematite, ferrihydrate, ZVI) (*77*). Previous works with hematite revealed NOM has an effect on arsenic speciation, which may influence the fate and transport of arsenic. Grafe et al. also reported the variable effects of humic acid and fulvic acid on the adsorption of As(III) and As(V) with goethite and ferrihydrite under different pH condition (*78,79*). And Ko et al. studied the effects of humic acid on the adsorption kinetics and speciation of arsenic (*80*). According to these studies, competition between NOM and arsenic for adsorption is a potentially important process in natural water.

In this section, NOM removal using NZVI will be discussed, obtained using different spectroscopic and microscopic techniques, as well as the interaction of NOM with arsenic.

Kinetics of NOM adsorption onto NZVI

The influence of NZVI concentrations (0.1–1.0 g L^{-1}) on the rate of NOM adsorption was investigated using 20 mg L^{-1} of humic acid (HA), obtained from the International Humic Substance Society (IHSS), at initial pH of 6. As can be seen in Figure 7, when the NZVI concentration is over 0.3 g L^{-1}, complete adsorption of HA occurred within 20 min.

The HA adsorption kinetics were then examined using pseudo-first-order reaction kinetics expressions. The pseudo-first-order rate constant (k_{obs}) for HA was approximately 0.7 min^{-1} at 1.0 g L^{-1} NZVI, which was significantly higher than for microscale ZVI (a few days) (*81*). This faster rate constant represents a rapid adsorption rate used to reach equilibrium. In addition, it was observed that HA removal increased with the increase in temperature (10–50 °C), thereby showing the endothermic nature of this process.

In this study, the equilibrium time was reduced as NZVI dosage was increased. This reduction can be explained based on the following: 1) there were sufficient surface sites for adsorption, and 2) there was a decrease of HA diffusion from the bulk solution to the NZVI surface, as the already adsorbed HA showed an unfavorable electrostatic interaction charge or a repulsion of negative ζ- potential (*82*).

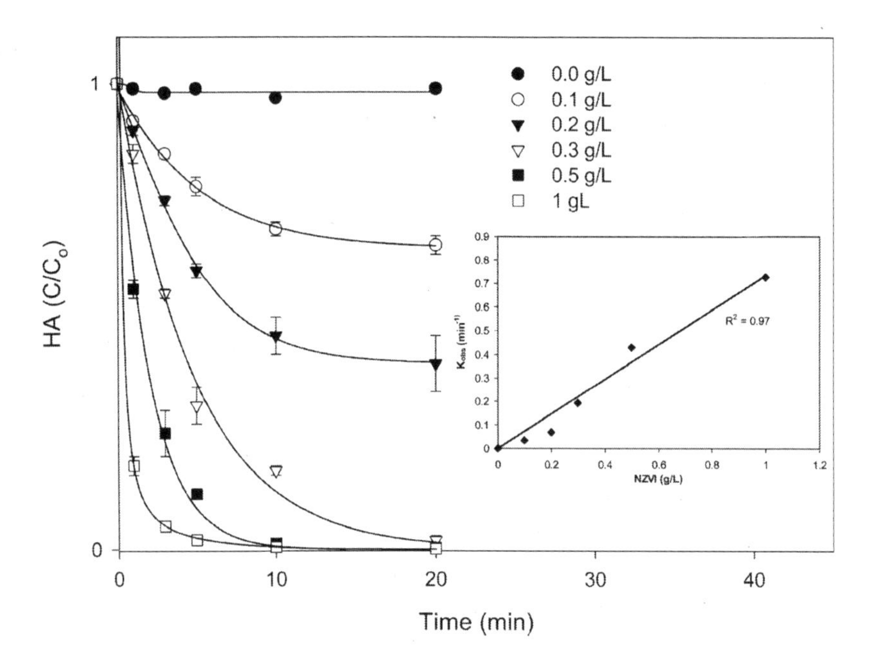

Figure 7. Kinetics of HA adsorption onto NZVI. The reaction is pseudo-first-order with respect to the total NZVI concentrations (48).

Influence of pH on HA adsorption

The effect of pH on HA adsorption onto NZVI is presented in Figure 8. The result shows that the maximum adsorption of HA is 80 mg g^{-1} in the pH range 3–9 and then decreases sharply at a pH over 10. These results of adsorption patterns of HA onto NZVI are consistent with the electrostatic interaction mechanism, as NZVI remains attractive to negatively charged HA as long as it has a positive charge on its surface. This attraction can be explained by the fact that NZVI became more positively charged as the pH decreased from its PZC at pH 7.7 (*19*).

The electrophoretic mobilities of HA, HA-treated NZVI, and pristine NZVI solutions were subsequently measured to determine PZC, at which the net surface charge is zero. The PZC of pristine NZVI and HA-coated NZVI was found to be around 7.7 and 6.2, respectively (Figure 9), and the PZC of HA-coated NZVI decreased due to the adsorption of HA on the positive-charged NZVI surface. These results are consistent with other organic matter sorption onto different iron oxides (*83,84*).

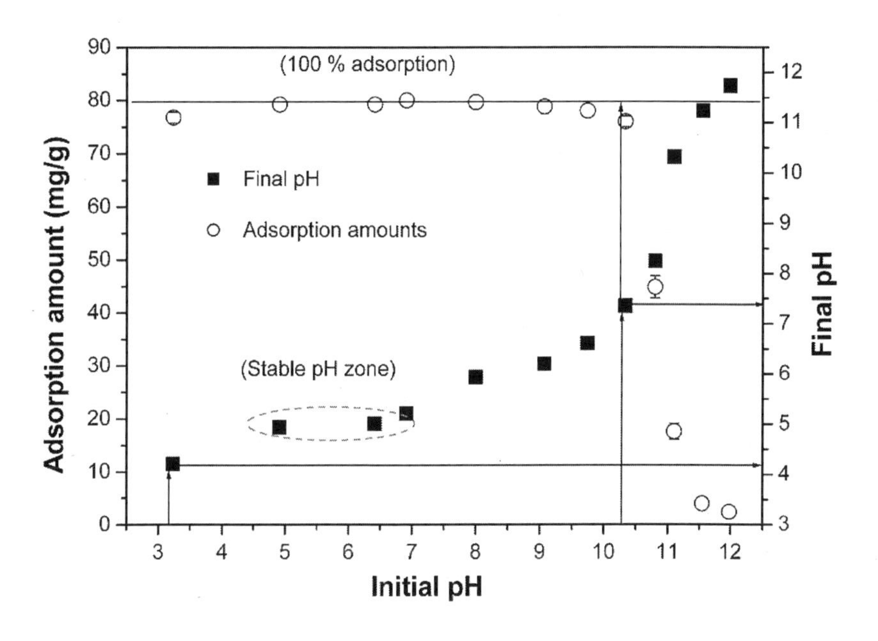

Figure 8. Effect of pH on HA adsorption onto NZVI. (48).

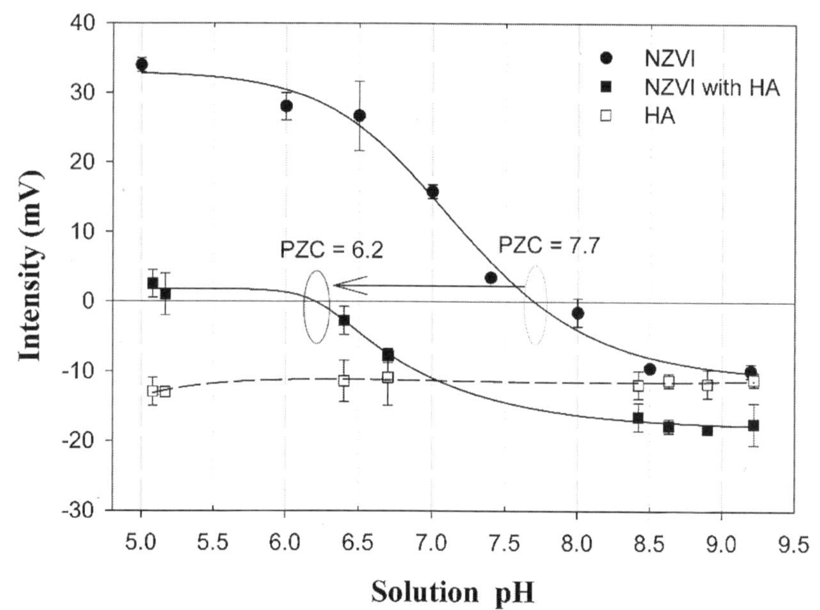

Figure 9. Electrophoretic mobility of HA treated and untreated NZVI with respect to pH (48).

The results of the FTIR spectra revealed that reaction between HA and NZVI made an inner-sphere complexation with carboxylate ion, aliphatic C-H, $C-H_2$, $C-H_3$, and carboxylic acid, consistent with previous results. In addition, the appearance of new strong bands at 1025 cm^{-1} implies that a complex chemical reaction occurred between NZVI and HA (*85*).

Effects of anionic and cationic compounds

In this study, the effects of different levels of individual anions (HCO^{3-}, SO_4^{2-}, NO_3^-, H_4SiO_4, and $H_2PO_4^{2-}$) on the adsorption of HA onto NZVI were investigated. The complete removal of HA was achieved in the presence of 10 mM NO_3^- and SO_4^{2-}, possibly due to the enhancement of the reactive sites of the NZVI species (*66*). However, HA removal efficiency decreased to 0 %, 18 % and 22 % in the presence of 10 mM $H_2PO_4^{2-}$, HCO_3^-, and $H_4SiO_4^0$, respectively. It can thus be concluded that competition between NZVI and $H_4SiO_4^0$, HCO_3^- and $H_2PO_4^{2-}$ reduce the adsorption capacity through inner sphere complex formation on NZVI from HCO_3^- and $H_2PO_4^{2-}$ (*59*).

In addition, HA adsorption in the presence of divalent cations (Ca^{2+}, and Mg^{2+}) is shown in Figure 10. Specifically, in presence of 2 mM Ca^{2+} and Mg^{2+}, HA adsorption capacity was significantly increased. This result can be explained as being due to the compression of the molecular volume of HA, the charge neutralization of both the adsorbate and adsorbent (*86*), and the bridging of Ca^{2+} and Mg^{2+} between HA molecules that neutralizes the repulsive forces between them. As a result, Ca^{2+} and Mg^{2+} can link NZVI with HA, forming an NZVI-metal-HA complex that can significantly enhance adsorption (*7*).

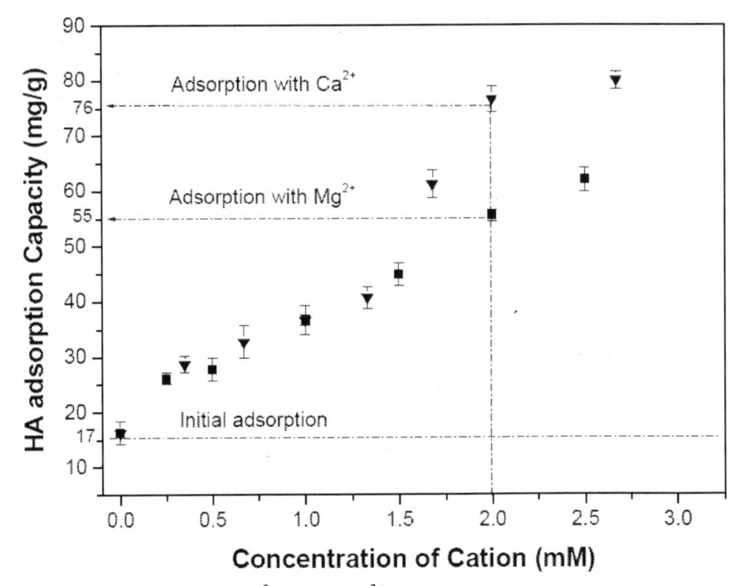

Figure 10. Effects of Ca^{2+} and Mg^{2+} on HA adsorption onto NZVI (48).

Long-term study

HA-treated NZVI was collected to investigate the long-term effects of HA adsorption onto NZVI after reaction for 1 to 90 days. The corresponding field emission scanning electron microscope (FE-SEM) images (Figure 11) show the different surface textures and different pore sizes with respect to time of adsorption and precipitation of HA onto NZVI.

Aggregation of particles was increased with reaction time due to the iron oxide/hydroxide precipitation. The FE-SEM images also clearly show that the very thin layer of NZVI disappeared over time to be replaced by a more stable phase structure, which followed the Gay-Lussac-Oswald ripening rule (*19*).

In the XRD data analysis of HA-treated NZVI for 1 day, there were no Fe^0 peaks, though a maghemite peak was observed; with time, several iron oxide/hydroxide peaks were observed.

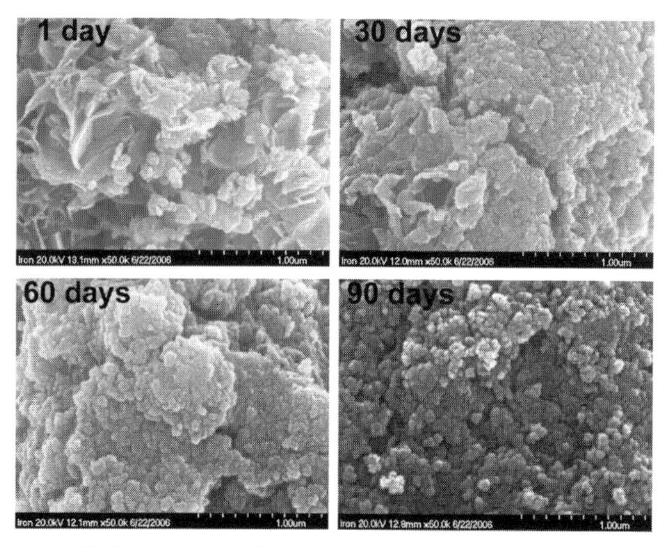

Figure 11. FE-SEM images of HA-treated NZVI and at 1, 30, 60, and 90 days (48).

Effects of HA on kinetics of Arsenic Removal

The effects of HA on kinetics of arsenic removal are presented in Figure 12. The surface normalized rate constant (k_{sa}) for As(III) and As(V) were found to be reduced in the presence of HA (20 mg L^{-1}) from 0.0168 to 0.0096 L m^{-2} min^{-1} and 0.024 to 0.0077 L m^{-2} min^{-1}, respectively. The observed reduction of surface-normalized rate constants (k_{sa}) potentially was a result of the following factors: 1) competition between arsenic species and HA for adsorption sites; 2) the coagulation of NZVI by HA, which decreased the available adsorption sites; and 3) the reduction of corrosion rates of NZVI by HA, which in turn reduced the formation of new arsenic sorption sites (*77*).

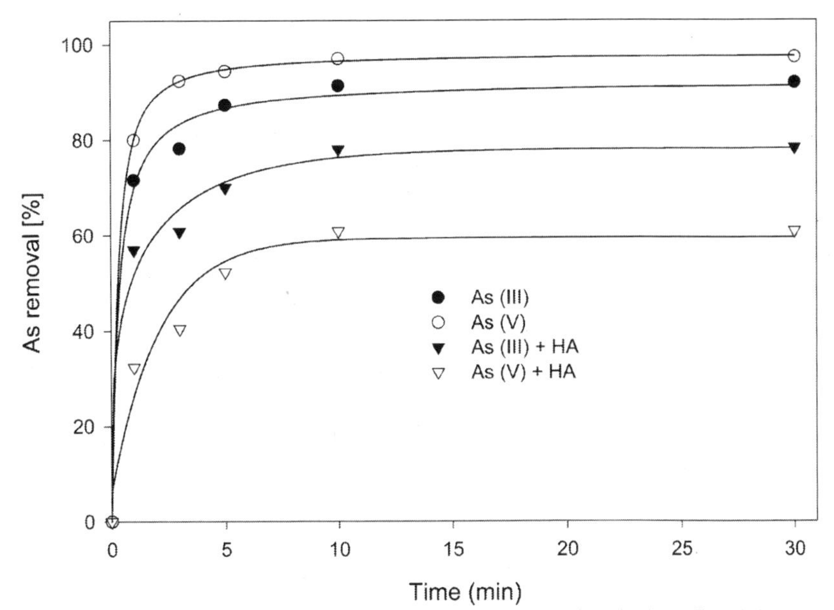

Figure 12. The effects of HA on adsorption kinetics of As(III) and As(V) using NZVI (48).

Summary

NZVI has a great potential for use in the remediation of pollutants already present in soil and groundwater, including arsenic. We have introduced evidence that As(III) and As(V) can be removed by adsorption onto NZVI within a minute time scale. Both As(III) and As(V) strongly sorb onto NZVI in a wide range of pH values; as As(III) and As(V) coprecipitate oxide/hydroxide corrosion products are involved.

Furthermore, several selected experiments were conducted to reveal the interaction between NOM and NZVI by considering groundwater chemistry (anion and cation effects) and the effects of NOM for arsenic (As(III) and As(V)) removal using NZVI were also investigated. It was subsequently determined that NOM can be completely removed by NZVI at several different pH values, and also that anionic and cationic compounds can affect the removal of NOM by NZVI. Furthermore, the results revealed that NOM has competitive effects with arsenic, and these effects should be considered during the application of adsorbents such as NZVI in contaminated water treatments.

For successful field applications of NZVI, a more detailed study of the surface reactions of NZVI with a variety of dissolved pollutants in the presence of HA should be further undertaken.

References

1. Ferguson, J. F.; Gavis, J. A review of the arsenic cycle in natural waters. *Water Res.* **1972,** 6, 1259-1274.
2. Goldberg, S.; Johnston, C. T. Mechanisms of arsenic adsorption on amorphous oxides evaluated using macroscopic measurements, vibrational spectroscopy, and surface complexation modeling. *J. Colloid Interface Sci.* **2001,** 234, 204-216.
3. Smedley, P. L.; Kinniburgh, D. G. A review of the source, behaviour and distribution of arsenic in natural waters. *Appl. Geochem.* **2002,** 17, 517-568.
4. WHO, *Guidelines for drinking-water quality vol 1 : Recommendations 2nd ed.* 1993.
5. Kaneco, S.; Itoh, K.; Katsumata, H.; Suzuki, T.; Masuyama, K.; Funasaka, K.; Hatano, K.; Ohta, K. Removal of natural organic polyelectrolytes by adsorption onto tobermorite. *Environ. Sci. Technol.* **2003,** 37, 1448-1451.
6. Davis, A. P.; Bhatnagar, V. Adsorption of cadmium and humic acid onto hematite. *Chemosphere* **1995,** 30, 243-256.
7. Murphy, E. M.; Zachara, J. M.; Smith, S. C. Influence of mineral-bound humic substances on the sorption of hydrophobic organic compounds. *Environ. Sci. Technol.* **1990,** 24, 1507-1516.
8. Summers, R. S.; Roberts, P. V. Activated carbon adsorption of humic substances : I. Heterodisperse mixtures and desorption. *J. Colloid Interface Sci.* **1988,** 122, 367-381.
9. Abu-Omar, M. M.; Appelman, E. H.; Espenson, J. H. Oxygen-transfer reactions of methylrhenium oxides. *Inorg. Chem.* **1996,** 35, 7751-7757.
10. Ogino, K.; Kaneko, Y.; Minoura, T.; Agui, W.; Abe, M. Removal of humic substance dissolved in water I. *J. Colloid Interface Sci.* **1988,** 121, 161-169.
11. Dror, I.; Baram, D.; Berkowitz, B. Use of nanosized catalysts for transformation of chloro-organic pollutants. *Environ. Sci. Technol.* **2005,** 39, 1283-1290.
12. Liu, Y.; Choi, H.; Dionysiou, D.; Lowry, G. V. Trichloroethene Hydrodechlorination in Water by Highly Disordered Monometallic Nanoiron. *Chem. Mater.* **2005,** 17, 5315-5322.
13. Liu, Y.; Majetich, S. A.; Tilton, R. D.; Sholl, D. S.; Lowry, G. V. TCE dechlorination rates, pathways, and efficiency of nanoscale iron particles with different properties. *Environ. Sci. Technol.* **2005,** 39, 1338-1345.
14. Liu, Y. Q.; Lowry, G. V. Effect of particle age (Fe^0 content) and solution pH on NZVI reactivity: H_2 evolution and TCE dechlorination. *Environ. Sci. Technol.* **2006,** 40, 6085-6090.
15. Wang, C.-B.; Zhang, W.-x. Synthesizing nanoscale iron particles for rapid and complete dechlorination of TCE and PCBs. *Environ. Sci. Technol.* **1997,** 31, 2154-2156.
16. Lowry, G. V.; Johnson, K. M. Congener-specific dechlorination of dissolved PCBs by microscale and nanoscale zerovalent iron in a water/methanol solution. *Environ. Sci. Technol.* **2004,** 38, 5208-5216.
17. Ponder, S. M.; Darab, J. G.; Mallouk, T. E. Remediation of Cr(VI) and Pb(II) aqueous solutions using supported, nanoscale zero-valent iron. *Environ. Sci. Technol.* **2000,** 34, 2564-2569.

18. Kanel, S. R.; Greneche, J. M.; Choi, H. Arsenic(V) removal from groundwater using nano scale zero-valent Iron as a colloidal reactive barrier material. *Environ. Sci. Technol.* **2006,** 40, 2045-2050.
19. Kanel, S. R.; Manning, B.; Charlet, L.; Choi, H. Removal of arsenic(III) from groundwater by nanoscale zero-valent iron. *Environ. Sci. Technol.* **2005,** 39, 1291-1298.
20. Choe, S.; Chang, Y.-Y.; Hwang, K.-Y.; Khim, J. Kinetics of reductive denitrification by nanoscale zero-valent iron. *Chemosphere* **2000,** 41, 1307-1311.
21. Choe, S.; Lee, S.-H.; Chang, Y.-Y.; Hwang, K.-Y.; Khim, J. Rapid reductive destruction of hazardous organic compounds by nanoscale Fe^0. *Chemosphere* **2001,** 42, 367-372.
22. Yang, G. C. C.; Lee, H.-L. Chemical reduction of nitrate by nanosized iron: kinetics and pathways. *Water Res.* **2005,** 39, 884-894.
23. Cullen, W. R.; Reimer, K. J. Arsenic speciation in the environment. *Chem. Rev.* **1989,** 89, 713-764.
24. Cummings, D. E.; Caccavo, F.; Fendorf, S.; Rosenzweig, R. F. Arsenic mobilization by the dissimilatory Fe(III)-reducing bacterium Shewanella alga BrY. *Environ. Sci. Technol.* **1999,** 33, 723-729.
25. Pichler, T.; Veizer, J.; Hall, G. E. M. Natural input of arsenic into a Coral-Reef ecosystem by hydrothermal fluids and its removal by Fe(III) oxyhydroxides. *Environ. Sci. Technol.* **1999,** 33, 1373-1378.
26. Smith, A. H.; Hopenhayn-Rich, C.; Bates, M. N.; Goeden, H. M.; Hertz-Picciotto, I.; Duggan, H. M.; Wood, R.; Kosnett, M. J.; Smith, M. T. Cancer risks from arsenic in drinking water. *Environ. Health perspective* **1992,** 97, 259-267.
27. Ungaro, F.; Ragazzi, F.; Cappellin, R.; Giandon, P. Arsenic concentration in the soils of the Brenta Plain (Northern Italy): Mapping the probability of exceeding contamination thresholds. *J. Geochem. Explor.* **2008,** 96, 117-131.
28. Han, F. X.; Su, Y.; Monts, D. L.; Plodinec, M. J.; Banin, A.; Triplett, G. E. Assessment of global industrial-age anthropogenic arsenic contamination. *Die Naturwissenschaften* **2003,** 90, 395-401.
29. Bhumbla, D. K.; Keefer, R. F. Arsenic mobilization and bioavailability in soils. In: JO Nriagu, editor. Arsenic in the environment. *A Wiley-Interscience Publication* **1994**, 51-82.
30. Christen, K. The arsenic threat worsens. *Environ. Sci. Technol.* **2001,** 35, 286A-291A.
31. Nickson, R.; McArthur, J.; Burgess, W.; Ahmed, K. M.; Ravenscroft, P.; Rahmann, M. Arsenic poisoning of Bangladesh groundwater. *Nature* **1998,** 395, 338-338.
32. Balasubramanian N.; Madhavan, K. Arsenic removal from industrial effluent through electrocoagulation. *Chem. Eng. Technol.* **2001,** 24, 519-521.
33. Brandhuber, P.; Amy, G. Alternative methods for membrane filtration of arsenic from drinking water. *Desalination* **1998,** 117, 1-10.

34. Chakravarty, S.; Dureja, V.; Bhattacharyya, G.; Maity, S.; Bhattacharjee, S. Removal of arsenic from groundwater using low cost ferruginous manganese ore. *Water Res.* **2002,** 36, 625-632.
35. Dixit, S.; Hering, J. G. Comparison of arsenic(V) and arsenic(III) sorption onto iron oxide minerals: Implications for arsenic mobility. *Environ. Sci. Technol.* **2003,** 37, 4182-4189.
36. Kartinen, E. O.; Martin, C. J. An overview of arsenic removal processes. *Desalination* **1995,** 103, 79-88.
37. Korngold, E.; Belayev, N.; Aronov, L. Removal of arsenic from drinking water by anion exchangers. *Desalination* **2001,** 141, 81-84.
38. Pierce, M. L.; Moore, C. B. Adsorption of arsenite on amorphous iron hydroxide from dilute aqueous solution. *Environ. Sci. Technol.* **1980,** 14, 214-216.
39. Twidwell, L. G.; Plessas, K. O.; Comba, P. G.; Dahnke, D. R. Removal of arsenic from wastewaters and stabilization of arsenic bearing waste solids: Summary of experimental studies. *J. Hazard. Mater.* **1994,** 36, 69-80.
40. Uheida, A.; Salazar-Alvarez, G.; Bjorkman, E.; Yu, Z.; Muhammed, M. Fe_3O_4 and γ-Fe_2O_3 nanoparticles for the adsorption of Co^{2+} from aqueous solution. *J. Colloid Interface Sci.* **2006,** 298, 501-507.
41. Park, H.; Myung, N. V.; Jung, H.; Choi, H. As(V) remediation using electrochemically synthesized maghemite nanoparticles. *J. Nanopart. Res.* **2009,** Accepted.
42. Huber, D. L. Synthesis, properties, and applications of iron nanoparticles. *Small* **2005,** 1, 482-501.
43. Çelebi, O.; Üzüm, Ç.; Shahwan, T.; Erten, H. N. A radiotracer study of the adsorption behavior of aqueous Ba^{2+} ions on nanoparticles of zero-valent iron. *J. Hazard. Mater.* **2007,** 148, 761-767.
44. Cao, J.; Elliott, D.; Zhang, W.-x. Perchlorate reduction by nanoscale iron particles. *J. Nanopart. Res.* **2005,** 7, 499-506.
45. Chen, S. S.; Hsu, H. D.; Li, C. W. A new method to produce nanoscale iron for nitrate removal. *J. Nanopart. Res.* **2004,** 6, 639-647.
46. Shu, H.-Y.; Chang, M.-C.; Yu, H.-H.; Chen, W.-H. Reduction of an azo dye Acid Black 24 solution using synthesized nanoscale zerovalent iron particles. *J. Colloid Interface Sci.* **2007,** 314, 89-97.
47. Wang, W.; Jin, Z.-h.; Li, T.-l.; Zhang, H.; Gao, S. Preparation of spherical iron nanoclusters in ethanol-water solution for nitrate removal. *Chemosphere* **2006,** 65, 1396-1404.
48. Giasuddin, A. B. M.; Kanel, S. R.; Choi, H. Adsorption of humic acid onto nanoscale zerovalent iron and its effect on arsenic removal. *Environ. Sci. Technol.* **2007,** 41, 2022-2027.
49. Joo, S. H.; Feitz, A. J.; Waite, T. D. Oxidative degradation of the carbothioate herbicide, molinate, using nanoscale zero-valent iron. *Environ. Sci. Technol.* **2004,** 38, 2242-2247.
50. Liu, Y.; Phenrat, T.; Lowry, G. V. Effect of TCE concentration and dissolved groundwater solutes on NUI-Promoted TCE dechlorination and H_2 evolution. *Environ. Sci. Technol.* **2007,** 41, 7881-7887.
51. Zhang, W.-x. Nanoscale iron particles for environmental remediation: An overview. *J. Nanopart. Res.* **2003,** 5, 323-332.

52. Zhang, W.-x.; Wang, C.-B.; Lien, H.-L. Treatment of chlorinated organic contaminants with nanoscale bimetallic particles. *Catalysis Today* **1998,** 40, 387-395.
53. Zhang, L.; Manthiram, A. Chains composed of nanosize metal particles and identifying the factors driving their formation. *Appl. Phys. Lett.* **1997,** 70, 2469-2471.
54. Zhang, W.; Quan, X.; Wang, J.; Zhang, Z.; Chen, S. Rapid and complete dechlorination of PCP in aqueous solution using Ni-Fe nanoparticles under assistance of ultrasound. *Chemosphere* **2006,** 65, 58-64.
55. Schrick, B.; Blough, J. L.; Jones, A. D.; Mallouk, T. E. Hydrodechlorination of trichloroethylene to hydrocarbons using bimetallic Nickel-Iron nanoparticles. *Chem. Mater.* **2002,** 14, 5140-5147.
56. Tee, Y.-H.; Grulke, E.; Bhattacharyya, D. Role of Ni/Fe nanoparticle composition on the degradation of trichloroethylene from water. *Indust. & Eng. Chem. Res.* **2005,** 44, 7062-7070.
57. Cheng, I. F.; Muftikian, R.; Fernando, Q.; Korte, N. Reduction of nitrate to ammonia by zero-valent iron. *Chemosphere* **1997,** 35, 2689-2695.
58. Huang, C.-P.; Wang, H.-W.; Chiu, P.-C. Nitrate reduction by metallic iron. *Water Res.* **1998,** 32, 2257-2264.
59. Su, C.; Puls, R. W. Nitrate reduction by zerovalent iron: Effects of Formate, Oxalate, Citrate, Chloride, Sulfate, Borate, and Phosphate. *Environ. Sci. Technol.* **2004,** 38, 2715-2720.
60. Westerhoff, P. Reduction of nitrate, bromate, and chlorate by zero valent iron (Fe^0). *J. Environ. Eng.* **2003,** 129, 10-16.
61. Sohn, K.; Kang, S. W.; Ahn, S.; Woo, M.; Yang, S.-K. Fe(0) Nanoparticles for nitrate reduction: Stability, reactivity, and transformation. *Environ. Sci. Technol.* **2006,** 40, 5514-5519.
62. Haag, W. R.; Hoign, J. g.; Bader, H. Improved ammonia oxidation by ozone in the presence of bromide ion during water treatment. *Water Res.* **1984,** 18, 1125-1128.
63. Weinberg, H. S.; Delcomyn, C. A.; Unnam, V. Bromate in chlorinated drinking waters: Occurrence and implications for future regulation. *Environ. Sci. Technol.* **2003,** 37, 3104-3110.
64. WHO, *Bromate in drinking-water. Background document for preparation of WHO guidelines for drinking-water quality.* 2003.
65. Xie, L.; Shang, C. Effects of copper and palladium on the reduction of bromate by Fe(0). *Chemosphere* **2006,** 64, 919-930.
66. Xie, L.; Shang, C. The effects of operational parameters and common anions on the reactivity of zero-valent iron in bromate reduction. *Chemosphere* **2007,** 66, 1652-1659.
67. Wang, Q.; Snyder, S.; Kim, J.; Choi, H. Aqueous ethanol modified nanoscale zerovalent Iron in bromate reduction: Synthesis, characterization, and reactivity. *Environ. Sci. Technol.* **2009**.
68. Cao, J.; Zhang, W.-X. Stabilization of chromium ore processing residue (COPR) with nanoscale iron particles. *J. Hazard. Mater.* **2006,** 132, 213-219.

69. Li, X.-q.; Zhang, W.-x. Sequestration of metal cations with zerovalent iron nanoparticles A study with High Resolution X-ray Photoelectron Spectroscopy (HR-XPS). *J. Phys. Chem. C* **2007,** 111, 6939-6946.
70. Henderson, A. D.; Demond, A. H. Long-term performance of zero-valent iron permeable reactive barriers: A critical review. *Environmental Engineering Science* **2007,** 24, 401-423.
71. Manning, B. A.; Fendorf, S. E.; Goldberg, S. Surface structures and stability of arsenic(III) on goethite: Spectroscopic evidence for inner-sphere complexes. *Environ. Sci. Technol.* **1998,** 32, 2383-2388.
72. Manning, B. A.; Hunt, M. L.; Amrhein, C.; Yarmoff, J. A. Arsenic(III) and Arsenic(V) reactions with zerovalent iron corrosion products. *Environ. Sci. Technol.* **2002,** 36, 5455-5461.
73. Farrell, J.; Wang, J. P.; O'Day, P.; Conklin, M. Electrochemical and spectroscopic study of arsenate removal from water using zero-valent iran media. *Environ. Sci. Technol.* **2001,** 35, 2026-2032.
74. Melitas, N.; Wang, J. P.; Conklin, M.; O'Day, P.; Farrell, J. Understanding soluble arsenate removal kinetics by zerovalent iron media. *Environ. Sci. Technol.* **2002,** 36, 2074-2081.
75. Su, C.; Puls, R. W. Arsenate and arsenite removal by zerovalent Iron: Kinetics, redox transformation, and implications for in situ groundwater remediation. *Environ. Sci. Technol.* **2001,** 35, 1487-1492.
76. Su, C. M.; Puls, R. W. Arsenate and arsenite removal by zerovalent iron: Effects of phosphate, silicate, carbonate, borate, sulfate, chromate, molybdate, and nitrate, relative to chloride. *Environ. Sci. Technol.* **2001,** 35, 4562-4568.
77. Redman, A. D.; Macalady, D. L.; Ahmann, D. Natural organic matter affects arsenic speciation and sorption onto hematite. *Environ. Sci. Technol.* **2002,** 36, 2889-2896.
78. Grafe, M.; Eick, M. J.; Grossl, P. R. Adsorption of arsenate (V) and arsenite (III) on goethite in the presence and absence of dissolved organic carbon. *Soil Sci Soc Am J* **2001,** 65, 1680-1687.
79. Grafe, M.; Eick, M. J.; Grossl, P. R.; Saunders, A. M. Adsorption of arsenate and arsenite on ferrihydrite in the presence and absence of dissolved organic carbon. *J Environ Qual* **2002,** 31, 1115-1123.
80. Ko, I.; Kim, J.-Y.; Kim, K.-W. Arsenic speciation and sorption kinetics in the As-hematite-humic acid system. *Colloids and Surfaces A: Physicochemical and Engineering Aspects* **2004,** 234, 43-50.
81. Dries, J.; Bastiaens, L.; Springael, D.; Kuypers, S.; Agathos, S. N.; Diels, L. Éffect of humic acids on heavy metal removal by zero-valent iron in batch and continuous flow column systems. *Water Res.* **2005,** 39, 3531-3540.
82. Deng, S.; Bai, R. B. Aminated polyacrylonitrile fibers for humic acid adsorption: Behaviors and mechanisms. *Environ. Sci. Technol.* **2003,** 37, 5799-5805.
83. Filius, J. D.; Lumsdon, D. G.; Meeussen, J. C. L.; Hiemstra, T.; Van Riemsdijk, W. H. Adsorption of fulvic acid on goethite. *Geochim. Cosmochim. Acta* **2000,** 64, 51-60.

84. Gu, B.; Schmitt, J. g.; Chen, Z.; Liang, L.; McCarthy, J. F. Adsorption and desorption of different organic matter fractions on iron oxide. *Geochim. Cosmochim. Acta* **1995,** 59, 219-229.
85. Cho, J. Natural organic matter (NOM) reflection by, and flux-decline of, nanofiltration (NF) and ultrafiltration (UF) membrane. *Ph.D. Thesis, University of Colorado at Boulder* **1998**.
86. Peng, X.; Luan, Z.; Chen, F.; Tian, B.; Jia, Z. Adsorption of humic acid onto pillared bentonite. *Desalination* **2005,** 174, 135-143.

Chapter 9

Nanostructured Multifunctional Materials for Environmental Remediation of Chlorinated Hydrocarbons

Tonghua Zheng[1], Jingjing Zhan[1], Jibao He[1], Bhanukiran Sunkara[1], Yunfeng Lu[2], Gary L. McPherson[1], Gerhard Piringer[1], Vladimir Kolesnichenko[3], Vijay T. John[1]

[1]Department of Chemical and Biomolecular Engineering, Tulane University, New Orleans, LA 70118
[2]Department of Chemical Engineering, University of California at Los Angeles. Los Angeles, CA 90095
[3]Department of Chemistry, Xavier University, New Orleans, LA 70125

Spherical silica particles containing nanoscale zero valent iron were synthesized through an aerosol assisted process. These particles are effective for groundwater remediation, with the environmentally benign silica particles serving as effective carriers for nanoiron transport. Incorporation of iron into porous submicron silica particles protects ferromagnetic iron nanoparticles from aggregation and may increase their subsurface mobility. Additionally, the presence of surface silanol groups on silica particles allows control of surface properties via silanol modification using organic functional groups. Aerosolized silica particles with functional alkyl moieties such as ethyl groups on the surface, clearly adsorb solubilized trichloroethylene (TCE) in water. These materials may therefore act as adsorbents which have coupled reactivity characteristics. Transport experiments indicate that these composite particles are in the optimal size range for effective transport through sediments. The nanoscale iron/silica composite particles with controlled surface properties have the potential to be efficiently applied for in-situ source depletion and in the design of permeable reactive barriers.

Introduction

Chlorinated hydrocarbons such as trichloroethylene (TCE) form a class of dense non-aqueous phase liquid (DNAPL) contaminants in groundwater and soil that are difficult to remediate. They have a density higher than water and settle deep into the sediment from which they gradually leach out into aquifers, causing long term environmental pollution. Remediation of these contaminants is of utmost importance for the cleanup of contaminated sites (*1-3*). In recent years, the reductive dehalogenation of such compounds using zerovalent iron (ZVI) represents a promising approach for remediation (*4-24*). The overall redox reaction using TCE as an example is

$$C_2HCl_3 + 4Fe^0 + 5H^+ \rightarrow C_2H_6 + 4Fe^{2+} + 3Cl^-$$

where gaseous products such as ethane result from complete reduction. The environmentally benign nature of ZVI and its low cost are attractive to the development of such remediation technologies. Compared to more conventional treatment processes, the in-situ direct injection of reactive zero-valent iron into the contaminated subsurface is a preferred method because it may more directly access and target the contaminants (*25, 26*). Nanoscale zero-valent iron particles (NZVI) often have higher remediation rates resulting from their increased surface area (*24,27-34*). More importantly, the colloidal nature of nanoiron indicates that these particles can be directly injected into contaminated sites for source depletion or, alternatively, be devised to construct permeable reactive barriers for efficient TCE remediation (*35-41*).

For successful in-situ source depletion of pure phase TCE, it is necessary for injected nanoscale ZVI to migrate through the saturated zone to reach the contaminant; one can consider successful strategies the environmental equivalent of targeted drug delivery. The transport of colloidal particles such as nanoiron through porous media is determined by competitive mechanisms of diffusive transport, interception by soil or sediment grains and sedimentation effects as shown through the now-classical theories of colloid transport (*38,42-45*). The Tufenkji-Elimelch model (*46*) which considers the effect of hydrodynamic forces and van der Waals interactions between the colloidal particles and soil/sediment grains is a significant advance in modeling transport of colloidal particles through sediment, and predicts optimal particle sizes between 200 nm – 1000 nm for zerovalent iron particles at typical groundwater flow conditions (*47,48*). At particle sizes exceeding 15 nm, however, ZVI exhibits ferromagnetism, leading to particle aggregation and a loss in mobility (*49*). The particles by themselves are therefore inherently ineffective for in-situ source depletion. One of the common methods to increase nanoiron mobility is to stabilize the particles by adsorption of organic molecules on the particle surface (*50-54*). The adsorbed molecules enhance steric or electrostatic repulsions between particles to prevent their aggregation. Techniques include the use of polymers, surfactants, starch, modified cellulose, and vegetable oils as stabilizing layers to form more stable dispersions (*47, 52-61*). These methods enhance steric or electrostatic repulsions of particles to prevent their aggregation and may be effective if the physically adsorbed stabilizers are retained during particle migration through sediments. Functionalization of ZVI nanoparticles with organic ligands is another alternative but such functionalization is not easy and it is unclear if the reactivity of ZVI is retained.

Figure 1 summarizes the objectives behind our recent work where we seek to develop multifunctional nanoscale materials for adsorption, reaction, transport and partitioning.

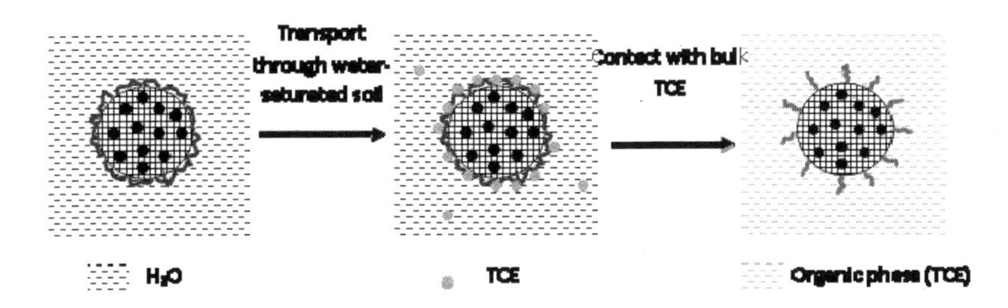

Figure 1: The design of functional composite particles for effective transport, reaction and partitioning.

On the left, we show the concept of entrapment of NZVI in porous submicron particles of functionalized silica. The functional groups are typically hydrophobic alkyl groups which, in aqueous solution, stay confined to the silica. The silica particles are designed to have the optimal size range for transport through sediment. As the particles travel through water- saturated sediment following groundwater flow streamlines, there is a significant adsorption of dissolved TCE onto the alkyl groups, thereby bringing the contaminant in close proximity to the NZVI (center). When the composite particles reach a site of bulk TCE, the alkyl groups extend out increasing the hydrodynamic radius of the particle thereby reducing its effective density (right). It is an objective to stabilize these particles either in the TCE bulk or at the water-TCE bulk interface.

The actualizations of these concepts are next presented. In order to separate the NZVI particles, we entrap them in a porous silica matrix, where the silica particles are typically submicron sized. The process of encapsulation is through an aerosol-assisted process (*62-64*). The postulated advantages behind the work are the following: [a] entrapment of ZVI into porous silica may make the ZVI less prone to aggregation, while maintaining reactivity; [b] silica is environmentally benign and entrapment of ZVI into porous silica reduces the safety concerns of nanoiron hazards of fire and explosion when exposed to air (*65*); [c] since the aerosol-assisted route to synthesis of porous colloidal silica is just a variation of is a variation of the spray drying process, scale up to produce large quantities of the material is feasible. [d] functionalization of silica is extremely simple and there are several methods of silica functionalization that could be exploited to allow maximum contact of ZVI with the contaminant (TCE) [e] alkyl groups are microbially degradable. Point [d] is especially relevant from two perspectives. First, it would be a significant advantage to target the delivery of ZVI so that the particles transport efficiently through the saturated zone and then effectively partition to the water-TCE interface upon encountering regions of bulk TCE. Second, if silica can be functionalized appropriately, the sparingly soluble pure phase TCE in water would partition to the silica, increasing local concentrations and accessibility to the ZVI nanoparticles.

Figure 2 illustrates the concepts of NZVI encapsulation in porous silica through the aerosol process. In this process, silica precursors such as tetraethyl orthosilicate (TEOS) and ethyl triethoxysilane (ETES) together with iron precursors are aerosolized with the aerosol droplets passing through a high temperature zone. During this process, silicates hydrolyze and condense in the droplet entrapping the iron species. The "chemistry in a droplet" process leads to submicron sized particles of silica containing iron nanoparticles which are then collected on a filter. Since the particles are essentially made with silica and iron they are environmentally benign. Of particular relevance also is the use of alkyl groups attached to the silica through the use of alkyl-silane precursors such as ETES. These groups introduce porosity into the silica. Additionally, these organic groups play an important role in that they serve as adsorbents for the TCE, thus bringing the organic contaminant to the vicinity of the iron species and facilitating reaction. We note that mixtures of ethyltrioxysilane (ETES) and tetraethylorthosilicate (TEOS) lead to particles where the degree of incorporation of the alkyl functionality can be adjusted.

Figure 3 shows the size distribution of the composite Fe/Ethyl-Silica particles, showing polydispersity that is inherent in the aerosol-assisted process. The inset shows a TEM indicating zerovalent iron nanoparticles decorating the silica matrix.

Figure 4 illustrates the reactivity characteristics of the composite particles when contacted with dissolved TCE. There is a significant drop in solution TCE concentration followed by a gradual decrease. The initial concentration drop is not due to reaction but to adsorption. This is clearly shown by the gaseous product evolution (ethane and ethylene) which is much more gradual. Additionally, when the composite particles are prepared without the alkyl functional groups, using just TEOS as the silica precursor, the sudden drop in TCE solution concentration is not observed (*66*).

$$CH_3CH_2-Si(OCH_2CH_3)_3$$

ETES: ethyltriethoxysilane

$$CH_3CH_2O-Si(OCH_2CH_3)_3$$

TEOS: tetraethyl orthosilicate

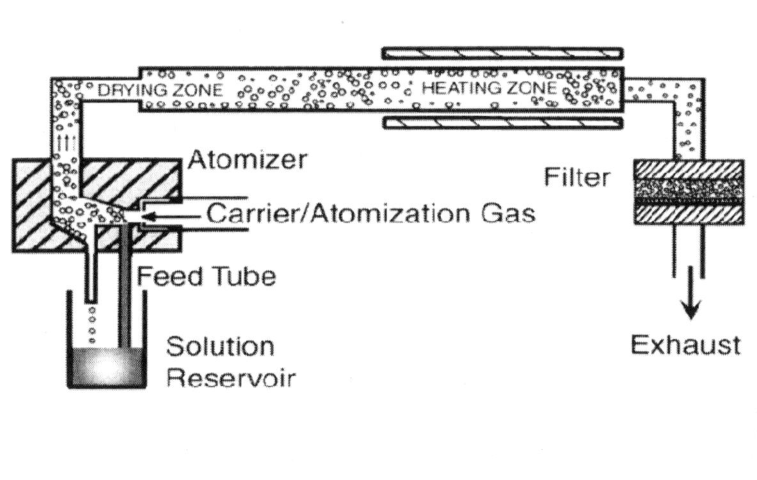

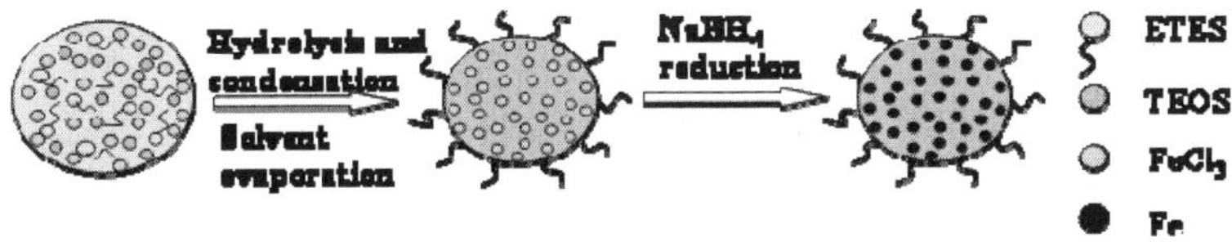

Figure 2: The encapsulation of NZVI in porous silica. The silica precursors are shown at the top, the aerosolizer in the middle and the "chemistry in a droplet" concept at the bottom.

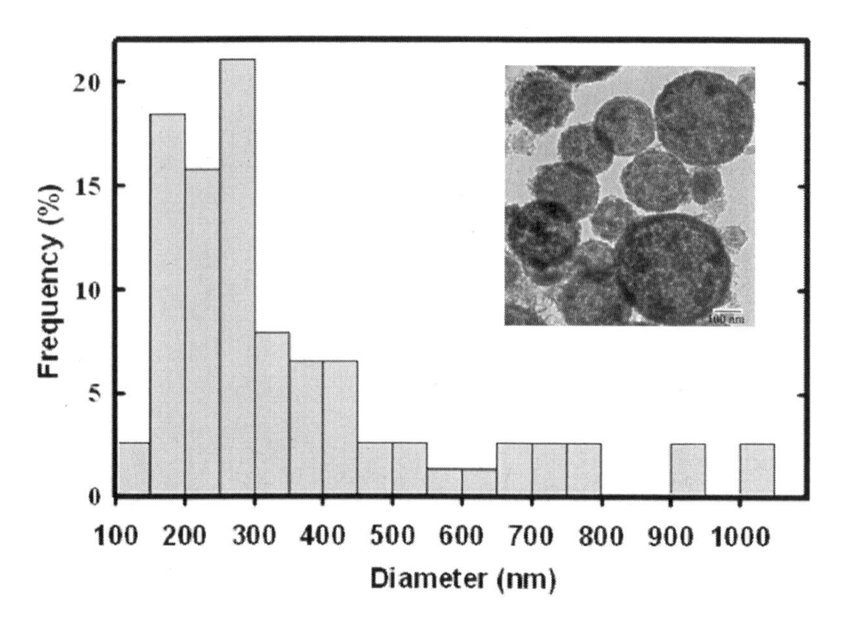

Figure 3: Particle size distribution and TEM of particles (inset)

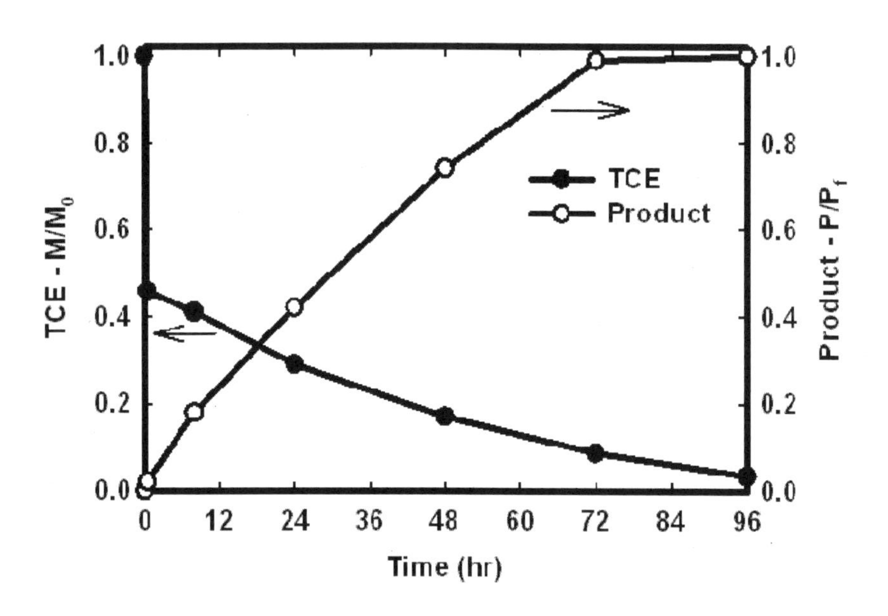

Figure 4: Reactivity Characteristics of the composite Fe/EthylSilica particles. The initial drop in solution TCE concentration is due to adsorption, bringing up the fact that these are adsorptive-reactive particles.

The adsorptive-reactive concept is extremely important in the design of multifunctional particles. Adsorption leads to high local concentrations in the vicinity of the reactive zerovalent iron, potentially facilitating reaction.

We also note that the reaction rate can be enhanced significantly upon deposition of small quantities of Pd through the incorporation of $PdCl_3$ in the precursor solution (*67*). The catalytic effect of Pd in dramatically enhancing reaction rates has been discussed in detail in the literature (*27,68,69*). The role of Pd is to dissociatively chemisorb hydrogen produced by redox reactions on Fe^0. Additionally, chlorinated hydrocarbons adsorb strongly on Pd, leading to the reduction of the chlorinated species through surface reaction with the adsorbed hydrogen).

The particle size range is extremely important from a transport perspective. Filtration theory predicts that the migration of colloidal particles through porous media such as soil is typically dictated by Brownian diffusion, interception and gravitational sedimentation (*70*). The Tufenkji-Elimelech model is perhaps the most comprehensive model to describe these effects in the presence of interparticle interactions (*46*), with the governing equation

$$\eta_0 = 2.4 A_S^{1/3} N_R^{-0.081} N_{Pe}^{-0.715} N_{vdW}^{0.052} + 0.55 A_S N_R^{1.675} N_A^{0.125} + 0.22 N_R^{-0.24} N_G^{1.11} N_{vdW}^{0.053}$$

where η_0 is the collector efficiency, simply defined as the probability of collision between migrating particles and sediment grains. The first term on the right characterizes the effects of particle diffusion on the collector efficiency, while the second and third terms describe the effects of interception and sedimentation. However, the Tufenkji-Elimelech equation does not provide the complete representation of particle transport, which also involves concepts such as bridging and attachment between the particles and the surfaces of soil grains, characterized through a "sticking coefficient" (*59*). For brevity, we limit the discussion of equation (1) to demonstrating the dependence of the collector efficiency on particle size as shown in Figure 5.

As seen in the figure, the collector efficiency is minimized at a particle size range 0.1 to 1 μm, which implies that this is the optimal size range for colloid particles to migrate through the soil, and is in the size range obtained through the aerosol-assisted process (Figure 3). Figure 5 also indicates an optical micrograph of commercially available ZVI nanoparticles, the reactive nanoiron particles (RNIP-10DS, which is uncoated or bare RNIP) from Toda Kogyo Corporation. While the intrinsic particle size of these particles is of the order 30-70 nm, aggregation to effective sizes over 10 μm make them ineffective for transport through soil (*49*).

We have carried out column and capillary transport experiments (Figure 6) on the Fe/Ethyl-Silica particles to determine the transport characteristics (*71*). As predicted by the TE equation, the Fe/Ethyl-silica particles elute efficiently through model sediment-packed columns, while the control bare RNIP remains aggregated at the head of the column.

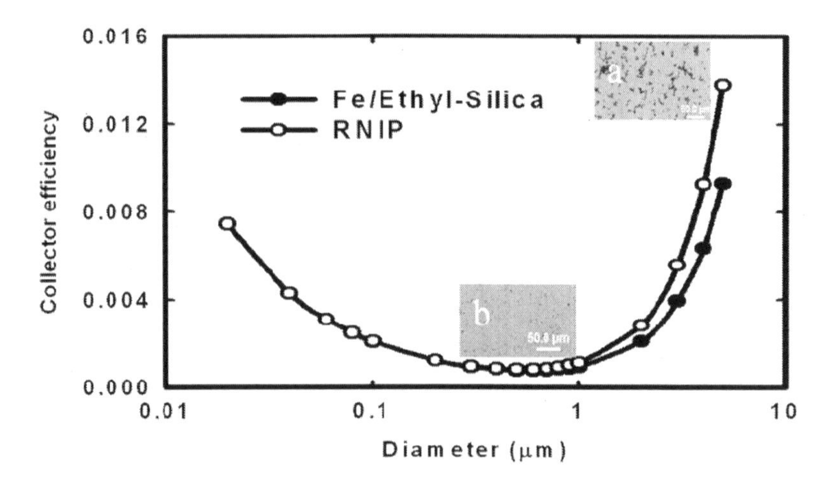

Figure 5: The TE equation applied to Fe-based system.

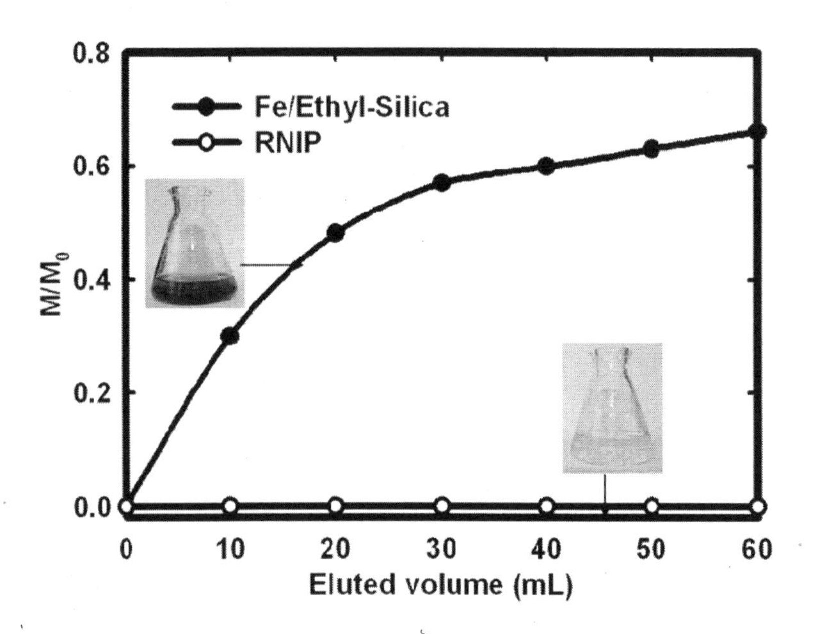

Figure 6: Column elution profiles of Fe/Ethyl-Silica particles compared to bare RNIP

Capillary studies confirm these findings (*71*). Figure 7 illustrates these studies where 1.5 mm horizontal capillaries are filled with sediment and particle transport is visualized through optical microscopy. Here it is clearly evident that the bare RNIP (panel (1) in the middle and panel (i) at the bottom) is retained at the capillary inlet, while visualization of Fe/Ethyl-Silica particles is clearly observed throughout the capillary during transport (panels (ii) in the middle and the bottom), and at the end of the capillary after elution is complete ((iii) in the middle). At the beginning of the experiment all system appear the same as (i) in the middle set of panels with a particle suspension at the inlet to the capillary.

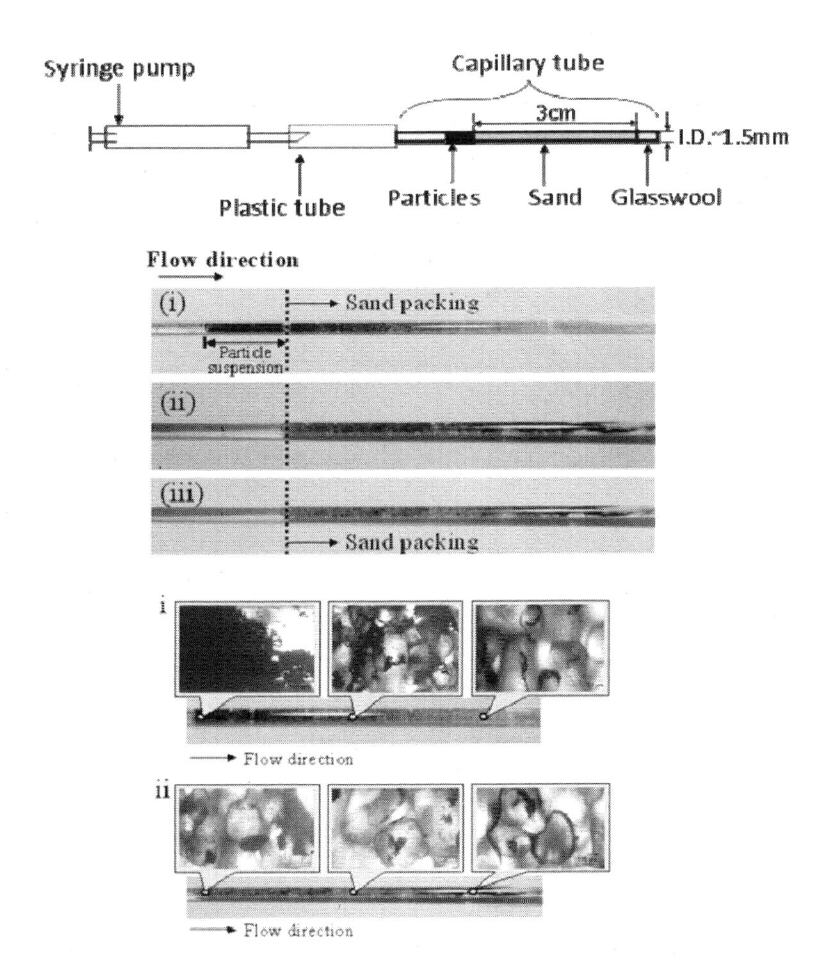

Figure 7: Capillary transport studies. The schematic of the setup is at the top, the middle panels illustrate the packed capillaries at varying stages of elution, and the bottom panels indicate the micrographs of the capillaries.

Figure 8 illustrates a microcapillary visualization experiment where a TCE droplet is injected using a micropipetter into a 200 μm capillary containing dispersed Fe/Ethyl-Silica particles in water. We see a stable aggregation of the particles on the TCE droplet interface.

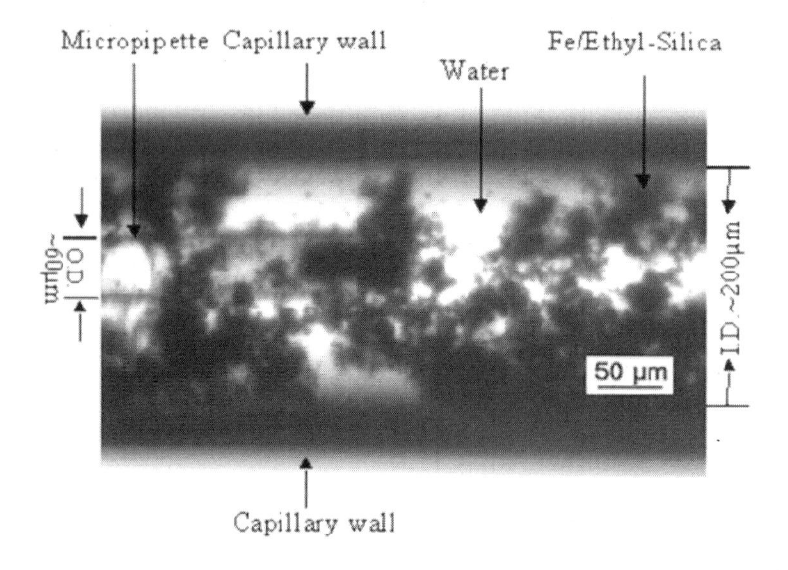

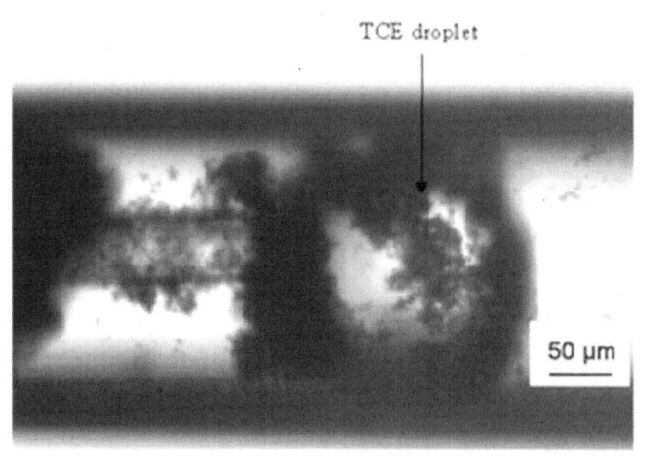

Figure 8: Visualization of particles partitioning to the interface of a TCE droplet when injected into a microcapillary

These results summarize our experiments with NZVI supported on novel Fe/Ethyl-Silica particles prepared through the aerosol assisted route. The disadvantage of making these materials a reality in environmental remediation is perhaps the cost of the silica precursor (ethyltriethoxysilane and tetraethyl orthosilicate) which becomes an overriding factor in developing applications to environmental remediation. In continuing research we are adapting carbon based materials to support ZVI nanoparticles. A brief description of our continuing research follows.

In 2001, there was a very interesting report published by Wang et al. in Carbon (*77*) illustrating a novel and simple method to synthesize spherical microporous carbons. These authors took sucrose in solution and subjected it to hydrothermal treatment at 200 C. At these conditions (12 bar vapor pressure), the sugar undergoes dehydration and the resultant material has the morphology of extremely monodisperse; carbon spheres. Upon pyrolysis of these materials, graphitic carbon spheres are obtained with sizes ranging from 200 nm to 6μm depending on the sugar concentrations used. These authors have been studying the materials for applications in Li-ion batteries as they make promising anode materials (*72*). We have been able to reproduce the morphologies of these materials as shown in the SEMs below (Figure 9). The particles on the left were prepared using a precursor sucrose solution of 0.15M concentration, while the particles on the right were prepared using a concentration of 1.5M.

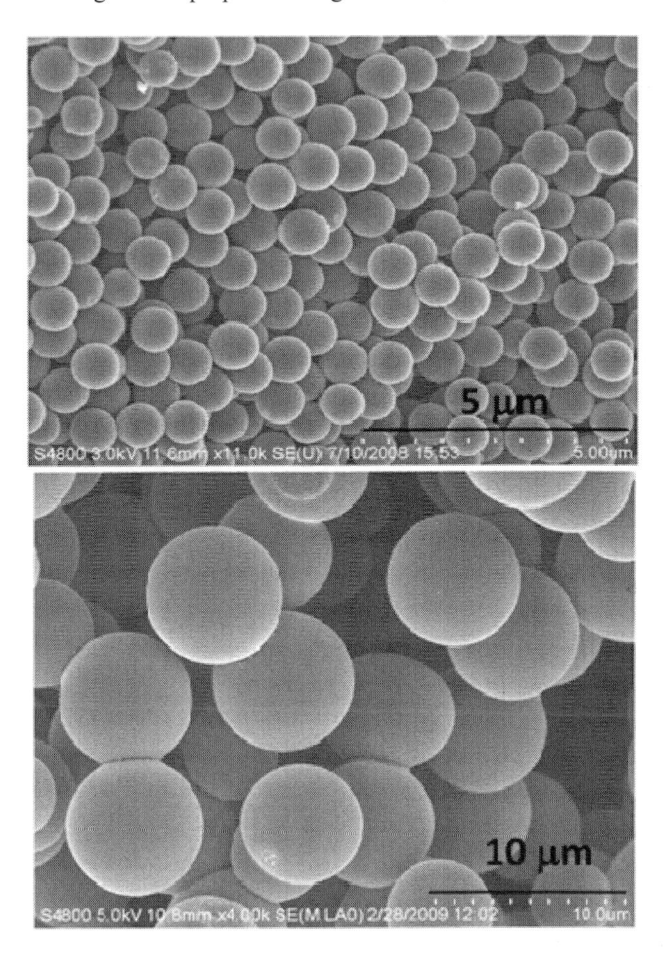

Figure 9: Synthesis of monodisperse carbon particles

These results become easily connected to the environmental problem of TCE remediation. Carbons are environmentally innocuous. The hydrothermal+pyrolysis process is simple and can be easily scaled up as a solution process. The carbon precursors are also inexpensive. Most importantly, the particle size can be tuned for optimal transport characteristics very easily by modifying precursor concentrations.

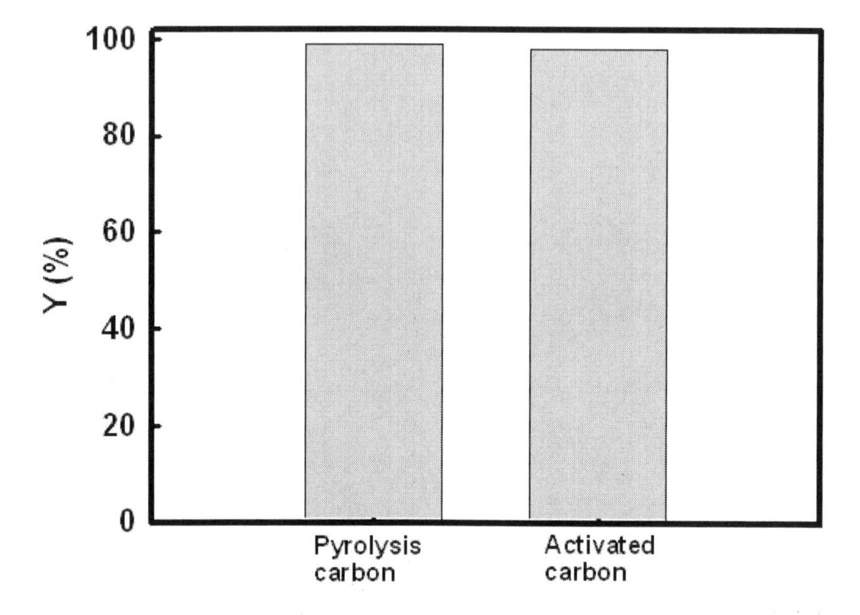

Figure 10: Adsorption capacties of porous carbon microspheres

Initial results inidcate that these carbon particles are microporous with surface areas in the order of 400 m^2/gm. They adsorb TCE to the same extent as activated carbons as shown in Figure 10 which illustrates a preliminary adsorption experiment done using 20 ml of 20 ppm TCE solution containing 0.2 g of the particles. Our continuing research seeks to impregrate these carbon microspheres with ZVI particles and stabilize these composites in solution using polymers (e.g. carboxymethyl cellulose). Complete studies will be reported in the literature.

Summary

The research described in this chapter is directed towards the development of zerovalent iron based multifunctional particles with the following characteristics (a) they are reactive to the reduction of chlorinated hydrocarbons (b) they transport through sediments and migrate to the sites of TCE contamination (c) they adsorb TCE lowering bulk dissolved TCE concentrations and bringing TCE to the proximity of the zerovalent reduction sites (d) they partition to the interface of bulk TCE. Fe/silica composite particles prepared through the aerosol-assisted route appear to have these characteristics. The aerosol route would not be difficult to be scaled up since it is just a variation of traditional spray drying processes. However, the precursor costs may make these materials unattractive for practical application. In continuing work, we are developing carbon based materials that exhibit these characteristics. The novelty of the carbon based approach is the methodology to make uniform sized particles in a simple manner.

References

1. Dowideit, P.; von Sonntag, C., Reaction of ozone with ethene and its methyl- and chlorine-substituted derivatives in aqueous solution. Environ. Sci. Technol. 1998, 32, 1112.
2. Cowell, M.; Kibbey, T.; Zimmerman, J.; Hayes, K., Partitioning of ethoxylated nonionic surfactants in water/NAPL systems: Effects of surfactant and NAPL properties. Environ. Sci. Technol. 2000, 34, 1583.
3. Nutt, M.; Hughes, J.; Wong, M., Designing Pd-on-Au bimetallic nanoparticle catalysts for trichloroethene hydrodechlorination. Environ. Sci. Technol. 2005, 39, 1346.
4. Matheson, L. J.; Tratnyek, P. G., Reductive dehalogenation of chlorinated methanes by iron metal. Environ. Sci. Technol. 1994, 28, 2045-2053.
5. Orth, W. S.; Gillham, R. W., Dechlorination of Trichloroethene in Aqueous Solution Using Fe0. Environ. Sci. Technol. 1996, 30, (1), 66-71.
6. Fennelly, J. P.; Roberts, A. L., Reaction of 1,1,1-trichloroethane with zero-valent metals and bimetallic reductants. Environmental Science & Technology 1998, 32, (13), 1980-1988.
7. Huang, C. P.; Wang, H. W.; Chiu, P. C., Nitrate reduction by metallic iron. Water Research 1998, 32, (8), 2257-2264.
8. Devlin, J. F.; Klausen, J.; Schwarzenbach, R. P., Kinetics of nitroaromatic reduction on granular iron in recirculating batch experiments. Environmental Science & Technology 1998, 32, (13), 1941-1947.
9. Gotpagar, J.; Lyuksyutov, S.; Cohn, R.; Grulke, E.; Bhattacharyya, D., Reductive dehalogenation of trichloroethylene with zerovalent iron: surface profiling microscopy and rate enhancement studies. Langmuir 1999, 15, 8412.
10. Li, T.; Farrell, J., Reductive dechlorination of trichloroethene and carbon tetrachloride using iron and palladized-iron cathodes. Environ. Sci. Technol. 2000, 34, 173.
11. Uludag-Demirer, S.; Bowers, A., Adsorption/reduction reactions of trichloroethylene by elemental iron in the gas phase: the role of water. Environ. Sci. Technol. 2000, 34, 4407.
12. Butler, E.; Hayes, K., Factors influencing rates and products in the transformation of trichloroethylene by iron sulfide and iron metal. Environ. Sci. Technol. 2001, 35, 3884.
13. Hu, H.; Goto, N.; Fujie, K., Effect of pH on the reduction of nitrite in water by metallic iron. Water Res. 2001, 35, 2789.
14. Doong, R.; Chen, K.; Tsai, H., Reductive dechlorination of carbon tetrachloride and tetrachloroethylene by zerovalent silicon-iron reductants. Environ. Sci. Technol. 2003, 37, 2575.

15. Su, C.; Puls, R., In situ remediation of arsenic in simulated groundwater using zerovalent iron: Laboratory column tests on combined effects of phosphate and silicate. Environ. Sci. Technol. 2003, 37, 2582.
16. Su, C. M.; Puls, R. W., Nitrate reduction by zerovalent iron: Effects of formate, oxalate, citrate, chloride, sulfate, borate, and phosphate. Environ. Sci. Technol. 2004, 38, 2715.
17. Tamara, M. L.; Butler, E. C., Effects of iron purity and groundwater characteristics on rates and products in the degradation of carbon tetrachloride by iron metal. Environ. Sci. Technol. 2004, 38, 1866.
18. Miehr, R.; Tratnyek, P.; Bandstra, J.; Scherer, M.; Alowitz, M.; Bylaska, E., Diversity of contaminant reduction reactions by zerovalent iron: role of the reductate. Environ. Sci. Technol. 2004, 38, 139.
19. Dries, J.; Bastiaens, L.; Springael, D.; Agathos, S.; Diels, L., Competition for sorption and degradation of chlorinated ethenes in batch zero-valent iron systems. Environ. Sci. Technol. 2004, 38, 2879.
20. Li, T.; Farrell, J., Electrochemical investigation of the rate-limiting mechanisms for trichloroethylene and carbon tetrachloride reduction at iron surfaces. Environ. Sci. Technol. 2001, 35, 3560.
21. Kenneke, J. F.; McCutcheon, S. C., Use of Pretreatment Zones and Zero-Valent Iron for the Remediation of Chloroalkenes in an Oxic Aquifer. Environ. Sci. Technol. 2003, 37, (12), 2829-2835.
22. Satapanajaru, T.; Shea, P.; Comfort, S.; Roh, Y., Green rust and iron oxide formation influences metolachlor dechlorination during zerovalent iron treatment. Environ. Sci. Technol. 2003, 37, 5219.
23. Arnold, W. A.; Roberts, A. L., Pathways and kinetics of chlorinated ethylene and chlorinated acetylene reaction with Fe(0) particles. Environ. Sci. Technol. 2000, 34, 1794.
24. Jeen, S.; Gillham, R.; Blowes, D., Effects of carbonate precipitates on long-term performance of granular iron for reductive dechlorination of TCE. Environ. Sci. Technol. 2006, 40, 6432.
25. Nyer, E. K.; Vance, D. B., Nano-scale iron for dehalogenation. Ground Water Monit. Remediat. 2001, 2, 41-46.
26. Schrick, B.; Hydutsky, B. W.; Blough, J. L.; Mallouk, T. E., Delivery Vehicles for Zerovalent Metal Nanoparticles in Soil and Groundwater. Chem. Mater. 2004, 16, (11), 2187-2193.
27. Wang, C. B.; Zhang, W. X., Synthesizing Nanoscale Iron Particles for Rapid and Complete Dechlorination of TCE and PCBs. Environ. Sci. Technol. 1997, 31, (7), 2154-2156.
28. Xu, Y.; Zhang, W., Subcolloidal Fe/Ag particles for reductive dehalogenation of chlorinated benzenes. Ind. Eng. Chem. Res. 2000, 39, 2238.
29. Lowry, G.; Reinhard, M., Pd-Catalyzed TCE dechlorination in groundwater: Solute effects, biological control, and oxidative catalyst regeneration. Environ. Sci. Technol. 2000, 34, 3217.
30. Schrick, B.; Blough, J. L.; Jones, A. D.; Mallouk, T. E., Hydrodechlorination of Trichloroethylene to Hydrocarbons Using Bimetallic Nickel-Iron Nanoparticles. Chem. Mater. 2002, 14, (12), 5140-5147.

31. Joo, S.; Feitz, A.; Waite, T., Oxidative degradation of the carbothioate herbicide, molinate, using nanoscale zero-valent iron. Environ. Sci. Technol. 2004, 38, 2242.
32. Tee, Y.; Grulke, E.; Bhattacharyya, D., Role of Ni/Fe nanoparticle composition on the degradation of trichloroethylene from water. Ind. Eng. Chem. Res. 2005, 44, 7062.
33. Dror, I.; Baram, D.; Berkowitz, B., Use of nanosized catalysts for transformation of chloro-organic pollutants. Environ. Sci. Technol. 2005, 39, 1283.
34. Liu, Y.; Majetich, S. A.; Tilton, R. D.; Sholl, D. S.; Lowry, G. V., TCE Dechlorination Rates, Pathways, and Efficiency of Nanoscale Iron Particles with Different Properties. Environ. Sci. Technol. 2005, 39, (5), 1338-1345.
35. Liu, Y.; Choi, H.; Dionysiou, D.; Lowry, G. V., Trichloroethene Hydrodechlorination in Water by Highly Disordered Monometallic Nanoiron. Chem. Mater. 2005, 17, (21), 5315-5322.
36. Wust, W. F.; Kober, R.; Schlicker, O.; Dahmke, A., Combined zero- and first-order kinetics model of the degradation of TCE and cis-DCE with commercial iron. Environ. Sci. Technol. 1999, 33, 4304.
37. Astrup, T.; Stipp, S.; Christensen, T., Immobilization of chromate from coal fly ash leachate using an attenuating barrier containing zero-valent iron. Environ. Sci. Technol. 2000, 34, 4163.
38. Elliott, D. W.; Zhang, W. X., Field Assessment of Nanoscale Bimetallic Particles for Groundwater Treatment. Environ. Sci. Technol. 2001, 35, (24), 4922-4926.
39. Yabusaki, S.; Cantrell, K.; Sass, B.; Steefel, C., Multicomponent reactive transport in an in situ zero-valent iron cell. Environ. Sci. Technol. 2001, 35, 1493.
40. Morrison, S., Performance evaluation of a permeable reactive barrier using reaction products as tracers. Environ. Sci. Technol. 2003, 37, 2302.
41. Casey, F.; Ong, S.; Horton, R., Degradation and transformation of trichloroethylene in miscible-displacement experiments through zerovalent metals. Environ. Sci. Technol. 2000, 34, 5023.
42. Shimotori, T.; Nuxoll, E.; Cussler, E.; Arnold, W., A polymer membrane containing Fe0 as a contaminant barrier. Environ. Sci. Technol. 2004, 38, 2264.
43. Yao, K.; Habibian, M.; O'Melia, C., Water and waste water filtration: Concepts and applications. Environ. Sci. Technol. 1971, 5, 1105.
44. Rajagopalan, R.; Tien, C., Trajectory analysis of deep-bed filtration with the sphere-in-cell porous media model. AIChE J. 1976, 22, 523.
45. Spielman, L., Particle capture from low-speed laminar flows. Annu. Rev. Fluid Mech. 1977, 9, 297.
46. Tufenkji, N.; Elimelech, M., Correlation Equation for Predicting Single-Collector Efficiency in Physicochemical Filtration in Saturated Porous Media. Environ. Sci. Technol. 2004, 38, (2), 529-536.
47. Hydutsky, B. W.; Mack, E. J.; Beckerman, B. B.; Skluzacek, J. M.; Mallouk, T. E., Optimization of Nano- and Microiron Transport through Sand Columns Using Polyelectrolyte Mixtures. Environ. Sci. Technol. 2007, 41, (18), 6418-6424.

48. He, F.; Zhao, D., Manipulating the size and dispersibility of zerovalent iron nanoparticles by use of carboxymethyl cellulose stabilizers. Environ. Sci. Technol. 2007, 41, 6216.
49. Phenrat, T.; Saleh, N.; Sirk, K.; Tilton, R. D.; Lowry, G. V., Aggregation and Sedimentation of Aqueous Nanoscale Zerovalent Iron Dispersions. Environ. Sci. Technol. 2007, 41, (1), 284-290.
50. Sayles, G. D.; You, G.; Wang, M.; Kupferle, M. J., DDT, DDD, and DDE Dechlorination by Zero-Valent Iron. Environ. Sci. Technol. 1997, 31, (12), 3448-3454.
51. Alessi, D. S.; Li, Z.; Synergistic, effect of cationic surfactants on perchloroethylene degradation by zero-valent iron. Environ. Sci. Technol. 2001, 35, 3713.
52. Quinn, J.; Geiger, C.; Clausen, C.; Brooks, K.; Coon, C.; O'Hara, S.; Krug, T.; Major, D.; Yoon, W. S.; Gavaskar, A.; Holdsworth, T., Field Demonstration of DNAPL Dehalogenation Using Emulsified Zero-Valent Iron. Environ. Sci. Technol. 2005, 39, (5), 1309-1318.
53. Saleh, N.; Phenrat, T.; Sirk, K.; Dufour, B.; Ok, J.; Sarbu, T.; Matyjaszewski, K.; Tilton, R. D.; Lowry, G. V., Adsorbed triblock copolymers deliver reactive iron nanoparticles to the oil/water interface. Nano Lett. 2005, 5, (12), 2489-94.
54. He, F.; Zhao, D., Preparation and Characterization of a New Class of Starch-Stabilized Bimetallic Nanoparticles for Degradation of Chlorinated Hydrocarbons in Water. Environ. Sci. Technol. 2005, 39, (9), 3314-3320.
55. Ditsch, A.; Laibinis, P. E.; Wang, D. I. C.; Hatton, T. A., Controlled Clustering and Enhanced Stability of Polymer-Coated Magnetic Nanoparticles. Langmuir 2005, 21, (13), 6006-6018.
56. Alessi, D. S.; Li, Z., Synergistic Effect of Cationic Surfactants on Perchloroethylene Degradation by Zero-Valent Iron. Environ. Sci. Technol. 2001, 35, (18), 3713-3717.
57. He, F.; Zhao, D.; Liu, J.; Roberts, C. B., Stabilization of Fe-Pd Nanoparticles with Sodium Carboxymethyl Cellulose for Enhanced Transport and Dechlorination of Trichloroethylene in Soil and Groundwater. Ind. Eng. Chem. Res. 2007, 46, (1), 29-34.
58. Kanel, S. R.; Goswami, R. R.; Clement, T. P.; Barnett, M. O.; Zhao, D., Two Dimensional Transport Characteristics of Surface Stabilized Zero-valent Iron Nanoparticles in Porous Media. Environ. Sci. Technol. 2008, 42, (3), 896-900.
59. Saleh, N.; Kim, H.-J.; Phenrat, T.; Matyjaszewski, K.; Tilton, R. D.; Lowry, G. V., Ionic Strength and Composition Affect the Mobility of Surface-Modified Fe0 Nanoparticles in Water-Saturated Sand Columns. Environ. Sci. Technol. 2008, 42, (9), 3349-3355.
60. Hoch, L. B.; Mack, E. J.; Hydutsky, B. W.; Hershman, J. M.; Skluzacek, J. M.; Mallouk, T. E., Carbothermal Synthesis of Carbon-supported Nanoscale Zero-valent Iron Particles for the Remediation of Hexavalent Chromium. Environ. Sci. Technol. 2008, 42, (7), 2600-2605.
61. Phenrat, T.; Saleh, N.; Sirk, K.; Kim, H.-J.; Tilton, R. D.; Lowry, G. V., Stabilization of aqueous nanoscale zerovalent iron dispersions by anionic polyelectrolytes: adsorbed anionic polyelectrolyte layer properties and their

effect on aggregation and sedimentation. J. Nanopart. Res. 2008, 10, 795-814.

62. Zheng, T.; Pang, J.; Tan, G.; He, J.; McPherson, G.; Lu, Y.; John, V. T.; Zhan, J., Surfactant templating effects on the encapsulation of iron oxide nanoparticles within silica microspheres. Langmuir 2007, 23, 5143.
63. Lu, Y.; Fan, H.; Stump, A.; Ward, T. L.; Rieker, T.; Brinker, C. J., Aerosol-assisted self-assembly of mesostructured spherical nanoparticles. Nature 1999, 398.
64. Zheng, T.; Zhan, J.; Pang, J.; Tan, G.; He, J.; McPherson, G.; Lu, Y.; John, V. T., Mesoporous carbon nanocapsules from enzymatically polymerized poly(4-ethylphenol) confined in silica aerosol particles. Adv. Mater. 2006, 18, 2735.
65. Li, A.; Tai, C.; Zhao, Z.; Wang, Y.; Zhang, Q.; Jiang, G.; Hu, J., Debromination of decabrominated diphenyl ether by resin-bound iron nanoparticles. Environ. Sci. Technol. 2007, 41, 6841.
66. Zheng, T.; Zhan, J.; He, J.; Day, C.; Lu, Y.; McPherson, G. L.; Piringer, G.; John, V. T., Reactivity Characteristics of Nanoscale Zerovalent Iron−Silica Composites for Trichloroethylene Remediation. Environ. Sci. Technol. 2008, 42, (12), 4494-4499.
67. Zheng, T.; Zhan, J.; He, J.; Day, C.; Lu, Y.; McPherson, G. L.; Piringer, G.; John, V. T., Reactivity Characteristics of Nanoscale Zerovalent Iron-Silica Composites for Trichloroethylene Remediation. Environ. Sci. Technol. 2008, 42, (12), 4494-4499.
68. Muftikian, R.; Fernando, Q.; Korte, N., A method for the rapid dechlorination of low molecular weight chlorinated hydrocarbons in water. Water Res. 1995, 29, 2434.
69. Schreier, C.; Reinhard, M., Catalytic hydrodehalogenation of chlorinated ethylenes using palladium and hydrogen for the treatment of contaminated water. Chemosphere 1995, 31, 3475.
70. Kuan-Mu Yao; Habibian, M. T.; O'Melia, C. R., Water and waste water filtration. Concepts and applications. Environ. Sci. Technol. 1971, 5, 1105-1112.
71. Zhan, J.; Zheng, T.; Piringer, G.; Day, C.; McPherson, G. L.; Lu, Y.; Papadopoulos, K.; John, V. T., Transport Characteristics of Nanoscale Functional Zerovalent Iron/Silica Composites for in Situ Remediation of Trichloroethylene. Environ. Sci. Technol. 2008, 42, (23), 8871-8876.
72. Wang, Q.; Li, H.; Chen, L. Q.; Huang, X. J., Monodispersed hard carbon spherules with uniform nanopores. Carbon 2001, 39, (14), 2211-2214.
73. Choi, H.; Agarwal, S.; Al-Abed, S. R., Adsorption and Simultaneous Dechlorination of PCBs on GAC/Fe/Pd: Mechanistic Aspects and Reactive Capping Barrier Concept. Environmental Science & Technology 2009, 43, (2), 488-493.
74. Choi, H.; Al-Abed, S. R.; Agarwal, S.; Dionysiou, D. D., Synthesis of reactive nano-Fe/Pd bimetallic system-impregnated activated carbon for the simultaneous adsorption and dechlorination of PCBs. Chemistry of Materials 2008, 20, (11), 3649-3655.

Field Simulation Studies

Chapter 10

Treatability Study for a TCE Contaminated Area using Nanoscale- and Microscale-Zerovalent Iron Particles: Reactivity and Reactive Life Time

Tanapon Phenrat[1,2], Daniel Schoenfelder[1], Mark Losi[3], June Yi[3], Steven A. Peck[4], and Gregory V. Lowry[1,2]

[1]Department of Civil and Environmental Engineering
[2]Center for Environmental Implications of Nanotechnology (CEINT)
Carnegie Mellon University, Pittsburgh, PA, USA
[3]Tetra Tech EC, Santa Ana, CA, USA
[4]BRAC Program Management Office, San Diego, CA, USA

Nanoscale zerovalent iron (NZVI) is a potentially attractive tool for *in situ* source zone remediation of chlorinated solvents. Microscale zerovalent iron (MZVI) is already widely used as a reactive media within permeable reactive barriers (PRBs) for treating plumes of chlorinated organics (e.g. trichloroethylene) in groundwater. Several types of NZVI and MZVI are commercially available, each made by different processes, having different surface properties, and therefore likely to perform differently as remedial agents. Treatability studies are conducted at sites proposed for treatment to ensure the suitability of a specific type of NZVI for that site's geochemical conditions. This laboratory study determined the rates of tricholoethylene (TCE) degradation, propensity to form chlorinated intermediates, the mass of TCE degraded per mass of ZVI added, and the effect of ZVI addition on the site geochemistry for five commercially available NZVI products and two MZVI products. The groundwater and aquifer materials used in this study were from a former naval air base in a marine location in California where ZVI products are proposed as remedial agents. The TCE dechlorination rates for the NZVI products were all faster than their MZVI counterparts. TCE half-life times ranged from as little as 1.2 hours to as high as 89 hours for NZVI (2 g/L), and ranged from 200 hrs to 2000 hrs using MZVI (2 g/L). The reactive lifetimes of NZVI ranged from 3 days to more than 60 days. Generally, the fastest reacting material had the shortest reactive lifetime. Acetylene, ethene, and ethane were the dominant reaction products, but chlorinated intermediates persisted in reactors where NZVI had fully reacted, accounting for as much as 5 mol% of the TCE degraded in one case, but typically less than 1-2 mol%. The ratios of mass of TCE reduced to mass of iron added at the end of the NZVI particle reactive lifetime ranged from as much as 1:17 to as little as 1:120. The,addition of NZVI or MZVI increased the pH from 7.3 to 7.8, but the high alkalinity of the site groundwater limited the pH increase. Differences in the TCE dechlorination rates and reactive life times between NZVI products were attributed to differences in their compositions (catalyzed vs. bare particles) and surface modifiers (polymeric modified vs. bare particles). Unfavorable geochemical conditions (high DO, neutral to acidic pH, and a high nitrate concentration) contributed to the relatively short reactive life times and low TCE dechlorination rates using NZVI products. Two NZVI products and two MZVI products were recommended as the most promising for application at this site.

Introduction

Zerovalent iron (ZVI) in the form of bulk iron filings has been widely used as permeable reactive barrier (PRB) to treat groundwater plumes of chlorinated organics *in situ* for a decades *(1-3)*. Recent advances in material science and nanotechnology gave rise to nanoscale zerovalent iron particles (NZVI), allowing greater utilization of Fe^0 in the particles and providing higher reactivity compared to the bulk micron sized iron filings (MZVI) due to their small size and higher surface to volume ratio *(4, 5)*. NZVI also provides the potential to install *in situ* reactive barriers deep in the subsurface using pneumatic injection *(6)*, gravity feed, or geoprobes*(4)*. Similarly, the small size of NZVI provides the potential to deliver it to the dense nonaqueous phase liquid (DNAPL) source area of a contaminated aquifer and accelerate the time to site closure *(4, 7, 8)*. Several types of MZVI and NZVI are commercially available, each made by different processes and having different surface properties*(4, 5, 9, 10)*. Some also have proprietary surface coatings to enhance their mobility in saturated porous media, making subsurface emplacement feasible *(11)*. Dechlorination by ZVI is a surface meditated reaction; therefore, different ZVI products with different surface properties can yield different performance in term of dechlorination rate and by-product formation. Site geochemistry (e.g. dissolved solutes, pH, natural organic matter, and competing oxidants) can also affect the efficacy of both NZVI *(12)* and MZVI *(13, 14)* for trichloroethylene (TCE) degradation. However, the fundamental reasons for these effects are not fully understood so treatability studies are conducted at sites proposed for treatment to ensure the suitability of a specific type of ZVI for that site's geochemical conditions.

This laboratory study evaluates NZVI and MZVI products for treating a contaminated site at a former naval air base in a marine location in California. The groundwater and soil is contaminated with trichloroethylene (TCE) and some chlorinated daughter products including dichloroethene (DCE) and vinyl chloride (VC). The contaminated zone with the highest aqueous TCE concentration detected (2400 µg/L) is referred to as source reduction area (SRA) where cleaned up using NZVI is proposed, while MZVI is proposed to treat the plume in an area called the plume interception area (PIA). This study compares the effectiveness of five types of commercially available NZVI for source zone treatment and two types of commercially available MZVI for the treatment of the plume. NZVI particles were selected for the SRA because their small size (<100 nm) makes them highly reactive and offers the potential for injection directly into the contamination source zone for *in situ* treatment. In contrast, two types of MZVI were evaluated as TCE plume interception because the larger particles size offers the potential for longer reactive lifetimes.

The objectives of this research were to determine the following for each product in a side-by-side study in aquifer sediment and groundwater (40% solids) from a TCE-impacted groundwater site at the a former naval air base in California.

- The overall effect of ZVI addition on the site geochemistry.
- The observed rate of TCE dechlorination for each ZVI.
- The propensity for formation of chlorinated byproducts.
- The mass of TCE reduced per mass of ZVI (or total ZVI product mass) added.

The change of pH and oxidation-reduction potential (ORP), TCE degradation and product formation kinetics, and reactive lifetime of the Fe^0 products were monitored in batch microcosms containing soil and groundwater from the site. The study also quantified the mass of TCE reduced per mass of ZVI product added as indicators of the cost of each ZVI product for application. The observed difference in TCE treatment efficiency for each ZVI product was discussed in terms of different particle properties and the effect of geochemical conditions of the site. Finally, a recommendation is made regarding the most appropriate types of NZVI and MZVI for the site.

Materials and Methods

Zero Valent Iron Products

The effectiveness of several NZVI products was evaluated, including four types of Reactive Nanoscale Iron Particles (RNIP) which are RNIP, MRNIP1, MRNIP2, and RNIP-R (Toda Kogyo Corp, Onada, Japan). Z-Loy^{TM}, which is a nanocomposite between a zerovalent metal exterior and a less-dense ceramic core, was also evaluated (On Materials). Two types of MZVI were evaluated including H-200 (Hepure Technologies, Wilmington, DE) and EHC (Adventus, Penticton, BC). Table 1 summarizes important physicochemical properties including particle size, surface area, structure/composition, and surface modifier of these ZVI products. RNIP are bare NZVI and H-200 is bare MZVI. MRNIP-1 and MRNIP-2 are coated with proprietary weak polyelectrolytes to enhance their mobility in the subsurface while RNIP-R is bare RNIP modified by a proprietary catalyst to increase reactivity and TCE dechlorination rate. EHC™ is an integrated ZVI and carbon source that is designed to stimulate reducing condition and both abiotic and biotic dechlorination.

Table I. Manufactures and Important Physicochemical Properties of ZVI Products

Particle Type	*Manufacture*	*Structure/ Composition*	*Size (nm)*	*Surface Area (m^2/g)*	*Modifier*
RNIP	Toda Koygo Corporation (Onada, Japan)	Fe^0 core surrounded by a Fe_3O_4 shell	20 to 70 nm in diam.	~15-25 m^2/g as measured by BET	*none*
MRNIP -1					Polyaspartate, weak polyelectrolytes
MRNIP -2					proprietary weak polyelectrolytes
RNIP-R		Fe^0 core surrounded by a Fe_3O_4 shell *coated with* a proprietary catalyst			*none*
Z-Loy™	OnMaterials	a zero-valent exterior and a less-dense ceramic core	<250 nm in diam.	~15m^2/g	proprietary modifier
H-200	ARS Technologies	Fe^0	45 to 140 microns (D_{10} to D_{90}) in diam.	~0.1 m^2/g	none
EHC™	Adventus	integrated carbon (72 to 82% fibrous organic material) and ZVI (17 to 28% iron) source	*N/A*	*N/A*	*N/A*

Groundwater Characterization

The samples of groundwater and aquifer materials from both the SRA and the PIA were collected from monitoring wells and core samples, respectively. The samples were preserved and prepared according to an appropriate standard method for each measurement of interest. The geochemical properties of the aquifer material and groundwater were measured including pH, dissolved oxygen (DO), oxidation reduction potential (ORP), dissolved organic carbon (DOC), the concentration and type of dissolved solids, and pollutant concentrations (TCE, DEC, and VC). pH and ORP were measured according to standard methods (Method 9045d and 2580)*(15)*. Groundwater was filtered prior to pH measurement. ORP was measured directly in the amended aquifer media as soon as the reactor was opened to minimize any effect of air (O_2). The ORP probe was calibrated using quinhydrone-amended pH buffer at pH=4 and 7 prior to use to ensure that the dynamic range of the probe was acceptable. The TOC of the aquifer material from the site was measured using an OI Analytical TOC Analyzer – Model 1010 with a Solids Module. The solids analyzer determined TOC by acidification of the sample then heating to 250 °C to remove inorganic carbon. The sample was then heated to 900 °C to combust the remaining TOC. The resulting carbon dioxide from the TOC was detected by a nondispersive infrared (NDIR) detector. Approximately 200 μg of aquifer material was added to a quartz sample cup. The sample was weighed and placed in the instrument for analysis. The measurements were performed in duplicate. Groundwater samples were analyzed for TCE and its chlorinated daughter products on aqueous samples using method SW 8260B.

Treatability Study Design

Unless otherwise specified, 50 mg/L of TCE was spiked into the microcosms with NZVI and aquifer materials from the source area reduction (SRA) . Whereas, 2 mg/L of TCE was spiked to the microcosms with MZVI and aquifer materials from the plume interception area (PIA). The concentration of ZVI was added at levels that simulate *in situ* field conditions or concentrations designated by the manufacturer. The concentration was selected so that the NZVI or MZVI could dechlorinate the TCE within the designated timeframe of the study (two months).

Microcosms and Analysis

Microcosms were 70 mL serum bottles containing 30 mL of deoxygenated groundwater with aquifer material and 40 mL headspace. The aquifer material contained 40% solids by mass. The microcosms were sealed with Teflon Mininert^{TM} tops to prevent the loss of TCE and the reaction products. For NZVI, TCE was added to the microcosms at least 48-hr prior to introducing NZVI to allow TCE adsorption equilibrium to establish between soil, water, and the gas phases. After equilibration, NZVI was injected into the reactors with a syringe. For MZVI, the particle sizes are too large to be injected into the reactors via syringe so particles were added to the reactors before introducing TCE. All reactors designated for TCE dechlorination were prepared in duplicate. Control reactors were run over the course of the study to ensure that TCE loss from other mechanisms was negligible. The reactors were slowly rotated on a roller-mixer to ensure that adequate mixing was achieved. The concentration of TCE and the byproducts of the reaction were regularly measured by GC/FID using established methods*(10)*. H_2 formation was measured by GC/TCD*(10)*. GC/FID and GC/TCD sampling took place until either the TCE had been completely degraded or the ZVI was no longer reactive as determined by lack of further change in the concentration of TCE and/or its reaction products. Following the complete degradation of TCE, the aquifer media was sparged with nitrogen, purged in a glovebox, and then an additional spike of TCE was added. This process was repeated until the ZVI was no longer reactive or until the duration of the study was completed (two months). When the ZVI ceased to be reactive, 20mL of hexane was added to the reactor and recovered to quantify TCE remaining in the system using GC/ECD.

To determine the impact of Fe^0 addition on the biogeochemistry of the site, pH and ORP were measured in similarly prepared reactors as those used to monitor TCE degradation. For pH and ORP, one measurement was made at approximately ½ of each material's reactive lifetime, and one measured at the end of each material's reactive lifetime. In select reactors that indicated the potential for biostimulation the counts for total *Eubacteria* and for *Dehalococcoides* were also measured. The general procedure for these measurements was as follows. The reactor was opened and the ORP was measured immediately by placing the ORP probe directly into the amended aquifer sediment according to Standard Method 2580*(15)*. After the ORP measurement, a 15-mL aliquot was removed from the reactor, filtered through a 0.45 micron syringe filter and placed into a 25-mL serum bottle where the pH was measured according to Standard Method 9045d*(15)*. In cases where samples were collected for analysis of Eubacteria and *Dehalococcoides* spp organisms (DHC), some aquifer media was removed quickly after opening the reactors and before the ORP measurement. The aquifer media was transferred to sterile centrifuge tubes using a sterilized spatula and immediately frozen for analysis. DNA was extracted from the amended and unamneded reactors following the procedures of Da Silva and Alvarez (2007) *(16)* however we used 0.625µM of each primer, 0.25 µM probe and the final reaction volume was 20 µl. Quantitative Polymerase Chain Reaction (qPCR) was performed to determine the effect of amendment on the total bacterial 16s gene copy number and using a suite of specific primers for *Dehalococcoides* spp as described in Ritalahti et al., 2006 *(17)*.

Results and Discussion

Geochemical Properties

The geochemical properties of groundwater and aquifer material can affect the reactivity and lifetime of Fe^0 with TCE. The properties of interest include pH, DO, ORP, DOC, the concentration and type of dissolved inorganic solids, and the presence of competing oxidants. Tables 2 and 3 summarize these geochemical characteristics of groundwater and soil samples in SRA and PIA.

DO/ORP.

High DO (9±3 mg/L) and a positive ORP measurement (+136 mV) in the SRA are indicative of an oxidizing environment. This oxidizing environment is expected to be unfavorable for Fe^0 application since DO can consume electrons from Fe^0 via the formation of water and superoxide radicals (Eqs. 1 and 2) and can promote particle passivation *(18)*. This H_2O and H_2O_2 production is a competitive reaction to the TCE dechlorination (Eqs. 3 and 4). Fe^{2+} (Eqs. 1 and 2) will be further oxidized to Fe^{3+} species in an oxic environment which subsequently precipitate as iron oxyhydroxides that may coat the Fe^0 surface and lower reaction rate*(18)*. In contrast, the PIA had low DO levels (<1 mg/L) and a negative ORP (-35 mV) which suggests a reducing environment favorable for Fe^0 application.

$$O_2 + 2Fe^0 + 4H^+ \longrightarrow 2Fe^{2+} + 2H_2O \qquad (1)$$

$$O_2 + Fe^0 + 2H^+ \longrightarrow Fe^{2+} + H_2O_2 \qquad (2)$$

$$Fe^{2+} + 2e^- \longrightarrow Fe^0 \qquad E_H^0 = -0.447\ \mathrm{V} \qquad (3)$$

$$TCE + n \cdot e^- + (n-3) \cdot H^+ \longrightarrow product + 3Cl^- \qquad (4)$$

pH

pH does not substantially affect the performance of PRBs using iron filings because PRBs normally consist of 10-50% by mass of iron, and the porewater within the barrier is buffered at a pH of 8-9 by the $Fe(OH)_2/H_2O$ or Fe_3O_4/H_2O equilibrium *(19)*. In contrast, the application of high surface area NZVI particles is normally at only 0.2-0.5 wt%. The high buffer capacity typical of most soil suggests that groundwater pH should not change much due to the injection of NZVI. The groundwater pH can therefore substantially affect the corrosion, electron utilization, and reactive lifetime of NZVI *(19)*. The SRA and PIA aquifer material both had a pH of <7. The pH of the aquifer is not expected to significantly change when NZVI is added because of the presence of significant amounts of bicarbonate alkalinity and a large buffering capacity (Table 3). At near neutral pH, the iron will corrode in the presence of water to form H_2 (Eq. 5). This reaction is a competitive reaction to TCE dechlorination. The higher the formation of H_2, the lower the electron utilization toward TCE dechlorination. However, certain kinds of ZVI materials, including ZVI modified with noble metals (Pd, Pt, Rh) *(20, 21)* such as RNIP-R in this study can activate H_2 and use it for dechlorination via hydrodechlorination (Eq. 6). The effect of a surface catalyst (e.g. RNIP-R) on TCE dechlorination is discussed later in this chapter.

$$H^+ + e^- \longrightarrow H^* \longrightarrow \frac{1}{2}H_2 \quad (5)$$

$$H_2(g) \longleftrightarrow 2H_{ad} \longleftrightarrow 2H^+_{(aq)} + 2e^- \quad (6)$$

Anionic Species

Due to geochemical cycling (dissolution and precipitation) of minerals in the subsurface, or to anthropogenic activities, groundwater normally contains various anionic species such as NO_3^-, Cl^-, SO_4^{2-}, HCO_3^-, and HPO_4^{-2}. Groundwater chemistry is well known to affect PRB performance through control of the Fe^0 corrosion rate *(2, 22)*, the dechlorination rate *(23-25)*, H_2 production *(2)*, microbial activity *(2, 26)*, formation of mineral precipitates on the surface of iron filing *(14, 23)*, and dissolution of the iron oxide layer on Fe^0 potentially leading to increased reactivity *(14)*. Similar effects of anionic species on NZVI performance have been reported *(12)*. At high concentration (~5 mM), NO_3^- deactivated RNIP after 3 days even though Fe^0 remained in RNIP. Presumably, at this high NO_3^- concentration, the surface reaction was shifted from cathodic control (i.e., reduction of TCE) to anodic control (i.e., release of Fe^{2+} and electrons) which facilitated the formation of a passivating FeOOH layer *(12)*. In contrast, anions such as Cl^-, SO_4^{2-}, HCO_3^-, and HPO_4^{-2} are not reducible by Fe^0, but these non reducible ionic species decreased the TCE dechlorination rate by up to a factor of seven compared to DI water. The order of their effect followed their affinity of anion complexation to hydrous ferric oxide, i.e. Cl^- < SO_4^{2-} < HCO_3^- < HPO_4^{2-} at pH 8.9 *(12)*. This implies that the inhibitory effect of these solutes on TCE degradation may be a result of reactive site blocking due to the formation of Fe-anion complexes on the RNIP surface.

Groundwater of both SRA and PIA contains relatively high concentrations of Cl^-, SO_4^{2-} and HCO_3^-. Therefore, an inhibitory effect is expected for TCE dechlorination using ZVI in this groundwater relative to DI water. Inhibition of TCE dechlorination in both SRA and PIA may be similar because the concentrations of each anion are similar in both zones. However, the concentration of NO_3^- in SRA (55 mg/L) is much greater than that in PIA (~3 mg/L). High NO_3^- concentration, in combination with high DO in the SRA create unfavorable conditions for TCE dechlorination. From the geochemical point of view, this unfavorable condition is not expected in the PIA due to its low DO and NO_3^- concentrations.

Dissolved Organic Carbon (DOC)

The major form of DOC in subsurface is natural organic matter (NOM). NOM is a natural charged macromolecule, carrying a net negative charge at natural pH due to the dissociation of carboxylic groups *(27)* . NOM consists of humic and fulvic acids. NOM was shown to adsorb on various kinds of colloids and nanoparticles *(28, 29)*. Similarly, carboxylic groups of NOM can specifically adsorb onto the iron oxide surface of ZVI. There are two common hypotheses regarding the impact of NOM on ZVI performance. First, NOM can enhance electron transfer and thus ZVI reactivity for pollutant degradation by acting as an electron shuttle *(13)*. NOM which consists of quinone groups with a standard potential E^0 of 0.23 V theoretically can transfer electrons from ZVI to a chlorinated ethene *(13)*. The enhanced dechlorination due to the presence of humic acid in a ZVI system was observed for PCE dechlorination but not TCE *(30)*. Second, adsorbed NOM decreases ZVI reactivity due to reactive site blocking *(13, 30, 31)*. Tratnyek et al.*(13)* reported that TCE degradation kinetics decreased by 21 and 39% in the presence of 20 and 40 mg/L, respectively, of Suwannee River organic matter, presumably due to reactive site blocking.

The DOC was 20mg/L and 10mg/L in the PIA and SRA, respectively. The presence of DOC can inhibit or enhance the reactivity of NZVI as mentioned above. However, the low concentration of DOC at this site suggests that the effects will be small. The high concentration of anionic species and pH should have greater impact on NZVI performance and life time.

NZVI Treatment in the Source Reduction Area

Effect of NZVI on geochemistry

As expected the addition of NZVI particles did not change the pH of the groundwater and aquifer materials, i.e., pH values for all microcosms were ~7.2-8. The addition of NZVI particles into the soil and groundwater mixture decreased the ORP from +80 to around -670 mV vs. Ag/AgCl at the reactive half life time of each NZVI materials. However, at the end of the particle reactive lifetime, the ORP increased to ~-360 mV vs. Ag/AgCl. The absence of a significant change in the pH of the microcosms suggests that corrosion of NZVI by water to form H_2 contributed to the relatively short lifetime of the NZVI*(32)*.

TCE Dechlorination

The observed TCE dechlorination rate constant and reactive lifetime for each NZVI (RNIP, RNIP-R, MRNIP-1, MRNIP-2, and Z-Loy™) in the SRA microcosms is given in Table 4. The range of NZVI reactive lifetimes was vast, ranging from 3 days to ~60 days with experimental TCE degradation half-life times that varied from ~1 hour to ~90 hours.

Z-Loy™ has the greatest TCE dechlorination rate constant (301 ±115x10^{-3} L g^{-1} hr^{-1}). The main by-product of TCE dechlorination using Z-Loy™ was ethane. Based on the very high TCE dechlorination rate and ethane as the major by-product, it is possible that TCE dechlorination using Z-Loy may involve a catalytic pathway (Eq.6). The high rate of reactivity comes at the cost of a relatively short reactive lifetime which was only 3-4 days. RNIP-R has the second greatest TCE dechlorination rate constant (56.4 ±14.7 x10^{-3} L g^{-1} hr^{-1}) due to the noble metal on its surface which presumably accelerates dechlorination via a catalytic pathway (Eq.6). The main by-products of TCE dechlorination using RNIP-R were ethene and ethane as expected for bimetallic NZVI particles. However, RNIP-R also has a relatively short reactive life time (~6 days). Bare RNIP provided moderate TCE dechlorination rate constants (17.4 ±5.5 and 22.1 ±6.7 x10^{-3} L$g^{-1}hr^{-1}$ for the first and second spikes, respectively). The reactive life time of RNIP is greater than 18 days. The major by-products are acetylene, ethene, and ethane, indicating β elimination as the major TCE dechlorination pathway *(33)*. All of the NZVI materials evaluated produced small amounts of chlorinated intermediates including 1,1-DCE, t-DCE, and c-DCE. GC measurements confirmed that 0.6 mol% to 5 mol% of the TCE degraded was present as chlorinated intermediates at their maximum. Z-Loy™ produced the highest quantity at 5 mol% whereas the other NZVI produced less than 1.8 mol%. In many cases, the chlorinated products appeared at the beginning of the reaction and were slowly degraded as long as the NZVI remained reactive.

The influence of the composition of NZVI on its performance is obvious when comparing the TCE dechlorination rate and life time of catalyzed NZVI

(RNIP-R and potentially Z-Loy™) with that of bare NZVI (RNIP) under the same conditions. The TCE dechlorination rates of RNIP-R and Z-Loy are ~3 to ~17 times greater than bare RNIP, presumably due to the dechlorination via catalytic pathway (Eq.6). However, the observed lifetimes of Fe^0 products with the highest reaction rates such as Z-Loy and RNIP-R were significantly less than bare RNIP which has a slower rate. In field application, these catalyzed NZVI products would have to be injected into the aquifer at more frequent intervals, but would provide more rapid degradation rates.

The influence of solution chemistry and the presence of aquifer material on NZVI performance in the SRA can be evaluated by comparing the TCE dechlorination rate and reactive life time of NZVI in SRA microcosms with that in DI water. The TCE dechlorination rate in SRA microcosms using RNIP is around 4 times slower than the dechlorination of RNIP in DI water at a similar pH ($81x10^{-3}$ L g^{-1} hr^{-1})*(12)*. In addition, the reactive life time of RNIP (18 days) in the aquifer materials and groundwater from the site is significantly shorter than that observed in DI water (~6 months at pH ~8.9)*(19)*. The relative short reactive life time and low TCE dechlorination rate are in good agreement with expectation when considering the high DO, near neutral pH, and high nitrate concentration in groundwater at this site.

In addition to the physical and chemical composition of the NZVI particles, surface modifiers can significantly affect NZVI performance. As shown in Table 4, the initial TCE dechlorination rates (1st spike) of MRNIP-1 (7 ±$2.6x10^{-3}$ L g^{-1} hr^{-1}) and MRNIP-2 (3.9 ±$1.8x10^{-3}$ L g^{-1} hr^{-1}) are substantially lower than that of bare RNIP. The surface modifiers of both MRNIP-1 and MRNIP-2 are charged polymers (polyelectrolytes), which adsorb onto the surface of NZVI. These adsorbed polymers can adversely affect reactivity of NZVI with TCE by 1) blocking the reactive sites of NZVI and 2) decreasing TCE availability for dechlorination by decreasing the surface concentration of TCE*(34)*. However, the relatively longer reactive life time of MRNIP-1 in comparison to RNIP suggests that surface modifiers might protect NZVI reactivity toward unfavorable conditions of solution chemistry in groundwater.

Polymeric surface modifiers, if bioavailable, may increase the microbial populations in aquifer materials where labile carbon can limit microbial growth. While the TCE dechlorination rate of bare RNIP decreased in the aquifer materials, the dechlorination rate of MRNIP-1 (7.4 ±$2.6x10^{-3}$ L g^{-1} hr^{-1}) increased in comparison to that in DI water (2.2 $x10^{-3}$ L g^{-1} hr^{-1}). As shown in Table 4, TCE dechlorination rate constants using MRNIP-1 increased from 7.4 ±$2.6x10^{-3}$ L g^{-1} hr^{-1} in the first spike to 17 ±5.1 $x10^{-3}$ L g^{-1} hr^{-1} in the fourth spike. Furthermore, MRNIP-1 has the longest reactive life time of any NZVI evaluated at ~ 2 months. The relatively long lifetime of MRNIP-1 suggests that its polyaspartate coating may have a biostimulating effect. The absence of significant chlorinated intermediates, however, would suggest that it is not biostimulating TCE reducers as one would expect to see a significant amount of chlorinated intermediates under these conditions. The samples of aquifer material in microcosms amended with MRNIP-1 and RNIP were analyzed for total Eubacteria and for *Dehalococcoides* (DHC) to determine if indeed the polyaspartate coating on MRNIP provides biostimulation. The polyaspartate coating increased Eubacteria 16s counts by nearly 2 orders of magnitude relative to the control reactors, suggesting that the polyaspartate coatings may in fact be serving as electron donor and stimulating microbial growth. However, no increase in the *Dehalococcoides* organisms was observed as the total counts for DHC remained below detection. A possible explanation for the increase of TCE dechlorination rates by MRNIP from the 1st spike to 4th spike is that microorganisms consumed polyaspartate on the surface of NZVI. This promotes the exposure of TCE to the bare, unblocked surface of MRNIP for dechlorination instead of the surface covered with adsorbed polymer layers. This is in good agreement with the fact that in the 4th spike, the TCE dechlorination rate of MRNIP-1 (17 ±5.1 x10^{-3} L g^{-1} hr^{-1}) is similar to that of bare RNIP (17.4 ±5.5 x10^{-3} L g^{-1} hr^{-1}).

Table IV. TCE reduction rate constants, primary reaction products, reactive lifetime, Fe^0 product required for TCE degradation measured in groundwater and aquifer solids slurry, and TCE Half-life times for 2 g/L concentration of each NZVI.

Material	*Mass added (g/L)*	*Measured Rate constant (L g-1 hr-1)*1000*	*Reaction products observed*	*Lifetime (days)*	*NZVI added/ TCE degraded (kg/kg)*	*TCE $t_{1/2}$ for 2 g/L (hr)*
MRNIP-1	10	7.4 ±2.6a 6.4 ±2.2 11.9 ±2.1 17 ±5.1	Acetylene, ethene and ethane	30-60	28	20-54
Z-Loy™ (w/ dispersant)[b]	3	301 ±115	Ethane, minor ethene	3-4	30	1.2
RNIP	4	17.4 ±5.5 22.1 ±6.7	Acetylene, ethene and ethane	≥18	27	16-20
MRNIP-2 (proprietary coating)	10	3.9 ±1.8 8.5 ±1.9	Acetylene and ethene	8-14	121	41-89
RNIP-R (proprietary catalyst)	3	56.4 ±14.7	Ethane and some ethane/ acetylene.	6	17	6
H-200	50	1.3-2.8	Ethene and ethane	~30	4500	125-270
EHC	10	0.12-0.22	Trace amounts of ethene and ethane	~21	1000b	1600-2900

[a] Errors are 95% confidence intervals to data fits of TCE and all products. Numbers and errors are average for two duplicates. Each reported rate constant is for one spike.

[b] Used 100 mg/L initial TCE concentration.

The ratio of NZVI added/ TCE degraded (kg/kg) for different NZVI products can be used as a preliminary criteria for evaluating the cost effectiveness of TCE dechlorination at a particular site condition by each NZVI product. This is calculated from the amount of TCE converted over the duration of the particle's reactive lifetime, assuming an initial Fe^0 content of 50%. The mass of NZVI added per mass of TCE degraded ranged from 17 kg/kg for RNIP-R to 121 kg/kg for MRNIP-2 (Table 4) while the ratio between NZVI added/ TCE degraded (kg/kg) for Z-Loy, RNIP, and MRNIP-1 is similar, i.e.~30. It also suggests that NZVI with the lowest Fe^0/TCE mass ratio are more efficient than those with higher Fe^0/TCE ratios. It should be noted that Z-Loy™ and RNIP-R were prepared using 100 mg/L TCE instead of 50 mg/L TCE which may have improved the selectivity for TCE for those NZVI. The ratio between NZVI added/ TCE degraded (kg/kg) of RNIP in aquifer material and groundwater is much higher than the ratio of 1:1 for RNIP added/ TCE degraded (kg/kg) calculated from a batch study of TCE dechlorination by RNIP in DI water *(33)*. This suggests that environmental conditions decrease the efficiency of the reaction, and that field application requires approximately 17-28 times more RNIP than the theory predicts from reactions in DI water.

ZVI for Plume Interception

Effect of ZVI on geochemistry

The addition of ZVI products into the aquifer material significantly lowered ORP levels of the microcosms to levels sufficient for dechlorination of TCE and its daughter products. The ORP of the ZVI in the PIA was measured at -523 to -662mV (vsAg/AgCl) which is appreciably lower than the control reactor, measured at -35 mV. The H-200 increased the pH of the aquifer material to 8.5 which suggests that the mass of iron added exceeded the buffering capacity of the soil and groundwater matrix. The EHC lowered the pH of the microcosm to 6.1, almost 1 pH unit lower than the control, suggesting that the organic material may be fermenting and forming H_2, which is by design for this product.

TCE dechlorination

EHC and H-200 degraded the TCE at slower rates than the NZVI products. For 2 g/L ZVI, half-life times for TCE ranged from 1600-2900 hours (66 to 120 days) for EHC to 125-270 hours (5 to 10 days) for H-200. The TCE degradation rate afforded by H-200 was higher than expected for a micron-sized iron with a specific surface area of $0.1 m^2/g$. Apparent reactive lifetimes for H-200 and EHC were 20 to 30 days. However, it is possible that the rates of degradation of TCE by these ZVI products just may have been too slow to measure after these times. Notably, the EHC produced very high levels of H_2 which may lead to degradation via biostimulation.

The mass of TCE degraded per mass of Fe^0 product added (Table 4) suggests that a significant amount of ZVI product would be required to degrade TCE *in situ*. The ratio ranged from 1:4500 for H-200 to 1:1000 for EHC. This estimate does not account for the slow long term degradation afforded by H-200 or the potential utilization of H_2 evolved from EHC for subsequent bioremediation of TCE. A longer-term study under flowing conditions or a well controlled field study is required to better assess the Fe^0 utilization efficiency of these materials.

The primary reaction products of H-200 and EHC in the PIA were ethane and ethene. In the H-200 microcosms, 1,1-DCE was initially detected at low levels and subsequently degraded. For EHC, no chlorinated intermediates were detected.

The organic acids in EHC suggest that it may have a biostimulating effect. This is the claim by the manufacturer Adventus. The absence of chlorinated intermediates, however, would suggest that it is not biostimulating TCE reducers as we would expect to see some chlorinated intermediates under these conditions. Samples of aquifer material in microcosms amended with EHC were analyzed for total Eubacteria and for *Dehalococcoides* (DHC) to determine if indeed EHC provides biostimulation. The addition of EHC increased Eubacteria 16s counts by more than 2.5 orders of magnitude relative to the control reactors, suggesting that EHC may be serving as electron donor and stimulating microbial growth. However, no increase the in the DHC organisms was observed as the total counts for DHC remained below detection.

Conclusions

Demonstration site

Based on the available biogeochemical data, the plume interception area appears to be the most promising demonstration site. In this area, dissolved oxygen is low and the iron added will be used to degrade TCE rather than to reduce dissolved oxygen. This can improve the efficiency of the Fe^0 with respect to the mass of TCE degraded per mass of Fe^0 product added. In general, sites with low dissolved oxygen, nitrate, and DOC are most favorable for treatment with NZVI or MZVI.

NZVI for Source Reduction

There are two NZVI products that appear to be viable alternatives at the site evaluated. These recommendations are based on the rate of TCE degradation, the mass of TCE degraded per mass of NZVI added, and on the low propensity to produce chlorinated byproducts. Z-Loy™ had a very high rate of TCE degradation, but has a relatively short (3-4 days) reactive lifetime. As such, multiple injections of Z-Loy™ at 4 to 7 day intervals appear to be a viable alternative for rapid reduction of the source mass. Of course, this rests on the assumption that the TCE "hot spot" is fairly well characterized such that the material will be placed directly into the source area. Missing this area will likely result in significant generation of H_2, but little TCE degradation since the reactive lifetime of Z-Loy™ is so short. MRNIP-1 (polyaspartate-modified RNIP) appears to be a good choice as well. Even though the rate of TCE degradation is not as rapid as that of Z-Loy™, MRNIP-1 provides a long term (~60 days) stable TCE reduction rate. Repeated applications of MRNIP-1 on an approximately monthly basis until the source has been depleted (based on monitoring) may be the best treatment strategy. In both cases, Z-Loy™ and MRNIP I, mobility testing in the aquifer media and groundwater from the site is required to assess the potential for either of these products to be mobile on the subsurface to enable adequagte emplacement.

ZVI products for plume interception

EHC and H-200 are viable ZVI products for the plume interception area. Neither appears to produce chlorinated intermediates to any level of concern over the time scale of the study. EHC, if it does result in biostimulation, may increase the likelihood of formation of chlorinated intermediates. Both result in a significant lowering of the redox potential (ORP) in site aquifer materials and groundwater and evolve H_2. Both also provide TCE dechlorination, albeit at a slower rate than for the NZVI products. H-200 provides fairly rapid TCE degradation for a micron sized iron with low specific surface area (0.1 m^2/g). At an application rate of 2 g/L, the TCE half-life time was between 125-270 hours (5 to 10 days). Thus to achieve 5 half lives using H-200 (this takes a 1ppm value down to 31ppb), a reactive zone of between 3 to 6 meters in length assuming 10 cm/d groundwater velocity (pore velocity) would be needed. This estimate scales linearly with groundwater velocity and with ZVI loading. Thus it seems that H-200 has the potential to serve as a plume interception material, i.e. an in situ PRB of sorts, however the reactive lifetime of H-200 was approximately 30 days so multiple injections may be needed on approximately monthly intervals to maintain the reactivity of the barrier.

References

1 Tratnyek, P. G.; Johnson, T. L.; Scherer, M. M.; Eykholt, G. R., Remediating Ground Water with Zero-Valent Metals: Chemical Considerations in Barrier Design *Ground Water Monit. Rem.* **1997,** *17*, (4), 108–114

2. Scherer, M. M.; Richter, S.; Valentine, R. L.; Alvarez, P. J. J., Chemistry and Microbiology of Permeable Reactive Barriers for In Situ Groundwater Clean Up. *Crit. Rev. Env. Sci. Technol.* **2000,** *30*, (3), 363-411.

3. Ebert, M.; Kober, R.; Parbs, A.; Plagentz, V.; Schafer, D.; Dahmke, A., Assessing degradation rates of chlorinated ethylenes in column experiments

with commercial iron materials used in permeable reactive barriers *Environ. Sci. Technol.* **2006,** *40*, (6), 2004-2010.

4. Lowry, G. V., Nanomaterials for Groundwater Remediation. In *Environmental Nanotechnology: Applications and Impacts of Nanomaterials*, Wiesner, M. R.; Bottero, J.-Y., Eds. McGraw-Hill: New York, 2007.
5. Zhang, W., Nanoscale iron particles for environmental remediation: An overview. *J. Nanopart. Res.* **2003,** *5*, 323-332.
6. Schnell, D. L.; Mack, J., Installation of Dispersed Iron Permeable Reactive Treatment Zones Using Pneumatic Injection. In *Chlorinated Solvent and DNAPL Remediation*, Henry, S. M.; Warner, S. D., Eds. American Chemical Society: Washington, DC, 2003; Vol. 837, pp 236-258.
7. Saleh, N.; Kim, H.-J.; Phenrat, T.; Matyjaszewski, K.; Tilton, R. D.; Lowry, G. V., Ionic Strength and Composition Affect the Mobility of Surface-Modified Fe0 Nanoparticles in Water-Saturated Sand Columns *Environ. Sci. Technol.* **2008,** *42*, (9), 3349-3355.
8. Phenrat, T.; Saleh, N.; Sirk, K.; Kim, H.-J.; Tilton, R. D.; Lowry, G. V., Stabilization of aqueous nanoscale zerovalent iron dispersions by anionic polyelectrolytes. *J. Nanopart. Res.* **2008,** *10*, 795-814.
9. Tratnyek, P. G.; Johnson, T. L.; Scherer, M. M.; Eykholt, G. R., Remediating Ground Water with Zero-Valent Metals: Chemical Considerations in Barrier Design *Ground Water Monit. Rem.* **1997,** *17*, (4), 108–114
10. Liu, Y.; Majetich, S. A.; Tilton, R. D.; Sholl, D. S.; Lowry, G. V., TCE dechlorination rates, pathways, and efficiency of nanoscale iron particles with different properties. *Environ. Sci. Technol.* **2005,** *39*, (5), 1338-1345.
11. Saleh, N.; Sirk, K.; Liu, Y.; Phenrat, T.; Dufour, B.; Matyjaszewski, K.; Tilton, R. D.; Lowry, G. V., Surface modifications enhance nanoiron transport and NAPL targeting in saturated porous media. *Environ. Eng. Sci.* **2007,** *24*, (1), 45-57.
12. Liu, Y.; Phenrat, T.; Lowry, G. V., Effect of TCE concentration and dissolved groundwater solutes on NZVI-promoted TCE dechlorination and H2 evolution. *Environ. Sci. Technol.* **2007,** *41*, (22), 7881-7887.
13. Tratnyek, P. G.; Scherer, M. M.; Deng, B.; Hu, S., Effects of Natural Organic Matter, Anthropogenic Surfactants, and Model Quinones on the Reduction of Contaminants by Zero-Valent Iron *Water Res.* **2001,** *35*, (18), 4435-4443
14. Agrawal, A.; Ferguson, W. J.; Gardner, B. O.; Christ, J. A.; Bandstra, J. Z.; Tratnyek, P. G., Effects of carbonate species on kinetics of dechlorination of 1,1,1-trichloroethane by zero-valent iron. *Environ. Sci. Technol.* **2002,** *36*, 4326-333.
15. Eaton, A. D.; Clesceri, L. S.; Rice, E. W.; Greenberg, A. E.; Franson, M. A. H., *Standard Methods for the Examination of Water & Wastewater.* American Public Health Association: New York, 2005.
16. Da Silva, M. L. B.; Alvarez, P. J. J., Assessment of anaerobic benzene degradation potential using 16S rRNA gene-targeted real-time PCR. *Environmental Microbiology* **2007,** *9*, (1), 72-80.

17. Ritalahti, K. M.; Amos, B. K.; Sung, Y.; Wu, Q.; Koenigsberg, S. S.; Lo¨ffler, F. E., Quantitative PCR Targeting 16S rRNA and Reductive Dehalogenase Genes Simultaneously Monitors Multiple Dehalococcoides Strains. *Applied and Environemntal Microbiology* **2006,** *72*, (4), 2765-2774.
18. Joo, S. H.; Feitz, A. J.; Waite, T. D., Oxidative Degradation of the Carbothioate Herbicide, Molinate. Using Nanoscale Zero-Valent Iron. *Environ. Sci. Technol.* **2004,** *38*, (7), 2242-2247.
19. Liu, Y.; Lowry, G. V., Effect of particle age (Fe0 content) and solution pH on NZVI reactivity: H2 evolution and TCE dechlorination. *Environ. Sci. Technol.* **2006,** *40*, (19).
20. Zhang, W.-X.; Wang, C.-B.; Lien, H.-L., Treatment of chlorinated organic comtaminants with nanoscale bimetallic particles. *Catal. Today* **1998,** *40*, 387-395.
21. Wang, C. B.; Zhang, W. X., Synthesizing nanoscale iron particles for rapid and complete dechlorination of TCE and PCBs. *Environ. Sci. Technol.* **1997,** *31*, (7), 2154-2156.
22. El-Naggar, M. M., Effects of Cl-, NO3-, and SO42- anions on the anodic behavior of carbon steel in deaerated 0.50 M NaHCO3 solutions. *Appl. Surf. Sci.* **2006,** *252*, 6179-6194.
23. Kohn, T.; Livi, K. J. T.; Roberts, A. L.; Vikesland, P. J., Longevity of Granular Iron in Groundwater Treatment Processes: Corrosion Product Development *Environ. Sci. Technol.* **2005,** *39*, (9), 2867-2879.
24. Su, C. M.; Puls, R. W., Nitrate reduction by zerovalent iron: Effects of formate, oxalate, citrate, chloride, sulfate, borate, and phosphate. *Environ. Sci. Technol.* **2004,** *38*, 2715–2720.
25. Klausen, J.; Vikesland, P. J.; Kohn, T.; Burris, D. R.; Ball, W. P.; Roberts, A. L., Longevity of granular iron in groundwater treatment processes: solution composition effects on reduction of organohalides and nitroaromatic compounds. *Environ. Sci. Technol.* **2003,** *37*, 1208-1218.
26. Van Nooten, T.; Springael, D.; Bastiaens, L., Positive Impact of Microorganisms on the Performance of Laboratory-Scale Permeable Reactive Iron Barriers *Environ. Sci. Technol.* **2008,** *42*, (5), 1680-1686.
27. Schwarzenbach, R. P.; Gschwend, P. M.; Imboden, D. M., *Environmental organic chemistry*. 2 ed.; John Wiley & Sons, Inc.: Hoboken, 2003; p 1198-1208.
28. Hyung, H.; Fortner, J. D.; Hughes, J. B.; Kim, J. H., Natural Organic Matter Stabilizes Carbon Nanotubes in the Aqueous Phase. *Environ. Sci. Technol.* **2007,** *41*, 179-184
29. Ramos-Tejada, M. M.; Ontiveros, A.; Viota, J. L.; Durán, J. D. G., Interfacial and rheological properties of humic acid/hematite suspensions. *J. Colloid Interface Sci.* **2003,** *268*, (1), 85-95.
30. Cho, H.-H.; Park, J. W., Sorption and reduction of tetrachloroethylene with zero valent iron and amphiphilic molecules. *Chemosphere* **2006,** *64*, 1047-1052.
31. Doong, R.-a.; Y.-l., L., Effect of metal ions and humic acid on the dechlorination of tetrachloroethylene by zerovalent iron. *Chemosphere* **2006,** *64*, 371-378.

32. Liu, Y.; Lowry, G. V., Effect of Particle Age (Fe0 content) and Solution pH on NZVI Reactivity: H2 Evolution and TCE Dechlorination. *Environ. Sci. Technol.* **2006,** *40*, 6085.
33. Liu, Y.; Majetich, S. A.; Tilton, R. D.; Sholl, D. S.; Lowry, G. V., TCE dechlorination rates, pathways, and efficiency of nanoscale iron particles with different properties. *Environ. Sci. Technol.* **2005,** *39*, 1338-1345.
34. Phenrat, T.; Liu, Y.; Tilton, R. D.; Lowry, G. V., Adsorbed Polyelectrolyte Coatings Decrease Fe0 Nanoparticle Reactivity with TCE in Water: Conceptual Model and Mechanisms. *Environ. Sci. Technol.* **2009,** *43*, (5), 1507–1514.

Chapter 11

Electrokinetically Enhanced Removal and Degradation of Subsurface Pollutants Using Nanosized Pd/Fe Slurry

Gordon C. C. Yang

Institute of Environmental Engineering, National Sun Yat-Sen University, Kaohsiung 80424, Taiwan

In this work a hybrid technology combining the injection of the slurry of palladized nanoiron (PNI) and electrokinetic (EK) remediation process was used to mimic the removal and degradation of trichloroethylene (TCE) and nitrate in the subsurface environment. Laboratory-prepared palladized nanoiron was characterized to have a mean diameter of 51.6 nm and a specific surface area of 101 m^2/g. PNI was further stabilized using 1 vol% polyacrylic acid (PAA) as a dispersant. The nanosized Pd/Fe slurry thus prepared was then used to evaluate the treatment efficiency of combined technologies of the injection of PNI slurry and EK remediation process in treating TCE- and nitrate-contaminated loamy sand in horizontal columns. Experimental results showed that electroosmosis is important to the transport of PNI slurry in the soil matrix and the subsequent hydrodechlorination of TCE, where electrophoresis and electromigration are important to migration of nitrate ions and PAA-modified PNI and subsequent chemical reduction of nitrate. The addition of PNI slurry to the anode reservoir yielded the lowest residual TCE concentration in soil, namely about 92.5% removal of TCE from soil. The residual TCE concentration in the cathode reservoir was about 8 mg/L. Although the addition of PNI slurry to the cathode reservoir could completely degrade TCE therein, its residual TCE in soil was up to 29.0%. In the case of nitrate decontamination, by injecting PNI slurry into the anode reservoir of the EK system, an efficiency of over 99% nitrate removal for the entire system was achieved. The cathode reservoir, however, was found to be the worst injection position.

Introduction

Electrokinetic (EK) remediation is capable of using electric currents to extract heavy metals, certain organic compounds, or mixed inorganic and organic species from soils and slurries *(1, 2)*. The principles of EK process is briefly described as follows: Electric fields applied across a saturated soil mass or alike would result in electrolysis of water, transport of species by ionic migration, electroosmosis, and diffusion. These transport processes are accompanied by sorption processes in the soil or a porous medium alike, precipitation and dissolution, and other aqueous-phase reactions in the pore fluid. Ionic migration (also known as electromigration) is the transport of charged ions in solution toward the oppositely charged electrodes. Namely, the anions will move toward the anode and the cations will move toward the cathode. Electromigration of ions is quantified by the effective ionic mobility, which can be defined as the velocity of the ion in the soil under the influence of a unit electrical potential gradient. In general, ionic migration is the most significant component of mass transport in electrokinetic remediation in most soils. The cause of electroosmotic flow (EOF) is an electrical double layer that forms at the solid/solution interface. When an electrical potential is applied to the system, the cations therein will might toward the cathode. Since these cations are solvated and clustered at the above-indicated interface, the migration of such cations would drag the rest of solution with them. The magnitude of the mass transport by electroosmosis in soils is often at least one order of magnitude less than that induced by electrical migration. In unenhanced EK process, the transport of the electrolysis products of water (i.e., H^+ and OH^-) would significantly affects the chemistry across the soil mass. The hydrogen ion movement toward the cathode (i.e., acid front) assists in desorption of species from clay surfaces and dissolution of the salts in the soil. The back migration and diffusion of the hydroxide ion generated at the cathode (i.e., base front) may lead to premature precipitation of the cations transported to this region. To resolve a problem of this kind, enhancement techniques are necessary to prevent this premature precipitation. EK process has been considered as an economic and effective method for soil and groundwater remediation. This method has also been listed by U.S. Environmental Protection Agency as one of best available and demonstrated technologies for the treatment of clayey soil contaminated by dense non-aqueous phase liquids *(3)*. EK has been known to have the following advantages: (i) various contaminants can be *in situ* removed from the contaminated porous media efficiently; (ii) there is little limitation on species to be treated; (iii) the direction of electroosmotic flow can be effectively controlled; (iv) it has no problem to remediate soils of low hydraulic conductivity; and (v) it could serve as a part of a treatment train technology, which is generally referred to the combination of different treatment technologies at one site for remediation (3).

In the past decade zero-valent iron (ZVI) has been proven to be capable of reductively degrading various chlorinated solvents (e.g., trichloroethylene, TCE) and a variety of other contaminants in aqueous phase *(4-8)*. In another study, it has been demonstrated that palladized iron (i.e., Pd/Fe bimetal) is preferable to elemental iron (e.g., iron fillings and microsized iron particles) for rapid and complete hydrodechlorination of low molecular weight chlorinated hydrocarbons in water *(9)*. Zhang et al. *(10)* further showed nanoscale bimetallic Pd/Fe particles outperformed microscale ZVI in the aspects of a higher dechlorination rate and a lesser amount of intermediate products such as dichloroethene and vinyl chloride. A study by Yang et al. *(11)* also demonstrated that the nanoscale bimetallic Pd/Fe slurry had a much better performance in TCE degradation than that of the nanoscale ZVI (NZVI) slurry.

Remediation of contaminated soil and groundwater by EK coupled with other technologies are commonly practiced. Examples include EK-iron wall process *(12)*, EK-Fenton process *(13, 14)*, EK-Fenton-biodegrdation process *(15)*, and EK-Fenton-iron wall process *(16)*. All the aforementioned processes employed iron fillings or microsized iron particles. Recently, Yang et al. *(17)* employed an electric field to enhance the chemical reduction of nitrates by the iron wall process. In this work the treatment performance of the combined technology coupling electrokinetic process and iron wall technology using 2.5 g of nanoiron (i.e., nanoscale zero-valent iron; NZVI) for nitrate removal and reduction was of main concern. The employment of 20 g of microsized iron particles as the permeable reactive material in the iron wall was merely for comparison purposes. In both cases, the iron wall was constructed in the soil compartment of 20 cm long at a position 5 cm away from the anode compartment and an electrical potential gradient of 1 V/cm was applied for 6 d. In the case of nanoiron wall consisted of 2.5 g of NZVI with a volume of 35.64 cm^3, the cumulative electroosmotic flow (EOF) in the cathode reservoir was determined to be 75 mL. The overall removal and destruction (i.e., removal + chemical reduction) efficiency was determined to be 99.58%, in which chemical reduction of nitrate accounted for 98.06%. On the other hand, 97.05% of overall removal and destruction efficiency, in which 95.08% contributed by chemical reduction of nitrate, was attainable using 20 g of microsized iron particles. Evidently, both forms of iron particles are capable of reducing nitrate very effectively and NZVI is superior to ZVI in this regard. Very recently, a novel process using EK to assist the transport of a nanoparticle slurry through a polluted porous medium has been developed by the present author *(18, 19)*.

In this work the aforementioned novel process (i.e., EK + injection of Pd/Fe nanoparticle slurry) was tested for the remediation of TCE- and nitrate-contaminated soils in a simulated subsurface environment. The objectives of this work were 3-fold: (1) to evaluate the effectiveness of employing EK as the driving force for transporting nanoscale palladized iron slurry (designated as "PNI slurry" hereinafter) in the soil matrix, (2) to study the effects of the injection position of PNI slurry relative to the anode compartment on removal of target contaminants in a simulated subsurface environment, and (3) to investigate the relvant reaction mechanisms in remediation of concerned contaminants.

Materials and Methods

Chemicals

Sodium borohydride ($NaBH_4$), iron(III) chloride 6-hydrate ($FeCl_3 \cdot 6H_2O$), palladium acetate ($Pd(CH_3CO_2)_2$), trichloroethylene (TCE; C_2HCl_3), and potassium nitrate (KNO_3) were all reagent grade. Poly acrylic acid (PAA), an anionic surfactant, with an MW of ca. 2,000 g/mol was selected in this work for stabilizing nanoparticles.

Preparation of Nanoscale Bimetallic Pd/Fe Particles and PNI Slurry

Nanoiron (i.e., nanoscale ZVI) was first prepared according to a chemical reduction method reported by Glavee et al. *(20)* Immediately after the nucleation of iron particles, 1 wt% of $Pd(CH_3CO_2)_2$ was added to the nanoiron suspension to lead to a chemical displacement of iron by palladium. In so doing, nanoscale bimetallic Pd/Fe would form according to the reaction equation given below:

$$Pd^{2+} + Fe^0 \rightarrow Pd^0 \downarrow + Fe^{2+} \qquad E = 1.43\ V \quad (1)$$

The metal displacement indicated above would be verified and further explored in the section of "Results and Discussion." To stabilize such nanoparticles in the suspension, 1 vol% of PAA was added to form PNI slurry. The intrinsic pH value of PNI slurry thus prepared was in the neighborhood of 9.0.

Soil Specimen

The soil specimen used in this work was obtained from a big public construction site in southern Taiwan. After collection, the plant roots and gravel in soil specimen were first discarded before air drying for a period of 3-7 d. Then the minus 10-mesh fraction (i.e., < 2.0 mm) of the soil specimen was collected and stored in big glass jars for later analyses and tests. The pre-treated soil specimen was determined to have the following characteristics: (i) texture: loamy sand; (ii) particle size distribution: 3.70% of finer than 2 μm, 20.77% in the range of 2-50 μm, and 75.53% in the range of 50-2,000 μm; (iii) pH: 7.9; (iv) BET (Brunauer-Emmett-Teller) surface area: 3.75 m2/g; (v) organic content: 2.36%; (vi) loss on ignition: 2.82%; (vii) cation exchange capacity: 3.63 meq/100g; and (viii) total iron concentration: 25,897.5 mg/kg.

Preparation of Soil Specimens Contaminated by TCE and Nitrate

A stock solution of TCE was first prepared. After the addition of TCE solution to the pre-treated soil specimen, they were homogeneously mixed. The TCE-contaminated soil thus obtained was fed into the soil columns. Such soil specimens were further subjected to 1-d aging in soil columns before they were tested in the EK system. In this work, the initial TCE concentrations in soil specimens were in the range of 160-181 mg/kg. Being prepared in a similar manner, the initial nitrate concentrations in soil specimens were kept constant at 7,316 mg/kg.

Experimental Set-Up of the EK Remediation System

To simulate the groundwater flow and *in situ* remediation of subsurface contaminantion, a bench-scale remediation system was designed to treat contaminated soil in the horizontal soil column by applying an electric field to enhance the transport of injected nanoscale bimetallic Pd/Fe slurry (i.e., PNI slurry) through the soil medium. The experimental set-up of the EK remediation system (see Figure 1) per se consists of a horizontal soil column (Pyrex glass; 20 cm L x 5.5 cm D), two electrode compartments (Pyrex glass; 5 cm L x 7.5 cm D), two graphite electrodes, and a DC power supply.

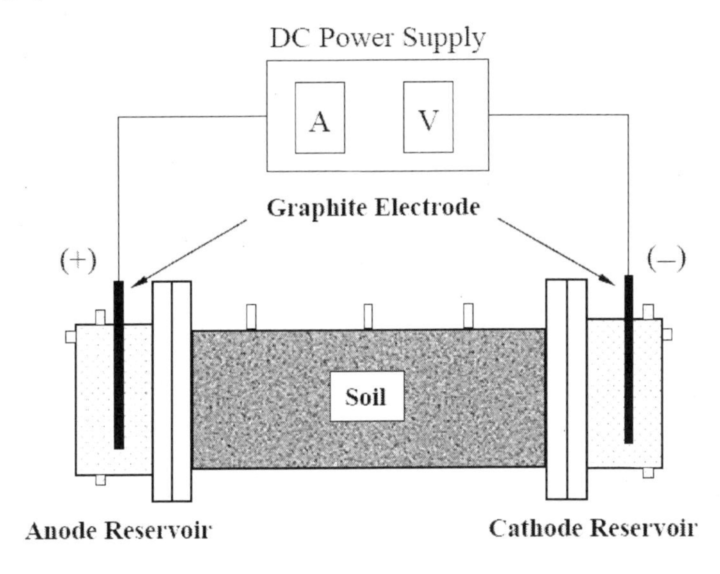

Figure 1. The experimental set-up of the electrokinetic remediation system employed in this work.

The electrode reservoirs were filled with de-ionized water or simulated groundwater. The simulated groundwater was synthesized according to the recipe adopted by Kao et al. *(22)*: Na^+, 2527.6 mg/L; Ca^{2+}, 44.1 mg/L; Mg^{2+}, 197.2 mg/L; NH_4^+, 10.7 mg/L; Cl^-, 98.9 mg/L; SO_4^{2-}, 98.6 mg/L; PO_4^{3-}, 1590.2 mg/L. In fact, this recipe is different from that of in the real groundwater, which generally would also contain K^+, HCO_3^-, and NO_3^-.

Decontamination by EK Coupled with the Injection of PNI Slurry

In this work, attempts were made to treat the above-indicated contaminated soil specimens by a hybrid process comprised of the injection of PNI slurry into the designated electrode compartment(s) and/or soil body of the remediation system and application of an external electric field. Unless otherwise specified, all tests were conducted under EK operating mode.

1. Remediation Tests for TCE-Contaminated Soil

In Tests 1 and 2, everyday 20 mL of PNI slurry was injected into the anode compartment and cathode compartment, respectively. In Test 3, one-half amount of PNI slurry was injected into each electrode compartment daily. In Test 4, only EK process was employed without injection of PNI slurry into any electrode compartment. Test 5 was a blank test, which had no application of the electric field and PNI slurry to the remediation system. In all tests, the following experimental conditions were employed: (1) a constant electrical potential gradient of 1 V/cm; (2) daily addition of 20 mL PNI slurry with a solid concentration of 2.5 g/L; and (3) a treatment time of 6 d.

2. Remediation Tests for Nitrate-Contaminated Soil

In the case of nitrate remediation, experimental conditions similar to the case for TCE remediation were employed except the concentration of nanoscale palladized iron in the slurry was 4.0 mg/L. Also different are the designations of test numbers. In Test 1, the remediation system was only subjected to EK alone without addition of PNI slurry. For the other tests, the systems were all subjected to the combined technology of coupling EK process with the injection of PNI slurry. In Tests 2 and 3, each day 20 mL of PNI slurry was injected into the anode reservoir and cathode reservoir, respectively in conjunction to the application of external electric field. In Test 4, daily injection of 10 mL PNI slurry into each of the injection ports 5 cm and 10 cm away from the anode reservoir was adopted. On the other hand, daily injection of 10 mL PNI slurry into each of the injection ports 5 cm and 10 cm away from the cathode reservoir was adopted in Test 5.

Equipment

In this work several apparatuses were employed for different purposes. The morphology examinations and elemental determinations of the bare nanoiron and palladized nanoiron were revealed by field emission scanning electron microscopy (FE-SEM) coupled with X-ray energy dispersive spectrometry (EDS) (JOEL-6700 and JOEL-6400, USA). A particle size analyzer (Malvern, Zetasizer Nano Series, UK) was employed for prepared nanoparticles to determine their size distributions. On the other hand, a separate particle size analyzer (Coulter, LS100, UK) was used for soil specimen. For determinations of BET (Brunauer-Emmett-Teller) surface area of fine particles, Micromeritics, ASAP 2010 Accelerated Surface Area and Porosimetry System (USA) was employed. Gas chromatograph with μECD detector (Aglient 6980, USA, capillary column of J&W P/N DB-5 with a tube diameter of 0.53 μm and a length of 30 m) was used for the analysis of TCE concentrations. In some instances, a purge and trap concentrator (Tekmar, 3000, USA) was also used to assist the TCE analysis. On the other hand, ion chromatography (Metrohm 861 Advanced Compact IC with 813 Compact Autosampler, Switzerland) was used for determinations of nitrate ion concentration.

Results and Discussion

Characterization of Nanoscale Bimetallic Pd/Fe Particles and PNI Slurry

The SEM micrograph clearly showed that the synthesized particles were spherical and aggregated in chain form. Detailed SEM-EDS (scanning electron microscopy-energy dispersive X-ray spectrometry) mapping results for elements of Pd and Fe in Pd/Fe nanoparticles can be found elsewhere *(21)*. In that work, nanoscale palladium was found to be evenly distributed in the test specimen mainly composed of nanoiron as the substrate. Prior to the addition of PAA dispersant, results of dynamic light scattering showed that the size of aggregated particles was ca. 500 nm. When PAA was added as the dispersant, however, PNI was pretty much dispersed through the mechanism of steric stabilization. The mean diameter as determined from the results of dynamic light scattering (see Figure 2) was 51.6 nm, in which particles of ca. 5 nm contributed about 20% of the specimen volume Thus, the palladized iron particles prepared in this work were truly in nanoscale. $N_{2(g)}$ adsorption/desorption results further showed that nanoscale Pd/Fe bimetal was mesoporous with a BET surface area of 101 m^2/g, which was upto 3-fold greater than those of reported by others *(10, 23-27)*.

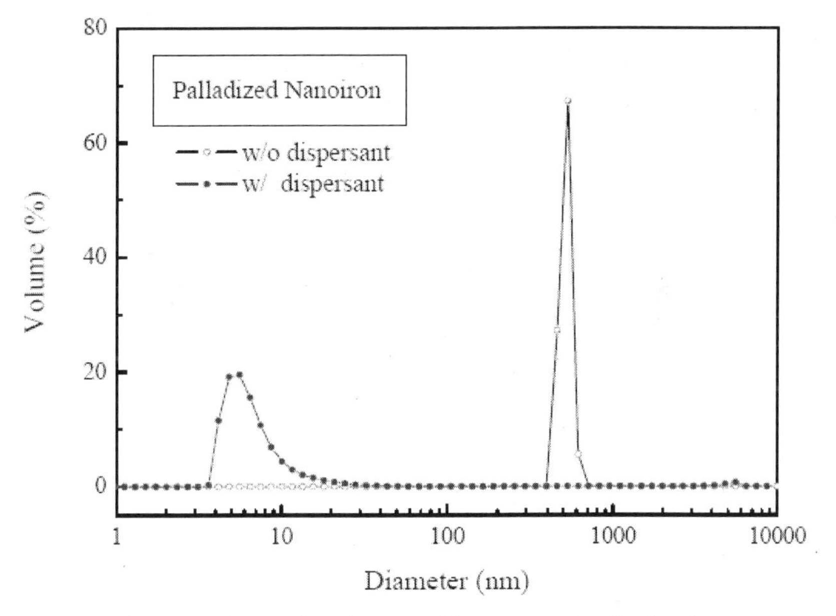

Figure 2. The particle size distribution of the prepared palladized nanoiron as determined by dynamic light scattering

Variations of the Cumulative Electroosmotic Flow (EOF) Quantities Under Different Test Conditions

Electroosmosis is an important phenomenon in EK process. Generally, the magnitude of EOF quantity would influence the extent of decontamination. Therefore, it is worthy of understanding the relevant EO parameters for different cases.

1. EO Parameters for TCE-Contaminated Soil

Experimental results showed that the direction of the EOF was from the anode toward the cathode for all tests. The 6-d accumulation of the EOF quantities for these tests ranged from 150 to 200 mL with a decreasing order shown below: Test 1 $\approx$ Test 3 > Test 4 > Test 2. The lowest EOF quantity for Test 2 might be due to the injection of PNI slurry into the cathode compartment resulting in a slightly greater hydraulic head than that of the anode compartment. This might render the penetration of the mixed fluid (i.e., the added PNI slurry and the intrinsic cathode reservoir water) into the neighboring soil body resulting in an offset for the EOF from the anode side. In addition, the PNI-originated iron ions transported by the aforementioned penetration of the mixed fluid into the soil body near the cathode reservoir accompanied by the intrinsic iron species in that fraction of the soil body would form $Fe(OH)_3$ precipitate in soil due to an EK-induced alkaline environment. As a result, it would further retard the mobilization of the EOF. Based on the experimental data obtained, the electroosmotic permeability coefficients for Tests 1-4 were calculated to be 1.78 x 10^{-5} $cm^2/V{\cdot}s$, 1.40 x 10^{-5} $cm^2/V{\cdot}s$, 1.81 x 10^{-5} $cm^2/V{\cdot}s$, 1.65 x 10^{-5} $cm^2/V{\cdot}s$, respectively.

2. EO Parameters for Nitrate-Contaminated Soil

Similar to the case of TCE-contaminated soil, the direction of the EOF was from the anode toward the cathode for all tests in the case of nitrate-contaminated soil. The 6-d accumulation of the EOF quantities for these tests were in the range of 130 to 170 mL with a decreasing order shown below: Test 1 > Test 2 > Test 3 > Test 5 > Test 4. The corresponding EO permeability coefficients were 1.56 x 10^{-5} $cm^2/V{\cdot}s$, 1.50 x 10^{-5} $cm^2/V{\cdot}s$, 1.37 x 10^{-5} $cm^2/V{\cdot}s$, 1.20 x 10^{-5} $cm^2/V{\cdot}s$, and 1.29 x 10^{-5} $cm^2/V{\cdot}s$.

3. Comaprison of EO Parameters for Different Contaminated Soils

In the case of daily injection of 20 mL PNI slurry into the anode reservoir, it would yield the greatest cumulative EOF quantities and EO permeability coefficients for both soil specimens contaminated by TCE and nitrate, namely Test 1 for the case of TCE-contaminated soil and Test 2 for nitrate-contaminated soil. Generally, the greatest EO permeability coefficients would result in the greatest removal efficiency of contaminant.

Variations of Contaminant Concentrations in the Electrode Reservoirs and Soil Body Under Different Test Conditions

1. Removal and Degradation of TCE

As the direction of EOF was from the anode reservoir toward the cathode reservoir, only TCE concentrations in the cathode reservoir and soil body are of concern. Figure 3 only shows residual TCE concentrations in the cathode reservoir during the treatment period for Tests 1-4 disregarding the blank test (i.e., Test 5). It was noticed that residual TCE concentrations of concern increased first and then decreased later. The probable reason is given below. During the early stage of the treatment, TCE in the soil body near the cathode compartment would be transported into the cathode reservoir by the EOF before the arrival of palladized nanoiron (PNI). Therefore, the TCE concentration in the cathode reservoir would accumulate in the first one or two day(s) of the treatment. After that short period of treatment time, presumably PNI had arrived at the cathode reservoir via the EOF and begun the reductive dechlorination of TCE therein. During the test period, it was also found that substantial amounts of TCE were found in the cathode reservoir in Tests 1 and 4, with Test 4 (i.e., applying only electric field without addition of PNI slurry) being the greatest in all tests.

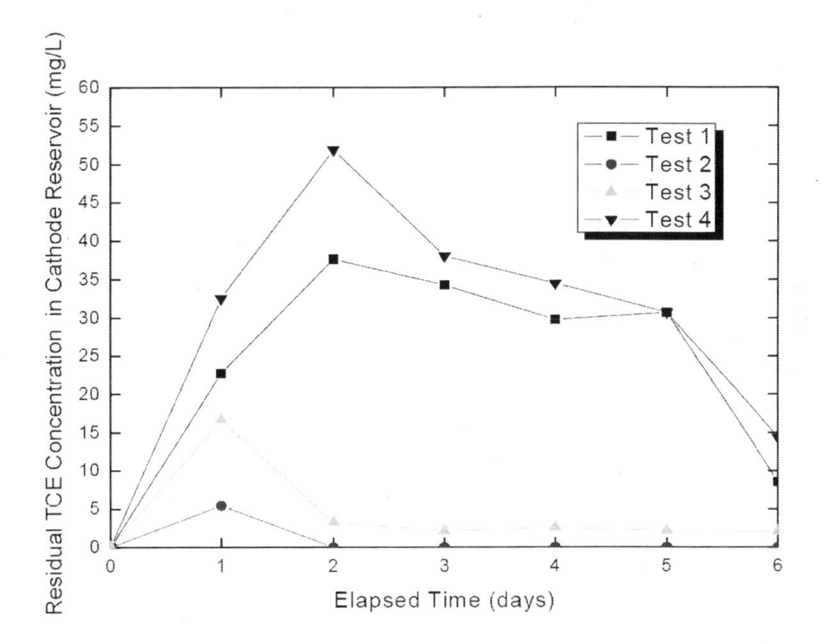

Figure 3. Variations of TCE concentration in cathode reservoir as a function of treatment time for various test runs

By examining the residual TCE concentrations in the soil bodies after six days of treatment, however, Test 1 yielded the best result of TCE remediation among various test runs. This is in accord with the fact that Test 1 had the greatest cumulative EOF quantity. In Test 4, TCE was only relocated from the soil body to the cathode reservoir without much degradation effect. On the contrary, a combined effect of reductive dechlorination by nanoscale Pd/Fe bimetal and enhanced transport of such nanoparticles through the soil body (from the anode compartment to the cathode compartment) by the EOF has removed and degraded TCE substantially, as shown in Test 1 (see Figure 4). By the same token, the experimental conditions employed in Test 3 (i.e., injecting PNI slurry into both electrode compartments) yielded the second best result for TCE remediation as compared with that of Test 2 (i.e., injecting PNI slurry into the cathode compartment only). The practice of injecting PNI slurry into the cathode compartment (e.g., Tests 2 and 3) also played a significant role in dechlorination of TCE. This is because that TCE removed from the soil body to the cathode reservoir would be hydrodechlorinated by the nanoscale bimetallic Pd/Fe particles therein. This would explain why Tests 2 and 3 yielded much lower residual TCE concentrations in the cathode reservoir as compared with Tests 1 and 4.

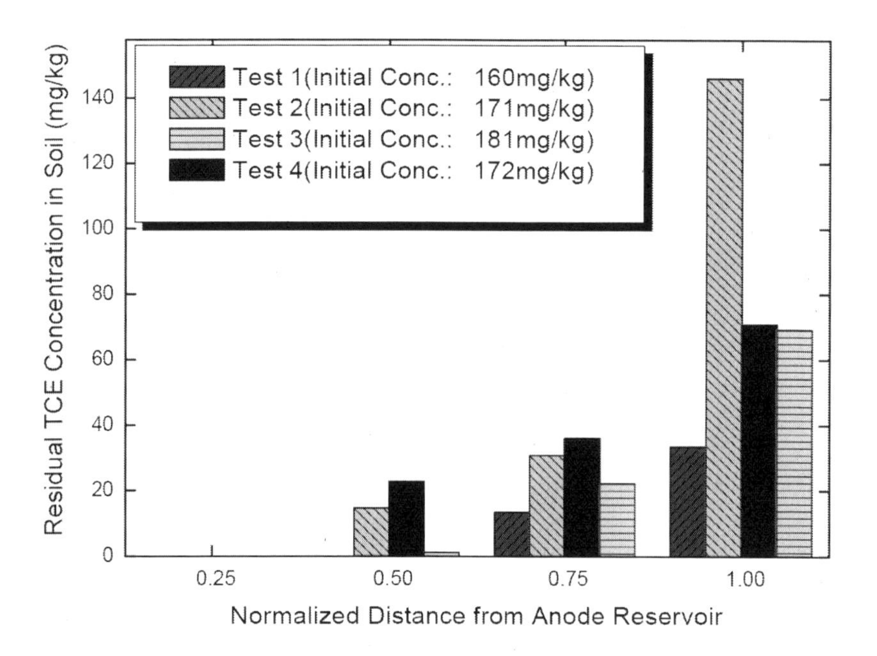

Figure 4. Distribution patterns of residual TCE concentrations in soil columns for various test runs

Figure 4 also shows that no TCE could be detected for the soil fraction near the anode compartment for Tests 1-4, whereas an increasing TCE content in the soil fractions toward the cathode compartment. This is also an indirect proof that the EOF has assisted the remediation of TCE via enhanced transport of PNI from the anode end toward the cathode end.

A comparison of residual TCE concentrations in soil for Tests 1-5 was made. The initial TCE masses in various tests ranged from 109.49 mg to 126.25 mg. After the 6-d treatment by the novel process of combining the injection of PNI slurry and EK, the residual TCE in soil were determined to be 7.56%, 29.04%, 12.99%, 18.94%, and 97.91% for Tests 1-5, respectively. Apparently, in Test 5 (i.e., the blank test; no application of the electric field and PNI to the remediation system), there was almost no removal of TCE from the soil body except the loss of TCE due to volatization. Among other tests, it was found that the injection of PNI slurry into the cathode compartment (i.e., Test 2) was the worst practice for removing TCE from the soil body. Again, this finding could be ascribed to a slightly greater hydraulic head in the cathode reservoir (due to the injection of PNI slurry) than that of the anode reservoir. Thus, there existed an offset for the EOF from the anode end to remove TCE in the soil body. On the contrary, the anode reservoir was found to be the best spot for the injection of PNI slurry because the transport of PNI into the soil body would be enhanced by the EOF toward the cathode end. That would explain why Test 1 yielded the greatest efficiency of TCE removal from soil. As for Test 4, though no PNI slurry was injected into any electrode compartment, application of an electric field to the system alone was still capable of removing TCE from the soil body to the anode reservoir. When one-half amount of PNI slurry was injected into each electrode compartment at the same time, namely Test 3, there would be no difference in hydraulic head. Under the circumstances, electrokinetics did play its role in generating the EOF, which would always bring along with PNI in the system. Therefore, the second greatest TCE removal efficiency was obtained in this case.

2. Removal and Reduction of Nitrate

Figure 5 shows the residual nitrate concentrations for various tests in the EK system having different injection positions of PNI slurry for the 6-d treatment of nitrate-contaminated soil. It was noticed that this figure was divided into three regions by two vertical dashed lines: (1) the left region denoted the anode compartment; (2) the middle region denoted the soil compartment having a length of 20 cm; and (3) the right region denoted the cathode compartment. In the region of soil compartment, A-05, A-10, A-15, and A-20 represented the soil fractions 0-5 cm, 5-10 cm, 10-15 cm, and 15-20 cm away from the anode compartment, respectively. Test 2 (i.e., injection of PNI slurry into the anode reservoir) was found to be superior to all other tests in terms of nitrate removal. This is in accord with the greatest cumulative EOF quantity disregarding Test 1 of no injection of PNI slurry. More specifically, nitrate removal efficiencies for the soil body and whole system (including the soil body and electrode reservoirs) were determined to be 99.5% and 99.2%, respectively in the case of Test 2. Presumably, electromigration of nitrate ions and transport of PAA-modified PNI by electrophoresis toward the anode had enhanced their reaction with PNI in the anode reservoir. In addition, a pH of about 3 or lower in the anode reservoir or the acid front might play a role of acid washing of PNI preventing the formation of passive layer of iron oxides on the PNI surface. On the other hand, Test 3 yielded very poor efficiencies of nitrate removal, 47.8% for the soil body and 8.03% for the whole system. This is ascribed to a rather alkaline environment (pH about 12) in the cathode reservoir, which would enhance the formation of iron oxides and/or iron hydroxides on the surface of PNI resulting in a much lower surface reactivity. It was also noticed that both Tests 4 and 5 yielded satisfactory efficiencies of nitrate removal. Test 4 was found to yield efficiencies of nitrate removal of 99.1% and 80.5% for the soil body and whole system, respectively. Test 5, on the other hand, yielded efficiencies of nitrate removal of 96.6% for the soil body and 92.7% for the whole system.

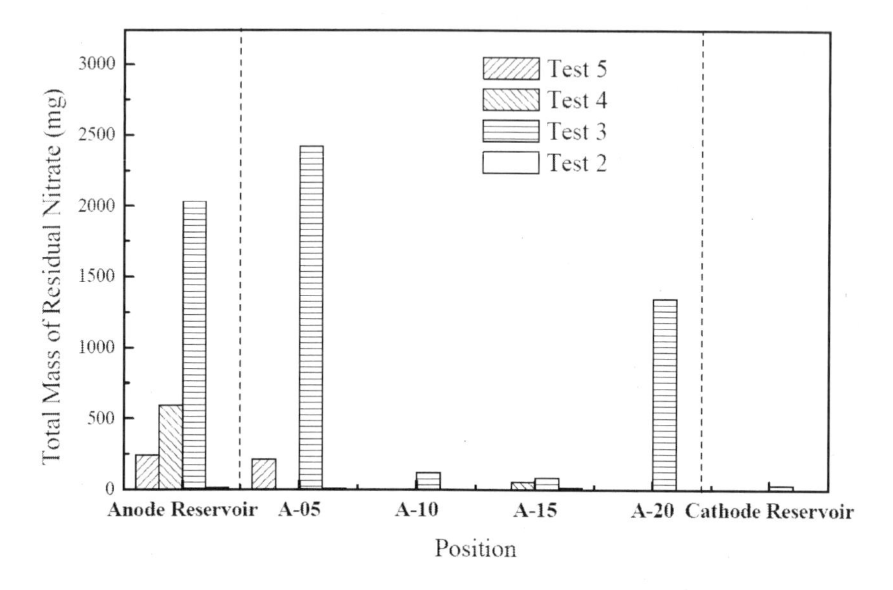

Figure 5. The distribution of the residual nitrate concentration in the system as a function of the injection position of nanosized Pd/Fe slurry.

Conclusions

In this work palladized nanoiron (PNI) was first synthesized by a solution chemistry method, then the addition of 1 vol% polyacrylic acid (PAA) was followed to prepare the slurry of PNI. PNI slurry thus prepared was injected into a simulated subsurface environment using electrokinetics (EK) as the driving force for delivering PNI through the soil body to treat trichloroethylene (TCE) or nitrate contamination therein. The research findings are summarized as follows:

1. The slurry of PNI could be considered as a phase of "mobile reactive nanoscale Pd/Fe bimetallic particles" as compared with a stagnant iron/nanoiron wall in the sense of conventional "permeable reactive barriers".
2. The injection position of PNI slurry was found to be critical to the overall treatment performance, with the injection into the anode reservoir being the best. This holds true for both TCE- and nitrate-contaminated soils.
3. The best TCE removal and degradation was obtained for the test having PNI slurry merely injected into the anode reservoir. In this case, only 7.56% of the initial TCE mass remained in soil after a treatment time of 6 d. This is ascribed to the greatest extent of transport of PNI slurry toward the cathode by the largest quantity of the electroosmotic flow (EOF) in this case.
4. By injecting PNI slurry into the anode reservoir of the EK system, an efficiency of over 99% nitrate removal and reduction for the entire system was achieved. Presumably, the driving force of moving nitrate ions by electromigration and negatively charged PAA-modified PNI toward the anode by electrophoresis must be the predominant migration mechanisms in this case. Chemical reduction of nitrate would occur mostly in the anode reservoir where PNI existed. The cathode reservoir was found to be the worst injection spot.
5. Evidently, a novel hybrid technology of injecting the nanoscale bimetallic Pd/Fe slurry coupled with the application of an electric field is an effective *in situ* remediation method for subsurface TCE and nitrate contamination. Presumably, it is applicable to the remeditation of other subsurface contaminants as well.

Acknowledgments

The author would like to express his gratitude to R.O.C. National Science Council (Project Nos. NSC 93-2211-E-110-006 & NSC 94-2211-E-110-014) and Center for Nanoscience and Nanotechnology, National Sun Yat-Sen University, Taiwan for their supports in this study. The experiments of this work were conducted by the author's former graduate students: Chih-Hsiung Hung, Der-Guang Chang, and Hsiu-Chuan Tu.

References

1. Acar, Y. B.; Alshawabkeh, A. N. *Environ. Sci. Technol.* 1993, 27 (13), 2638.
2. Acar, Y. B.; Gale, R. J.; Alshawabkeh, A. N.;. Marks, R. E.; Puppala, S.; Bricka, M.; Parker, R. *J. Hazard. Mat.* 1995, 40 (2), 117.
3. U.S. EPA, http://www.cluin.org (2000).
4. Gilham, R. W.; O'Hannessin, S. F. *Ground Wate,* 1994, 32 (6), 958.
5. Matheson, L. J.; Tratnyek, P. G. *Environ. Sci. Technol.* 1994, 28 (12), 2045.

6. Burris, D. R.; Campbell, T. J.; Manoranjan, V. S. *Environ. Sci. Technol.* 1995, 29 (11), 2850.
7. Roberts, A. L.; Totten, L. A.; Arnold, W. A.; Burris, D. R.; Campbell, T. J. *Environ. Sci. Technol.* 1996, 30 (8), 2654.
8. Johson, T. L.; Scherer, M. M.; Tratnyek, P. G. *Environ. Sci. Technol.* 1996, 30 (8), 2634.
9. Muftikian, R.; Fernando, Q.; Korte, N. *Water Res.* 1995, 29 (10), 2434.
10. Zhang, W. X.; Wang, C. B.; Lien, H. L. *Cata. Today*, 1998, 40 (4), 387.
11. Yang, G. C. C.; Hung, C. H.; Chang, D. G. Proc. Intern. Sym. Environ. Nanotechnol. 2004: Taipei, Taiwan, 2004; pp. 263-268.
12. Chew, C.F.; Zhang, T. C. *Water Sci.Technol.* 1998, 38 (7), 135.
13. Yang, G. C. C.; Long, Y. W. *J. Hazard. Mat.* 1999, 69 (3), 259.
14. Yang, G. C. C.; Liu, C. Y. *J. Hazard. Mat.* 2001, 85 (3), 317.
15. Yang, G. C. C.; Chen, J. T. Proc. EREM 2001 (3rd Symp. Status Report Electrokinetic Remedi.): Karlsruhe, Germany, 2001; pp. 18.1-18.10.
16. Hung, Y. C. M.S. Thesis, Inst. Environ. Eng., Nat. Sun Yat-Sen Univ., Kaohsiung, Taiwan, 2002.
17. Yang, G. C. C.; Lee, H. L.; Hung, C. H. *J. Chinese Inst. Environ. Eng.* 2004, 14 (4), 255.
18. Yang, G. C. C. *A Method for Transporting Nanoparticles-Containing Slurry through Porous Media,* R.O.C. Patent Pending, 2005.
19. Yang, G. C. C. *Method for Treating a Body of a Polluted Porous Medium,* U.S. Patent No. 7334965 B2, 2008.
20. Glavee, G. N.; Klabunde, K. J.; Sorensen, C. M.; Hadlipanayis, G. C. *Inorg. Chem.* 1995, 34 (1), 28.
21. Yang, G. C. C.; Hung, C. H.; Tu, H. C. *J. Environ. Sci. Health A* 2008, 43 (8), 945.
22. Kao, C. M.; Chen, S. C.; Su, M. C. *Chemosphere* 2001, 44 (5), 925.
23. Choe, S.; Lee, S.-H.; Chang, Y.-Y.; Hwang, K.-Y.; Khim, J. *Chemosphere.* 2001, 42 (4), 367.
24. Numri, J. T.; Tratnyek, P. G.; Sarathy, V.; Baer, D. R.; Amonette, J. E.; Pecher, K.; Wang, C.; Linehehan, J. C.; Matson, D. W.; Penn, R. E.; Driessen, M. D. *Environ. Sci. Technol.* 2005, 39 (5), 1221.
25. Liu, Y.; Choi, C.; Dionysiou, D.; Lowry, G. V. *Chem. Mat.* 2005, 17 (21), 5315.
26. Liou, Y. H.; Lo, S. L.; Lin, C. J.; Kuan, W. H.; Weng, S. C. *J. Haz. Mat.* 2005, B127 (1-3), 102.
27. Kanel, S. R.; Greneche, J.; Choi, H. *Environ. Sci. Technol.* 2006, 40 (6), 2045.

Technology Demonstrations and Field Applications

Chapter 12

Status of nZVI Technology

Lessons Learned from North American and International Implementations

Michael J. Borda, Ph.D., Ramesh Venkatakrishnan, Ph.D., P.G., Florin Gheorghiu, C.P.G., P.G.

Golder Associates Inc., Mt. Laurel, New Jersey, USA, 08054

With nearly ten years of experience, Golder Associates Inc. (Golder) is a leader in the manufacture and implementation of nano-scale zero-valent iron (nZVI) for environmental remediation applications under licensing agreement with Lehigh University. Golder has designed and implemented nZVI injections in the United States, Canada and across Europe at twenty sites, either as the lead consultant or in partnership with Universities and other contractors. In addition, the United States Environmental Protection Agency (USEPA) and state regulatory agencies have participated actively in providing comments and feed-back on proposed nZVI injections, resulting in the further development and understanding of this maturing technology. Golder's global experience has led to realization of the state-of- the-technology including: determination of the importance of a well-developed Site Conceptual Model (SCM); verification of the need to include surface modifiers to enhance the mobility of nZVI in the subsurface; verification of the need to include a catalyst for *in situ* treatment using mechanically crushed material; and, determination of the enhanced treatment potential from combination nZVI/enhanced bioremediation applications. The following chapter expands on these advancements and looks forward to the future needs of this maturing technology.

Introduction

The use of zero-valent iron (Fe^0) to treat sites impacted by chlorinated solvents is a well-established technology in the environmental remediation industry and is considered an accepted technology. Research over the last fifteen years began to focus on the possibility that using smaller particles of zero-valent iron (ZVI) may be effective at treating chlorinated solvents more rapidly *in situ*, as well as, potentially treating more recalcitrant chlorinated compounds due to increased reactivity of particles as a function of size. Nano-scale zero-valent iron (nZVI) entered the sector of contaminated site remediation over a decade ago, and research is actively conducted around the world to better understand the advantages of using nano-scale particles versus their micro-scale counterparts. nZVI consists of sub-micron ($<10^{-6}$ m) particles of zero-valent iron. The rapid destruction of a wide range of contaminants is based on surface-promoted redox processes where the contaminant serves as an electron acceptor and nZVI as the electron donor. Treatment can be accomplished either *in situ* or *ex situ*. The effectiveness of nZVI has been demonstrated in laboratory (bench-scale) and field (pilot-scale) treatability tests and also full-scale remedial projects. It is effective for complete reductive dechlorination of chlorinated organic compounds, e.g., chlorinated ethenes and chlorinated ethanes in contaminated soils, sediments, groundwater and wastewater.

Technological Feasibility

Geologic and Hydrogeologic Considerations

Site geology and hydrogeology play an important role in remedy selection. Implementing nZVI requires a good understanding of subsurface heterogeneities, groundwater flow direction, contaminant distribution, and contaminant migration pathways. These factors define the conceptual site model (CSM) as illustrated in Figure 1.

Subsurface conditions influence nZVI mobility. Presence of low permeable materials (clays, silts and fine sands) may prevent nZVI implementation because of significant limitation on nZVI mobility. Presence of materials with higher porosity and large pore spaces (coarse sand and gravel) is favorable for nZVI mobility. Increased nZVI mobility occurs under higher groundwater flow velocities characteristic of fracture flow.

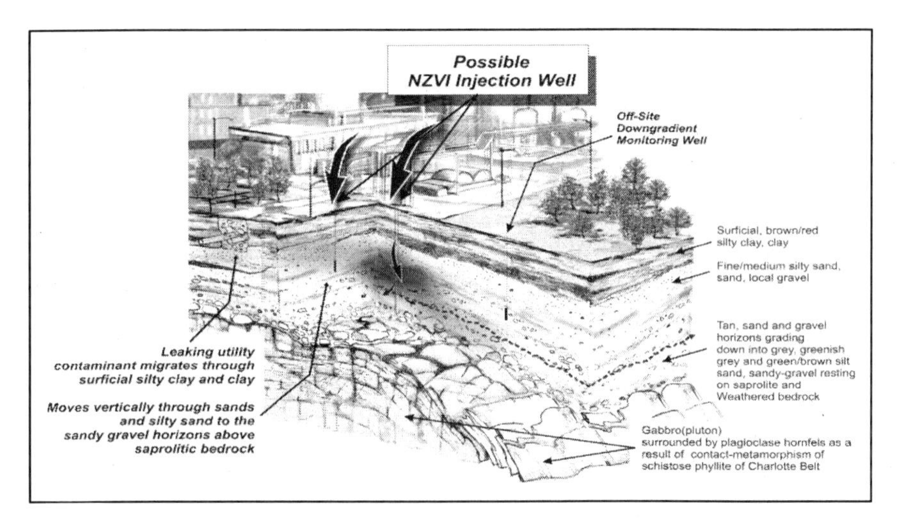

Figure 1. Example of a Site Conceptual Model showing geologic and hydrogeologic and contaminant information along with conceptual design information of an nZVI injection remedy.

The design of nZVI injection is highly dependent on subsurface geologic and hydrogeologic heterogeneities. Well screen location, and nZVI mass injection should target the permeable sections of the subsurface stratigraphy. It is important that the boreholes in a proposed nZVI remedial area be logged in detail and stratigraphic correlations be made. Detailed stratigraphic correlations allow for placement of monitoring points downgradient of the injection area with screens set at the appropriate level. Water level monitoring and detailed understanding of groundwater flow, groundwater fluctuations, and variability in groundwater gradients and flow direction are also important for the same reasons discussed above. Tracer tests and detailed monitoring with multi-parameter data loggers is recommended prior and during nZVI injection. Continuous data logger monitoring allows for refinements of the hydrogeologic conditions in the nZVI test area and help the interpretation of the geochemical changes following nZVI injections. Not understanding the complexities of local stratigraphy and groundwater flow migration may result in monitoring of zones not affected by nZVI injections.

Injection well drilling, installation, and development are important for the successful implementation of an nZVI remedy. The mobility of nZVI particles can be obstructed by the presence of clay lenses that could generate a smearing zone, improper well screen design, incomplete well development, and other factors. An nZVI remedy should take into account the need for injection well rehabilitation and well replacements.

Geochemical Considerations

Along with developing a concise Site Conceptual Model and consideration of the important geological and hydrogeological conditions at a given site, a critical step in evaluating the technological feasibility of using nZVI as a remedial technology is the determination of nZVI dosage. This determination relies on a thorough knowledge and evaluation of the site geochemistry and has broad implications for the effectiveness of the remedy, the mass of nZVI necessary for treatment and the overall remedial cost. The current method that Golder employs for this evaluation is the calculation of the stoichiometric natural reductant demand (NRD), i.e., the concentration of electrons (e^-) necessary (donated from nZVI) to overcome the electron demand from naturally occurring redox species in the system. These naturally occurring species include, but are not limited to: iron (Fe); manganese (Mn); sulfate (SO_4^{2-}); nitrate (NO_3^-); and, dissolved oxygen (DO).

These types of calculations require standard water quality data, knowledge of the approximate treatment zone volume, hydraulic parameters including: conductivity (k); hydraulic gradient (i); and, total porosity (n_t) and contaminant data. The stoichiometric electron demand provided by nZVI is assumed to be 2 electron equivalents per mole (e^- equiv/mol), although a treatment efficiency factor is typically used to address the fact that the transfer of electrons from nZVI is inefficient. Typically, sulfate is the redox sensitive species of most concern as the electron demand is 8 e^- equiv/mol and can radically change the mass of nZVI necessary for treatment, and therefore, the remedial cost.

For example, if one assumes a treatment volume of 125,000 cubic feet (ft^3) of contaminated groundwater with a total porosity of 30% the resulting total pore volume is 37,500 ft^3. A sulfate concentration in groundwater of 1 mg/L produces an electron demand of 88 e^- equiv requiring approximately 12.5 kilograms (kg) of nZVI to satisfy the electron demand. An increase in sulfate concentration to 10 mg/L requires an order of magnitude more nZVI, or approximately 125 kg and a concomitant order of magnitude cost increase. It is critical to determine the total NRD compared to the electron demand from contaminants to evaluate the cost impact of side-reactions. However, there may be occasions where the express purpose of the remedial action is to radically change the system's geochemistry and this can be achieved using nZVI.

An example of an injection condition where nZVI can be effectively placed in the subsurface is a condition where the aquifer is predominantly sand and gravel with limited silt and no clay. Under these conditions large quantities of nZVI material can be injected and potentially overcome NRD concerns. In one example, a plume extended approximately 13,000 ft downgradient of three source zones and is approximately 3,000 ft wide and up to 180 ft deep in an unconsolidated multi-layer soil formation (predominantly sand and some gravel). The dissolved plume consists mainly of dissolved concentrations of trichloroethene (TCE) with low concentrations of cis-1,2-dichlorethene (cis-1,2-DCE) and traces of vinyl chloride (VC). The technology was implemented at a pilot scale on an approximately 100,000 ft^2 area with an approximate depth of 130 ft. Successive injections of a total 4,500 kg of bi-metallic nano-particles (BNPs) was achieved under a forced gradient from a set of eight injection wells screened at four different depths. A food-grade organic dispersant (soy protein) was used as an nZVI surface modifier to increase nZVI mobility in the subsurface.

The baseline conditions at the pilot test site suggested the maximum volatile organic compound (VOC) concentrations were encountered in the deepest part of the aquifer, with TCE mean concentrations around 250 µg/L and mean values of 10 µg/L and less than 1 µg/L corresponding to the middle and upper part of the saturated zone respectively. Prior to nZVI injections, the aquifer was under aerobic conditions with DO concentrations of about 10 mg/L and an oxidation-reduction potential (ORP) of about +150 mV. Nitrate and sulfate concentrations were low, with values lower than 1 mg/L for nitrate and around 6 mg/L for sulfate. pH was slightly acidic with a mean value around 6.5. Hydrogen concentrations were measured upgradient of the treatment zone with concentrations around 3.6 nanomolar (nM).

A total of 4,550 kg of nZVI, coated with soy protein was injected in three distinct injection phases. An injection system using pneumatic packers was used in order to isolate injection sections within the injection wells (IWs). Chase water was injected after each nZVI slurry batch injection to increase nZVI mobility and dispersion. The mass of nZVI injected in each well screen varied based on distribution of TCE concentrations at each location. Preparation of the nZVI slurry was performed on site using high energy mixers.

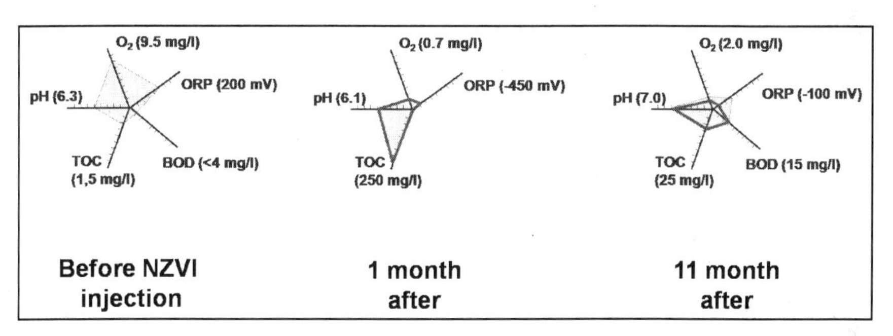

Figure 2. Graphical representation of the geochemical changes that occur after the injection of nZVI into a groundwater system.

Significant changes occurred in the geochemical conditions post-injection as illustrated in Figure 2. Redox potential decreased significantly following nZVI injection with mean values at -450 mV in IWs and -400 mV in the treatment area and up to 15 ft downgradient of IWs. Dissolved oxygen concentration decreased to lower than 1 mg/L in the injection wells and to lower than 2 mg/L in the treatment zone. pH slightly increased following nZVI injections. Almost one year after nZVI injections, redox potential slightly increased in IWs with mean values around -350 mV. Dissolved oxygen concentrations also slightly increased in injection wells with values lower than 3 mg/L. Within the treatment zone, hydrogen concentrations were high, with values in the order of 4,600 nM. Methane was also detected and methane production was inferred to be related to degradation of the soy protein used for nZVI injection.

Four months after the completion of the injections, TCE concentrations had decreased by more than 90% and were still decreasing. After one year, TCE reduction reached more than 95% in the treatment area. As an example of the impact of low transmissivity layers, it was found that a significant mass of contaminant was trapped below a discontinuous silty clay lens making it difficult to achieve contact with nZVI material. To account for these non-homogeneous conditions in the treatment zone and to increase residence time in high hydraulic regime zones, a groundwater recirculation system with dissolved hydrogen injection was implemented one year after nZVI injections. nZVI injections combined with groundwater recirculation in high hydraulic regime zones confirmed that this approach can be used at this site to intercept the TCE plume and achieve the site-specific regulatory criteria of 5 μg/L.

Advances in Delivery

One of greatest challenges of this technology is the successful delivery of nZVI to the impacted zone in the subsurface. Processes that diminish this ability begin to occur immediately after the production of the material and continue during product transport through to the time of delivery into the subsurface. During the processing of nZVI material (Golder currently uses a top-down process to generate nano-scale material from a macro-scale ZVI stock) it has been observed that the primary particle size range of nZVI produced is approximately 50 nanometers (nm) to 100 nm. Particles then begin to agglomerate, based on surface charge and on magnetic properties of ZVI, forming secondary particle sizes that may approach the micron-scale (*1*). Based on this size information, it is expected that the particles being used do not exhibit the "extraordinary" behavior that is ascribed to "true" nano-size particles (*1*). One of the critical questions that still surround the particle size, reactivity and delivery issues with nZVI is the impact of the transition between primary particle size and secondary particle size, i.e., do particles with a secondary particle size of approximately one micron (μm) still exhibit increased reactivity and mobility like a material with a primary particle size of 100 nm? Much of the focus at Golder has been on using food-grade additives to limit the degree of agglomeration to attempt to deliver primary particle size materials to the subsurface.

Importance of Surface Modification

The reactivity of nZVI toward contaminants of concern (COCs) has been thoroughly researched with successful results, leaving the delivery of nZVI to impacted zones in the subsurface as the most critical path to successful site remediation. To this end, Golder has tested and currently utilizes several surface modifiers to diminish the attractive forces between nano-particles which cause agglomeration and limit mobility in the subsurface. Recent research suggests that agglomeration, rather than interaction with aquifer materials, may be a significant cause to the limited mobility observed in nZVI applications (*2*). This has resulted in a series of research papers evaluating different types of surface modifiers including: surfactants (Tween-20); poly acrylic acid (PAA); carboxymethyl cellulose (CMC); cellulose acetate; starch; and, oil emulsions (*3*). Golder along with Lehigh University have evaluated 12 additives for use during injections including: Pluronic® P65 (BASF) surfactant; Pluronic® F68 surfactant pastille; Pluronic® L121 surfactant; Lauryl Sulfate; Polyethylene-block-poly-(ethylene glycol); PAA; oxalic acid; Sodium Acid Pyrophosphate; Soybean Milk and related soy-based powders; Pthalic acid; Poly (acrylic acid-co-acrylamide); and, Polysorbate 80. At several project sites (unconsolidated sediments), Golder has produced surface-modified nZVI particles using Soy Protein to establish a negative surface charge on particles and increase the degree of particle-particle repulsion allowing for gravity-fed and pressurized injection radii of influence (ROI) on the order of 10 ft to 20 ft.

Along with using surface modifiers, a critical factor that has been observed in enhancing the particle mobility is the use of freshly produced nZVI particles. Because the processes that cause nZVI to become immobile (e.g., agglomeration) are kinetically controlled, using material that is less than one week old has been observed to significantly impact the quality of the product. nZVI is processed in dense slurry, approximately 800 g/L, which promotes the agglomeration of charged particles. It is critical to deliver the slurry and dilute it to injection slurry concentration (typically 1 g/L to 10 g/L) quickly to ensure that agglomeration is slowed and more dispersed slurry is used.

Enhancing Delivery with Hydraulic Fracturing

Hydraulic fracturing can be used to initiate new fractures in a bedrock aquifer to increase the transmissivity and enhance the ability to inject reagents into the subsurface. A typical hydraulic fracturing program consists of sealing off a short segment (1 ft to 10 ft) of a borehole at a desired depth (using inflatable packers), injecting fluid (typically water) into the isolated zone at a sufficient rate to raise the hydraulic pressure rapidly and bring about hydraulic fracturing of the borehole wall (*4*). Hydraulic fracturing occurs when the fluid pressure in the isolated portion of the borehole reaches a critical level, called the breakdown pressure (P_c). At this breakdown pressure the rock fractures causing hydraulic fluid loss and a drop in pressure.

To illustrate the importance of increasing transmissivity in impacted bedrock aquifers the following example describes a Site where significant quantities of nZVI were successfully delivered to a previously low transmissivity formation using hydraulic fracturing techniques. The test began by setting a single packer setup at 39.5 ft bgs and inflating to 500 pounds per square inch (psi) and pressurizing the test zone at a constant flow rate until a maximum pressure of approximately 260 psi was reached. This was followed by fracture initiation. Hydraulic jacking tests were performed and consisted of several constant pressure steps designed to define the fracture re-opening or "jacking" pressure. The maximum interval pressure after several pressurization events of 64 psi at a flow rate of 2.5 gallons per minute (gpm) was reached indicating the creation of a jacked feature (re-opened fracture). Based on the response from several cycles, connection of the jacked feature in the borehole with a feature or zone of higher hydraulic conductivity away from the borehole (i.e., in the vicinity of a monitoring point approximately 40 ft away) has been observed. The recovery hydraulic conductivity was calculated using the program FlowDim© and was observed to be approximately 1×10^{-4} cm/sec. The original hydraulic conductivity measured in this area prior to hydraulic fracturing of the bedrock was observed to be $<10^{-6}$ cm/s based on hydraulic testing.

Approximately 0.05% wt/wt palladium acetate was added to a dense slurry of nZVI material and diluted to the appropriate slurry density for effective injection. The total nZVI injected was 90 kg mixed in 2,500 gallons of water, or an average nZVI concentration of 9.5 g/L. The injection rate was initially on the order of 6 gpm with a back-pressure in the nZVI injection line of 60 psi (near the jacking pressure). The nZVI slurry injection rate increased with time to approximately 12 gpm with a decrease in the back-pressure to 50 psi. This decrease in back-pressure likely represents the re-opening of fractures created during the initial hydraulic fracturing cycle or further connection to high transmissivity areas at a distance from the injection location.

The slurry arrival was observed in the multi-parameter graphs at a monitoring point approximately 100 feet downgradient from injection well by the sudden increase in conductivity and pH. ORP also showed a decrease with stabilization at -440 mV. A longer slurry arrival time was observed at a location 50 ft upgradient of the injection location potentially due to the low hydraulic conductivity (7×10^{-5} cm/s) of the surrounding formation. The slurry arrival was observed to coincide with a decrease in specific conductivity and DO. ORP also shows a decrease with stabilization at about -180 mV. It has been interpreted that the continued injection allowed the propagation of the initial fractures, allowing hydraulic connection between the injection well and upgradient and downgradient wells.

Post-nZVI injection groundwater sampling indicated the arrival of nZVI particles at the monitoring wells. This was initially observed during the second sampling event two months after nZVI injection. Additional observations were made during the third sampling event which documented nZVI particles in purge water:

- Injection well – Large amount of particles present
- MW located 60 ft downgradient – Large amount of particles present
- MW located 45 ft upgradient – Small to trace amount of particles present
- MW located 100 ft downgradient – Small to trace amount of particles present

The results collected to date are consistent with previous pilot test results and the results of previously-published laboratory jar tests. Significant percent reduction in the pre-injection baseline concentration was observed in the injection well and observation wells. Maximum concentration reduction occurred three months after nZVI injection. The concentration treated at the injection well is about one order of magnitude higher that the concentrations previously treated during pilot-testing. The Pilot-Scale test was successful in meeting the following objectives:

- Identify the bedrock fracturing pressure (i.e., 260 psi for a 41 ft deep well)
- Identify the fracture re-opening (jacking) pressure and nZVI slurry injection pressure (i.e., 50 to 60 psi for a 41 ft deep well)
- Demonstrate the feasibility of injecting larger iron quantities (i.e., around 90 kg per injection well)
- Identify the ROI of the injection (100 ft based on slurry arrival at monitoring well locations and visual observation of nZVI particles)
- Identify the time to concentration rebound for the pilot test implemented in an area with the highest groundwater concentration

Reactivity Issues

There are numerous examples of using nano-scale materials for Site Remediation including, nonionic amphiphilic polyurethane (*5*) or alumina-supported noble metals (*6*), however, the most promising and most well-studied are those which contain nZVI (*1*). This interest in nZVI material is driven by the increased capacity the material has from a reactivity standpoint when compared to its larger micro-scale counterpart, ZVI (*7, 8,* and *9*). This interest has lead to the rapid development of this technology over the past decade (*1*).

Enhanced reactivity of nZVI is often thought to be the result of larger overall surface area, greater density of reactive sites on the particle surfaces, and/or higher intrinsic reactivity of the reactive surface sites (*1*). This rich chemistry has resulted in several results that differentiate nZVI from macro-scale ZVI including: degradation of contaminants that do not react with larger particles of ZVI (e.g. polychlorinated biphenyls [*10*]); more rapid degradation of contaminants that ZVI has already been shown to treat (e.g. chlorinated ethylenes [*11,12*]); and, more favorable products from contaminants that are rapidly degraded by larger materials but that yield undesirable byproducts (e.g. carbon tetrachloride [*13*]) (*1*). A general trend graph is shown in Figure 3 containing a compilation of results from pilot-scale and full-scale field implementations of nZVI illustrating the optimal treatment envelop.

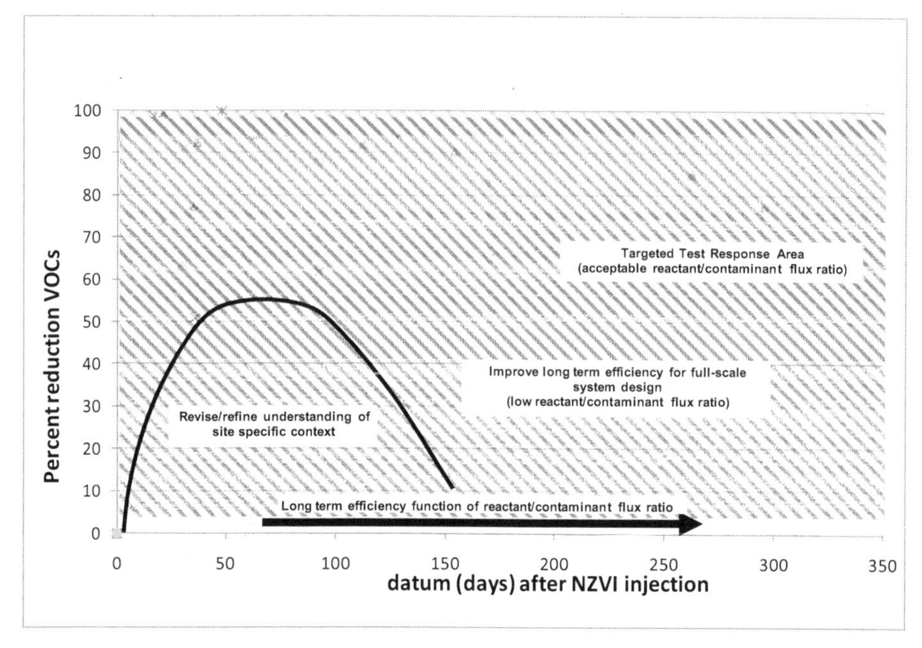

Figure 3. Graph drawn from data obtained in the injection wells and monitoring wells used during pilot-scale tests undertaken by Golder Associates Inc. at locations around the world. Target test response area illustrates the optimum treatment conditions that can be achieved using nZVI technology with 60% to 100% treatment efficiency and long-term (<1 yr) treatment times. Rapid rebound of contaminants after treatment (illustrated by the black curve) suggests that the CSM be refined. Modified from Mace et al. 2006 (14).

To further enhance this reactivity difference from ZVI particles, the addition of surface catalysts, generating BNPs, has found favor in the nZVI field. BNPs consist of sub-micron ($<10^{-6}$ m) of zero valent iron (Fe^0) with a trace coating of noble metal catalyst (typically, palladium). The effectiveness of BNP has been demonstrated in laboratory (bench-scale) and field (pilot-scale) treatability tests. BNP is effective for complete reductive dechlorination of chlorinated volatile organic compounds (cVOCs) such as TCE, polychlorinated biphenyls (PCBs), and chlorinated pesticides/herbicides in contaminated soils, sediments, groundwater, and wastewater. Typically, the injection concentration of BNP is < 0.1% by weight of the injected BNP.

Importance of Surface Catalysts

Precipitated nZVI has reactivity unparalleled by its mechanically crushed counterpart. However, by exploiting the chemistry of noble metals the reactivity of mechanically crushed nZVI can be restored. This process involves the addition of a small concentration of palladium (Pd) to the mechanically crushed material. Palladium irreversibly adsorbs to the nZVI surface and exhibits catalytic activity as the non-reactive palladium is in contact with the highly reactive ZVI surface causing enhanced corrosion of the Fe^0. It has been shown in the literature that iron oxidation is retarded when iron is in contact with a more redox active metal (e.g., zinc) and enhanced when iron is in contact with a less redox active metal (e.g., Pd) (15). The combination of iron and a less active metal forms a galvanic cell where iron acts as the anodic site and is preferentially corroded. Because the less active metal is protected, it exhibits catalytic properties and continues to enhance the further corrosion of iron. This enhanced corrosion of iron can then be used to more efficiently degrade chlorinated aliphatic hydrocarbons (CAHs), e.g., chlorinated ethenes (*16, 17*).

Long-Term Treatment

Coupling nZVI with Enhanced Bioremediation

During the largest pilot-scale nZVI injection performed to date (~4,500 kg nZVI injected) a thorough microbiological assessment was performed illustrating the transition of abiotic CAH degradation to enhanced anaerobic bioremediation of CAHs. Microbial counts, polymerase chain reaction (PCR) and denatured gradient gel electrophoresis (DGGE) analyses were conducted before, and after nZVI injections. Before treatment implementation, bacterial analysis indicated predominance of aerobic bacteria and a low bacterial population density. Traces of *Dehalococcoides* and *Dehalobacter* were detected. PCR and DGGE analysis confirmed that, following nZVI injections, population density became much less diverse. With time, population density increased to its original state, but presented a different set of dominant bacterial species. Microbial counts confirmed that aerobic heterotrophic bacterial population remained relatively stable, but iron-reducing bacteria, sulfate-reducing bacteria and nitrifying bacteria increased significantly in the treatment zone. In general, bacterial population increased by five orders of magnitude compared to the upgradient bacterial population. *Dehalococcoides* was detected in the treatment zone, but its distribution was non-homogeneous over the treatment zone.

This transition is driven by the redox condition of the aquifer due to nZVI reactivity (ORP values changed from +200 mV to -500 mV upon injection) coupled with the addition of complex source of soluble and sparingly soluble carbon (Soy Protein) which acts as an electron donor. This transition and continued degradation of CAHs continued for over one year and achieved a groundwater concentration target of 5 parts per billion (ppb) for TCE without build-up of intermediate degradation products. This suggests that a combined remedy of nZVI injection and long-term enhanced bioremediation may be a strong candidate technology for a number of CAH impacted sites.

The abiotic treatment of CAHs using nZVI is driven by the direct contact of nZVI particles with the CAH and consequent electron transfer. With particle mobility being a significant issue and further agglomeration and sedimentation of nZVI particles placing an upper limit of the functional treatment time of this technology, direct-contact abiotic nZVI treatment may not be a sustainable remedial option. To effectively continue abiotic treatment, successive injections are necessary and can become costly. In addition, aquifer clogging and limited electron transfer efficiency can begin to render multiple injections unsuccessful. In contrast to abiotic nZVI treatment, the presence and reactivity of nZVI in the subsurface causes broad changes in the geochemistry which is not direct-contact driven. Changes in ORP to more reducing conditions, stripping of DO and removal of terminal electron acceptors (TEAs) can occur over a significantly larger area than direct-contact abiotic treatment is occurring. Typically under these scenarios the capacity of the system for intrinsic bioremediation is high; however, the systems are typically carbon limited. The addition of an electron donor (e.g., soy protein) can stimulate the indigenous microorganisms and result in successful enhanced bioremediation over a significantly larger treatment area than the initial nZVI impact. A phased approach may be applicable and sustainable as nZVI can be injected in source areas to abiotically treat high dissolved concentration of CAHs and to condition the aquifer to enhance bioremediation. Further treatments of electron donor may continue treatment for significantly longer time-frames than nZVI alone allowing for less frequent nZVI injections to condition the aquifer.

Next Steps

Looking Forward at the Future of nZVI Technology

The current use of additives to enhance the mobility of nZVI particles in the subsurface represents a facet of the technology where significant advancements may still exist. Although, PAA and soy-based products have been approved and used in the United States and in Canada, progress should be made to develop better surface modifiers with even less impact to the natural system. Future development of "green" polymers may advance this ability to improve the mobility of nZVI and achieve better contaminant targeting in the subsurface with more sustainable technology. These advancements will also be strongly tied to the further development of coupling nZVI with enhanced bioremediation as the additives used to enhance mobility also represent soluble forms of carbon that can enhance bioremediation.

Along with the use of additives, mechanically enhancing mobility shows a great deal of promise for the further development of this technology. Golder has used hydraulic fracturing to enhance mobility in bedrock; however, the use of fracturing technologies in unconsolidated sediments may represent a potential opportunity to enhance the dispersion of nZVI in a number of systems.

As mentioned previously, further research on the reactivity and mobility of agglomerated nano-scale primary particle size materials with a macro-scale secondary particle size compared to macro-scale primary particle size materials may change the paradigm that smaller particles are necessary for environmental applications. Similar mobility and reactivity may be achieve with larger particle size materials (~500 nm to 1 μm) with the continued use of surface modifiers and catalysts resulting in significantly lowered remedial costs and potentially eliminating perceived risk.

Although nZVI has shown significant progress on sites with unconsolidated sediments it has performed well in formations that trend towards larger particle size (i.e., gravel and sand). Typically, *in situ* treatments have suffered from limited applicability in complex geology and also where non-aqueous phase liquid (NAPL) is present. A potential remedial option that has shown promise is the use of electro kinetics to deliver nZVI into DNAPL and dissolved phase contaminants zones with low transmissivity. The electro kinetic movement of articles can overcome the limitations of conventional injections and physical flow to achieve rapid and uniform contact of nZVI with targeted compounds.

Finally, it is important to note that during the early development of this technology a number of myths became widely prevalent and continue to influence the opinion of regulatory agencies and policy makers. These include the concept that nano-scale materials travel indefinitely in groundwater and can treat elevated concentrations of contaminants over highly dispersed areas. This provoked many people to demand more information on the fate and transport of particles and created a great deal of interest in the human health and environmental risk of using this technology. Over ten years of applying nZVI one of the most critical design issues of this technology is particle mobility. In fact, as stated above, a number of additives have been used to enhance the mobility of particles to achieve even modest dispersion in the subsurface. The future of this technology is strongly tied to the ability of researchers and professionals applying this technology to continue to better understand and report information observed during nZVI work. It is also important that this technology finds its place as a remedial tool which under the appropriate site conditions can be highly successful as a treatment technology. However, this is not the remedial panacea that it was first thought to be and is not widely applicable at all CAH-impacted sites. A thorough Site Conceptual Model must be developed and all geologic, hydrogeologic and geochemical information must be considered to properly evaluate the feasibility of this technology. Finally, rigorous field-scale pilot testing must be performed to determine the applicability of the technology on a small scale under actual site conditions. Only then can a successful full-scale application be designed and implemented.

References

1. Tratnyek, P.G. and Johnson, R.L. Nanotechnologies for Environmental Cleanup. *Nanotoday*. **2006**. Vol. 1. No. 2, 44-48.
2. Phenrat, T., Saleh, N., Sirk, K., Kim, H., Matyjaszewski, K., Titlton, R., Lowry, G.V. Stabilization of Aqueous Nanoscale Zerovalent Iron Dispersions by Anionic Polyelectrolytes: Adsorbed anionic polyelectrolyte layer properties and their effect on aggregation and sedimentation. *J Nanoparticle Res*. **2008**. *10*. 795-814.

3. Kanel, S.R., Goswami, R.R., Clement, T.P., Barnett, M.O. and Zhao, D. Two Dimensional Transport Characteristics of Surface Stabilized Zero-valent Iron Nanoparticles in Porous Media. *Environ. Sci. Technol.* **2008**. *42*, 896-900.
4. ASTM Designation D 4645-87. Standard Test Method for Determination of the in situ Stress in Rock Using the Hydraulic Fracturing Method. **1992**
5. Tungittiplakorn, W., Cohen C, Lion LW, Engineered polymeric nanoparticles for the bioremediation of hydrophobic contaminants. *Environ. Sci. Technol.* **2005**. *39*. 1354
6. Nutt, M. O., Hughes JB, Wong MS, Designing Pd- on-Au bimetallic nanoparticles for trichloroethylene hydrodechlorination. *Environ. Sci. Technol.* **2005**. *39*. 1346
7. Zhang, W., Nanoscale iron particles for environmental remediation: An overview. J. Nanoparticle Res. **2003**. *5*. 323
8. Tratnyek, P. G., Scherer, M., Johnson, T., and Matheson, L., In: Chemical Degradation Methods for Wastes and Pollutants: Environmental and Industrial Applications, Marcel Dekker, New York, NY **2003**. 371
9. Interstate Technology and Regulatory Council (ITRC), Permeable Reactive Barriers: Lessons Learned/New Directions, ITRC **2005**.
10. Lowry, G. V., and Johnson, K. M., Congener-Specific Dechlorination of Dissolved PCBs by Microscale and Nanoscale Zerovalent Iron in a Water/Methanol Solution. *Environ. Sci. Technol.* **2004**. *38*. 5208
11. Liu YQ, Majetich SA, Tilton RD, Sholl DS, Lowry GV, TCE dechlorination rates, pathways, and efficiency of nanoscale iron particles with different properties. *Environ. Sci. Technol.* **2005**. *39*. 1338
12. Song, H., and Carraway, E. R., Reduction of chlorinated ethanes by nanosized zero-valent iron: Kinetics, pathways, and effects of reaction conditions. *Environ. Sci. Technol.* **2005**. *39*. 6237
13. Nurmi, J. T., Tratnyek PG, Sarathy V, Baer DR, Amonette JE, Pecher K, Wang CM, Linehan JC, Matson DW, Penn RL, Driessen MD., Characterization and properties of metallic iron nanoparticles: spectroscopy, electrochemistry, and kinetics. *Environ. Sci. Technol.* **2005.** *39*. 1221
14. Mace, C., Desrocher, S., Gheorghiu, F., Kane, A., Pupeza, M., Cernik, M., Kvapil, P., Venkatakrishnan, R., and Zhang, W-X. Nanotechnology and Groundwater Remediation: A Step Forward in Technology Understanding. *Remediation*. Spring **2006**
15. Crow, D. *Principles and Applications of Electrochemistry*. Chapman and Hall. London. **1988**
16. Grittini, C., Malcomson, M., Fernando, Q. and Korte, N. Rapid Dechlorination of Polychlorinated Biphenyls on the Surface of a Pd/Fe Bimetallic System. *Environ. Sci. Technol.* **1995**. *29*. 2898.
17. Muftikian, R., Fernando, Q., Korte, N. A Method for the Rapid Dechlorination of Low Molecular Weight Chlorinated Hydrocarbons in Water. *Water Res*. **1995**. *29*. 2434.

Chapter 13

Iron Nanoparticles for In Situ Groundwater Remediation of Chlorinated Organic Solvents in Taiwan

Yu-Ting Wei[1], Shian-Chee Wu[1], Chih-Ming Chou[2], De-Huang Huang[3] and Hsing-Lung Lien[2]

[1]Graduate Institute of Environmental Engineering, National Taiwan University, Taipei, Taiwan, ROC
[2]Department of Civil and Environmental Engineering, National University of Kaohsiung, Kaohsiung, Taiwan, ROC
[3]Chinese Petroleum Corporation, Kaohsiung, Taiwan, ROC

A 200-m^2 pilot-scale field study successfully demonstrated that palladized nanoscale zero-valent iron (NZVI) is capable of remediating groundwater contaminated with a variety of chlorinated organic solvents including vinyl chloride (VC), dichloroethanes and dichloroethylenes in southern Taiwan. Major contaminant is VC that has a concentration ranging from 10 to 5000 μg/L. The concentration distribution is depth-dependent at the site where contaminant concentrations increased with depth. A total iron mass of about 20 kg on-site synthesized NZVI (Pd 0.05 wt%) suspended in 8,500 L water was injected via gravity into the sandy aquifer. Thirteen multi-level monitoring wells allowing to collect samples from three different depths (6, 12, 18 m) were installed. For a monitoring period of 3 months, a spatial and temporal decrease in VC concentrations was observed. The degradation efficiency was greater than 90% at both upper and middle layers but was about 20-85% at the bottom layer. Oxidation-reduction potential (ORP) measurements indicated a homogeneous reducing condition (ORP -450 ~ -280 mV) was achieved in the testing field. Analysis of total iron concentrations found iron was mainly trapped at the upper layer. NZVI-enhanced biodegradation was observed.

Introduction

The use of nanoscale zero-valent iron (NZVI) for the remediation of groundwater impacted by a variety of contaminants including chlorinated hydrocarbons and heavy metals has received much research attention over the past decade *(1)*. The NZVI technology has been demonstrated to be suitable for in-situ treatment of contaminant "hot-spots" given its high reactivity and flexible deployment in the field *(2-6)*. Taiwan has long been known as an industrial island. Major economic activities rely on advanced electronic devices and petroleum manufactures. In this study, a groundwater contaminated site was selected from a vinyl chloride monomer (VCM) manufacturing plant. In this paper, we present the first field test for groundwater remediation by NZVI technology in Taiwan. The study focuses on a pilot-scale field demonstration of injecting the palladium-catalyzed and surfactant-dispersed NZVI to control the contaminated plume.

The VCM manufacturing plant is located in southern Taiwan. High concentrations of VC (4562 μg/L), 1,1-dichloroethylene (430 μg/L), cis-1,2-dichloroethylene (1151 μg/L) and trichloroethylene (682 μg/L) in groundwater were detected from the monitoring well nearby the plant (Figure 1). The NZVI pilot test was conducted in a small area (10 meters by 20 meters) south of the VCM plant in downstream groundwater direction. The unconfined aquifer, composed of medium to coarse sand and few silt, lies approximately 4 to 18 meters below ground surface (m bgs).

Materials and Methods

Test Area Design

Three injection wells and thirteen nested multi-level monitoring wells were installed on a 200-m^2 pilot. In the downstream direction of each injection well, four additional multi-level monitoring wells were installed. The positions of the four nested monitoring wells were approximately one, two, three, and five meters from the injection well. The injection wells were all eighteen-meter deep with fifteen-meter screens. In addition, every nested monitoring well included three separate wells which were approximately six, twelve and eighteen-meter deep with three-meter screens (Figure 2). There was one nested monitoring well located upstream for the purpose of background monitoring. Initially, about 1,000 liters of on-site synthesized NZVI were injected into well IW-3 by gravity. Another 7,500 liters of NZVI suspension was injected into IW-1 via gravity after ten days. The total iron mass was about 20 kilograms companioned with 100 g of palladium catalyst (0.05% of total iron mass).

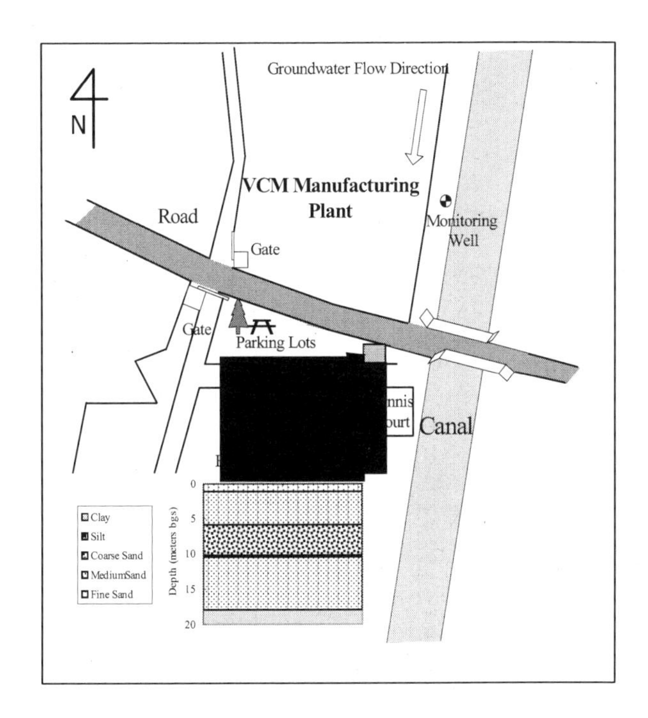

Figure 1. Site map and geologic cross section of the NZVI plot.

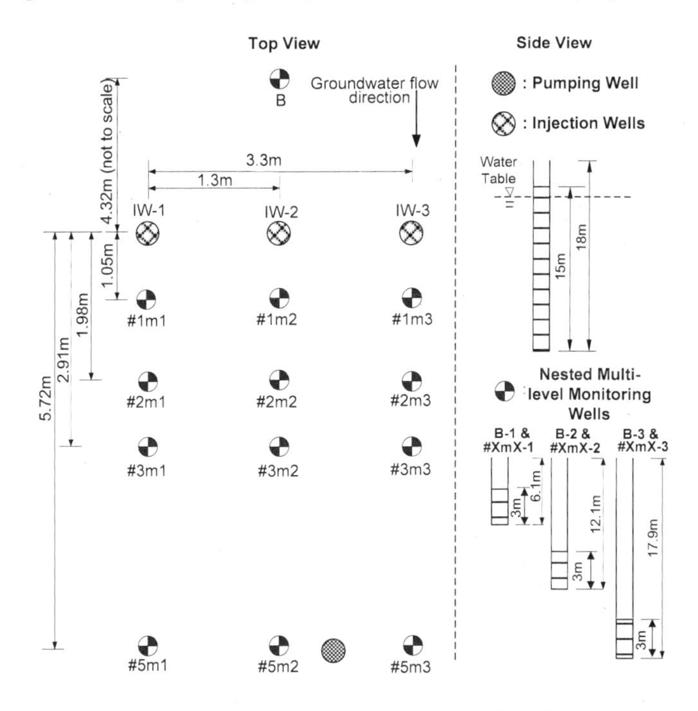

Figure 2. Injection and monitoring locations within the test area.

Production of On-Site Synthesized NZVI

A semi-continuous reactor system was used to produce NZVI on site (Figure 3). The reactor system is designed to load on a trailer for convenient mobility. The on-site synthesized NZVI was prepared by slowly adding ferrous sulfate solution into sodium borohydride (>98.5%, Beckman Coulter, Inc.) solution containing nonionic surfactant (Taiwan NJC Corp., industrial-grade) at the concentration of 5,000 mg/L in a 1,000-liter tank. After the reaction was completed, palladium acetate was mixed with the NZVI suspension. The NZVI mixture was then pumped into a storage tank for the injection later. Because of the high reactivity, dry iron nanoparticles tend to explode in contact with air. Nevertheless, the on-site synthesized NZVI preserved in aqueous solutions can be safely handled without the danger of explosion. TEM analysis indicated that the on-site synthesized NZVI has the particle size in the range of 80-120 nm with a specific surface area of 29.3 m^2/g.

Figure 3. A semi-continuous reactor system for on-site synthesis of NZVI.

Analytic Methods

Volatile organic compounds were measured by GC/MS (Angilent 6890/5973 with a DB-624 capillary column) using a purge and trap sampling equipment (OI Analytical. Model 4560). Methane, ethane and ethene were measured from the headspace of serum vials containing water samples after equilibration. The headspace was analyzed for the target gases by GC/FID (HP 5890 with GS-GASPRO column). Dissolved oxygen (DO), oxidation-reduction potential (ORP) and pH were measured by a portable equipment (YSI 650 MDS-6600 V2-4 Sonde, YSI Inc.). Total iron was measured by atomic absorption spectrophotometer (Perkin Elmer Aanalyst 800) after acid digestion.

Results and Discussion

Injectability

Injectability of NZVI is evaluated by the total iron concentration in the aquifer. The background concentration of total iron in the testing site was about 10 mg/L. As shown in Figure 4a, the total iron concentration measured at the upper layer increased significantly at the whole testing site after NZVI was injected. The iron concentration was in a range of 40-370 mg/L. In general, the iron concentration decreased with increasing distance downstream from the injection well. Figure 4b shows the iron concentration at the testing site after NZVI was injected for 60 days. A dramatic decrease in iron concentrations with time was observed. This suggests that iron either was consumed through the corrosion or transported through the groundwater flow. In terms of the depth, it was found that the iron concentration decreased in the order: upper layer > middle layer > lower layer. It is believed that with the gravity injection, because of the aquifer heterogeneity, much of the NZVI first seeped through channels in the unsaturated zone, causing NZVI to accumulate more in the upper layer as compared to those in the lower layer.

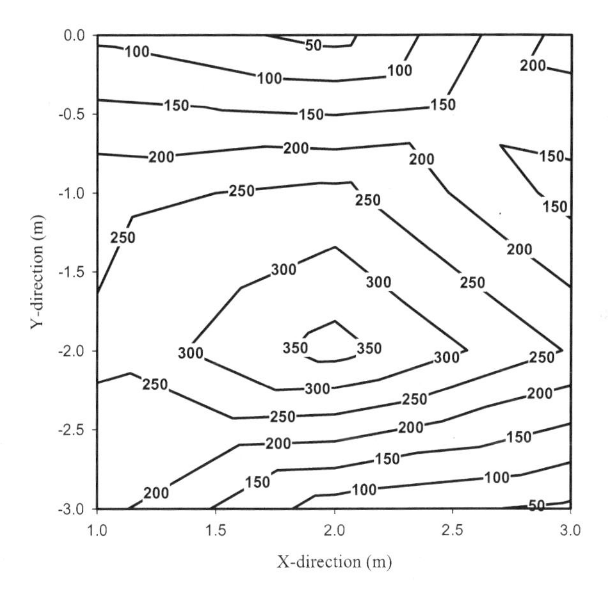

Figure 4a. Total iron concentration measured at the upper layer after NZVI injection.

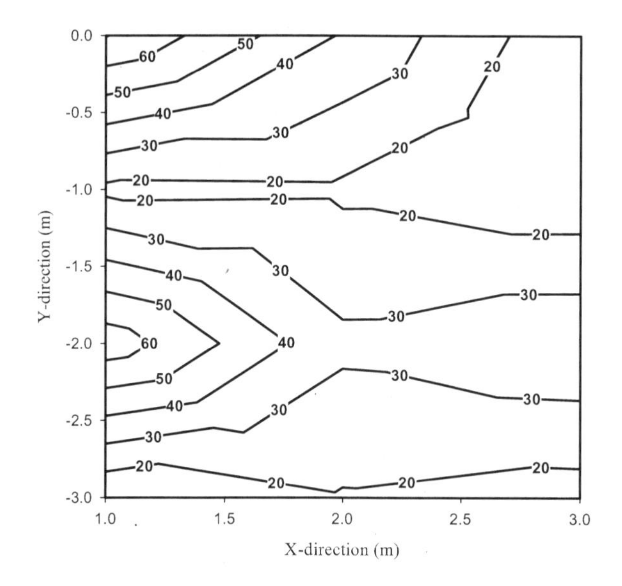

Figure 4b Iron concentration at the testing site after NZVI was injected for 60 days.

Mobility

As it has been observed in field tests of the NZVI technology, the pH and E_h profiles at given monitoring locations over time can be used as a convenient indicator for the NZVI reactivity and to track the migration path of the nanoparticles *(2-4)*. In this study, our data of the iron concentration shown in Figure 4a suggests that the NZVI is mobile. Furthermore, as illustrated in Figure 5, the ORP decreased from about -100 to -400 mV at the central area of the testing site during the 30 days. The E_h profiles shown in Figure 5 suggest the NZVI gradually migrated downstream. This is consistent with the previous studies indicating the ORP can serve as a convenient indicator for the mobility of NZVI *(2-4)*. Overall, the data from this study suggest that NZVI is an effective means of achieving highly reducing conditions in the subsurface environment.

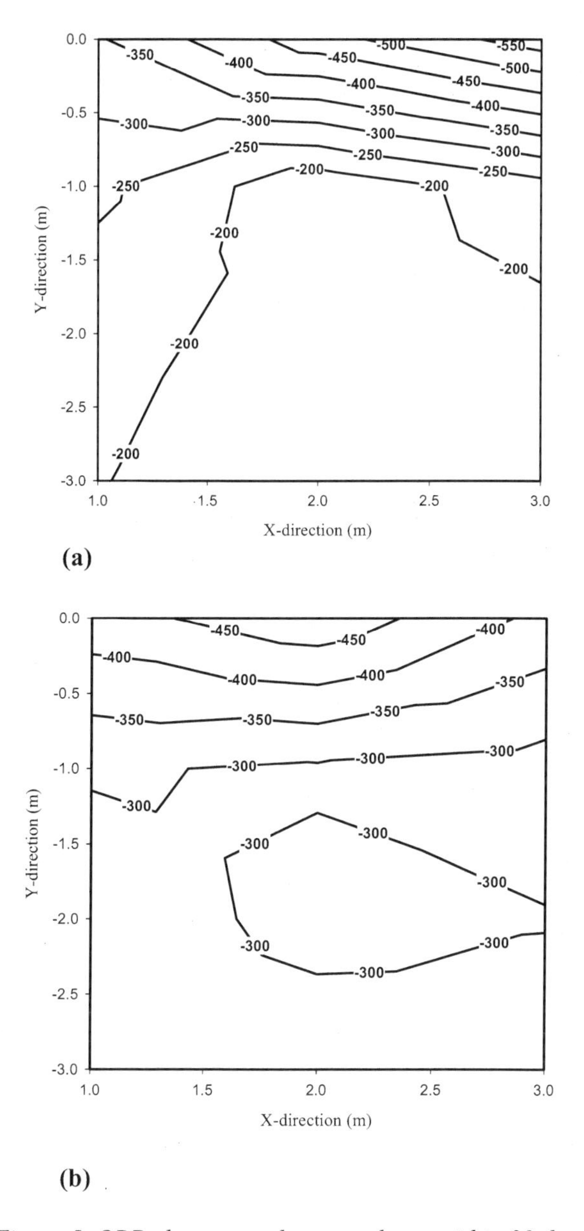

Figure 5. ORP changes at the upper layer within 30 days.

Effectiveness

The concentrations of VC monitored in various times for the upper, middle and lower layers are totally summarized in Figure 6a. It is clear that the VC concentration steadily decreased as the test date progressed, with few exceptions. Furthermore, the decrease in VC concentrations corresponded to a decrease of ORP (Figure 6b). This is in agreement with the previous studies suggesting that ORP can act as an indicator for the NZVI reactivity (4). The average removal efficiency determined at most of the monitored wells was 50-99%. The lowest removal efficiency was about 20%, which was found at the bottom layer of the monitoring well #5M1.

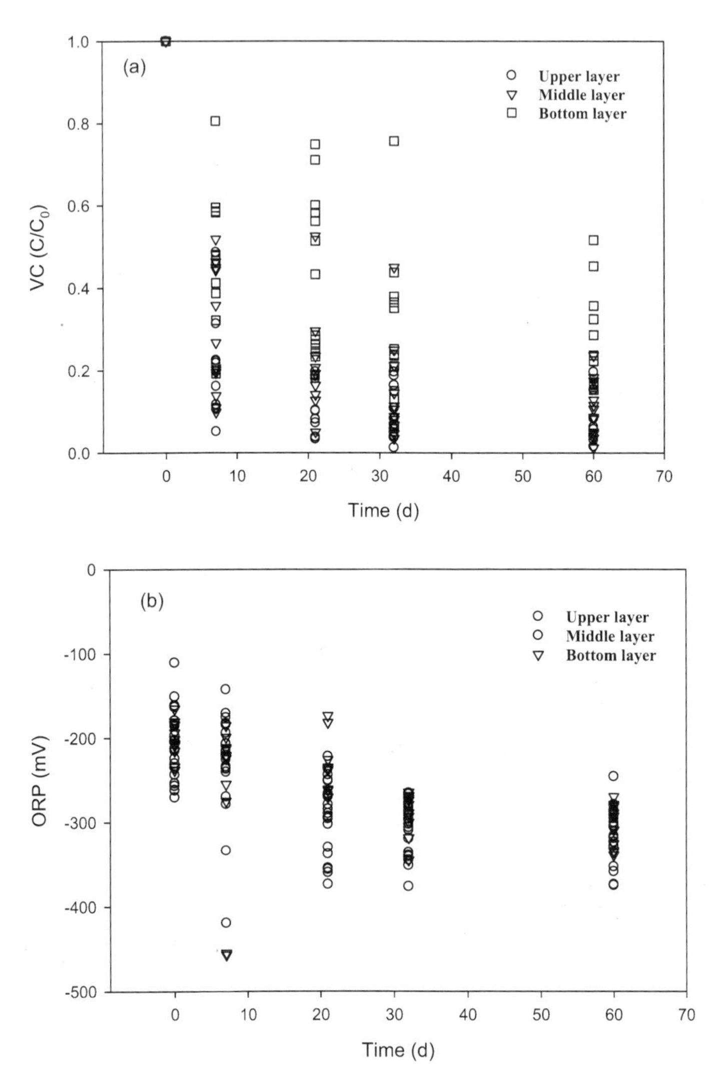

Figure 6. Trends of (a) VC concentration and (b) ORP during a 60-day operation at the testing site.

Microbial Interaction

Figure 7 shows the concentrations of hydrocarbons and total organic carbon (TOC) measured along the downstream direction in the bottom layer. An unexpected high concentration of ethylene was determined in the testing site. The TOC concentration is used to reflect the influence of added biodegrable surfactants. An unexpected high concentration of ethene was determined in the testing site. Currently, the cause is still unclear. Nevertheless, methane was observed at 25 days after the injection.The gradual increase of methane concentration suggests methanogenesis took place at the testing site. The methanogenic conditions require the ORP value lower than -240 mV, which can be established in this specific testing site. A small amount of ethane was also found. The formation of ethane is likely due to the reduction of VC to ethane by NZVI, and the other being the presence of 1,1-dichloroethane that would reduce to chloroethane and ultimately to ethane. A gradual increase in TOC concentration followed by a subsequent leveling off provides further evidence to support enhanced biodegradation occurring in the testing site because the added biodegrable surfactant may serve as the carbon source to stimulate microbial growth (7-8).

Conclusion

This paper presents a successful pilot-scale field study for applying on-site synthesized nanoscale zero-valent iron to remediate groundwater contaminated with chlorinated organic compounds. A total amount of 20 kg palladized NZVI was injected into the groundwater via gravity at a 10 m × 20 m testing site. The VC degradation efficiency determined at most of the monitoring wells was 50-99%. High concentrations (up to 20 mg/L) of methane and ethylene were detected. Though the cause of which is still unclear at the current stage, it is likely that enhanced bioremediation was involved at the testing site because of its strongly reducing conditions. An increase in VC degradation efficiency corresponded to a decrease of ORP values, which is in agreement with the previous studies suggesting that ORP can serve as a proper indicator for the NZVI reactivity.

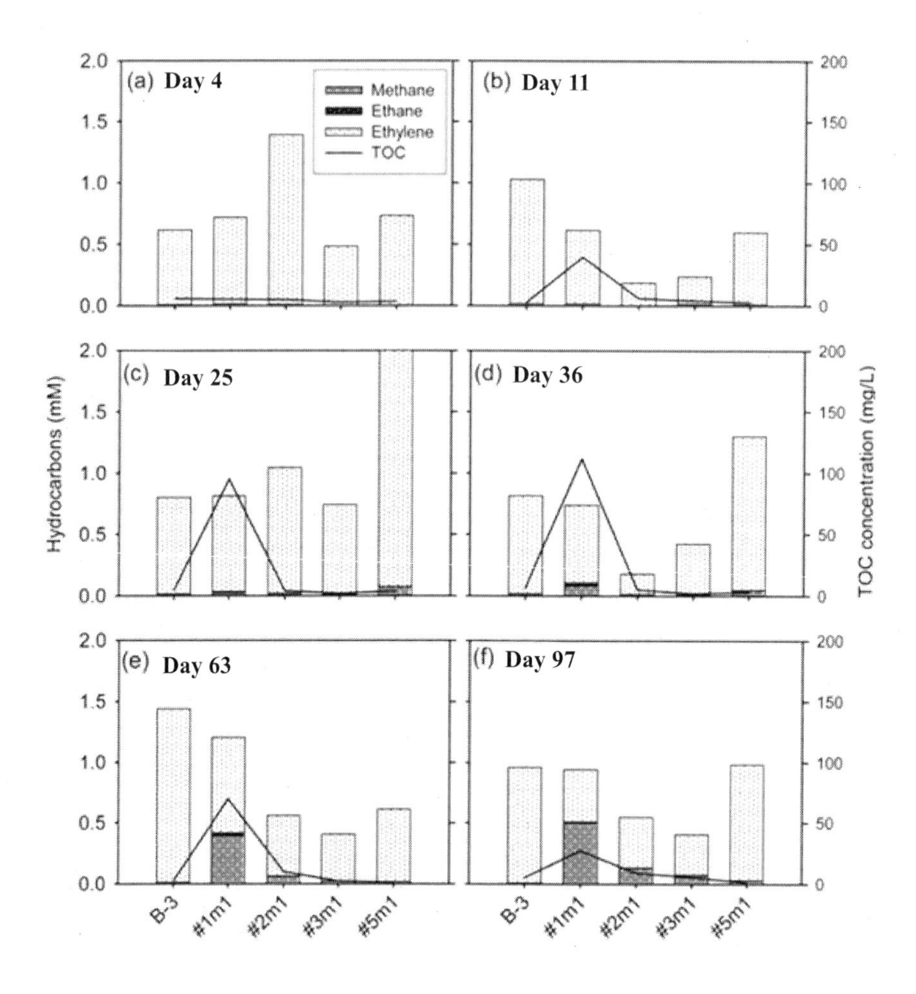

Fig. 7 TOC and molar of methane, ethane and ethylene found in the bottom layers.

Acknowledgements

The authors would like to thank the National Science Council (NSC), Taiwan, R.O.C. for the financial support under Grant no. NSC 95-2221-E-002-162-MY2 and NSC95-2221-E-390-014-MY2. We also like to thank Chinese Petroleum Corporation for its on-site assistance.

References

1. Lien, H-L.; Elliott, D. W.; Sun, Y-P.; Zhang, W. X. Recent progress in zero-valent iron nanoparticles for groundwater remediation. *J. Environ. Eng. Manage.* **2006**, 16, 371-380.
2. Zhang, W. X. Nanoscale iron particles for environmental remediation: An overview. *J. Nanopart. Res.* **2003**, 5, 323-332.
3. Elliott, D. W.; Zhang, W. X. Field assessment of nanoscale bimetallic particles for groundwater treatment. *Environ. Sci. Technol.* **2001**, 35, 4922-4926.
4. Glazier, R.; Venkatakrishnan, R.; Gheorghiu, F.; Walata, L.; Nash, R.; Zhang, W. Nanotechnology takes root. *Civil Engineering.* **2003**, 73, 64-69.
5. Jung, B. M.; Sakulchaicharoen, N.; O'Carroll, D. M.; Herrera, J. E.; Sleep, B. E. Characterization of iron nanoparticles stabilized for enhanced delivery to TCE source zones. 237th ACS National Meeting, Salt Lake City, UT, United States, March 22-26, 2009.
6. Klimkova, S.; Cernik, M.; Lacinova, L.; Nosek, J. Application of nanoscale zero-valent iron for groundwater remediation: Laboratory and pilot experiments. *NANO* **2008**, 3, 287-289.
7. Ramsburg, C.; Abriola, L.; Pennell, K.; Loffler, F.; Gamache, M.; Amos, B.; Petrovskis, E. Stimulated microbial reductive dechlorination following surfactant treatment at the Bachman road site. *Environ. Sci. Technol.* **2004**, 38, 5902-5914.
8. Low, A.; Schleheck, D.; Khou, M.; Aagaard, V.e; Lee, M.; Manefield, M. Options for in situ remediation of soil contaminated with a mixture of perchlorinated compounds. *Bioremediation J.* **2007**, 11, 113-124.

Chapter 14

Practical Applications of Bimetallic Nanoiron Particles for Reductive Dehalogenation of Haloorganics: Prospects and Challenges

Teik-Thye Lim and Bao-Wei Zhu

School of Civil and Environmental Engineering, Nanyang Technological University, Republic of Singapore

This chapter provides a brief review of the potential of using Pd/Fe nanoparticles to dehalogenate haloaliphatics and chloroaromatics, and discusses the recent findings on reactivities of Pd/Fe nanoparticles in various aqueous systems. The prevailing method of synthesis and characteristics of Pd/Fe are described. The role of Pd in the bimetallic particles in enhancing the reductive dehalogenation of chlorinated benzenes and halogenated methanes are discussed. The influences of various matrix species such as inorganic anions and amphiphilic molecules on the dechlorination process and kinetics are examined. Finally, challenges facing field application of the Pd/Fe nanoparticles are deliberated for the future research to address.

Introduction

Halogenated organic compounds (HOCs) including aliphatics and aromatics, are widely used as solvents in degreasing, cleaning and extracting, as synthetic intermediates for plastics, dyes and pesticides, and as fire retardant chemicals, moth repellents and deodorants. They are known toxic chemicals, and some are potential carcinogens. Due to their intensive usage in industries and households, and recalcitrance in natural environments, they are widespread contaminants found in soil, sediment, groundwater and surface water. Their continual releases into these environmental compartments have resulted in their increasing bioaccumulation in biota, threatening human and ecosystem health.

To date, there are only a handful of effective technologies to degrade HOCs. One promising degradation pathway for HOCs is through abiotic reductive transformation that certainly reduces the degrees of halogenation and often enhances biodegradability of their intermediates and end products. In the last two decades, various types of zero-valent iron (ZVI) particles have been intensively investigated for their reactivities towards HOCs and other toxic oxidized organics and inorganics. Initially, the granular form of ZVI were introduced in permeable reactive barrier (PRB) systems to treat groundwater plume contaminated with chlorinated organics (*1, 2*). To improve over the PRB treatment system, nanoscale ZVI particles (nZVI) have been synthesized, and their field applications in treating contaminant source zone have been well documented (*3*). The nZVI particles offers several advantages compared to the granular ZVI for in-situ treatment of contaminated subsurface, such as higher reactivity due to greater density of reactive surface sites (and possibly with higher intrinsic reactivity too), low material cost, higher mobility, and flexible delivery into deep or stratified contaminated source zones in aquifer. In addition, the nZVI can be dispersed easily, and this allows its ex-situ applications in slurry or expanded bed reactors.

To further enhance the reactivity of nZVI making the nanoparticles capable of hydrodechlorinating some persistent haloaromatics, the nanoscale iron-bimetallic composite particle, has been synthesized in recent years. The bimetallic nanoiron particle (BNIP) has a noble metal (e.g., Pd, Pt, Ni, or Ag) deposited on the nZVI surface. These noble metals have lower hydrogen overpotentials compared to iron. In such BNIP composite system, the noble metal serves as catalyst to catalyze the reductive transformation of the oxidized organics, while the nZVI functions as electron donor or in some cases as catalyst too through its iron oxide shell (*4*). Among these catalytic metals, Pd and Ni are the most commonly used in BNIP in the past years. The BNIPs show enhanced reactivities towards various HOCs, including chloroaromatics (*5-10*). The various toxic organics that can be reductively transformed by the BNIP or bare (monometallic) nZVI are presented in Figure 1. In particular, the Pd/Fe particles with a Pd percentage as low as 0.01% (w/w) has been found still exhibiting catalytic hydrodechlorination of chloroaromatics that would otherwise inert towards the bare nZVI (*9*).

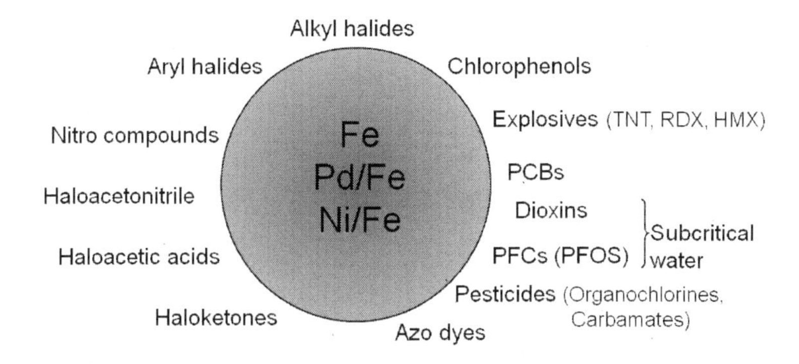

Figure 1. Reductively degradable organics with nZVI or BNIPs.

Figure 2 shows a schematic illustrating the role of Pd in the Pd/Fe particle during reductive transformation of a chlorinated benzene. In the composite, Pd serves as hydrogen collector and subsequently catalyzes the hydrodechlorination reaction (*11*). It has also been suggested that a galvanic couple can form between the two metals in the bimetallic composite particles, and it is critical for the generation of the activated atomic hydrogen species (H^*) (*6, 7*). Pd, the lower hydrogen overpotential metal in the iron/water system, could form a galvanic cell with iron. Thus, the corrosion rate of anodic Fe increased in the Pd/Fe system, resulting in higher hydrogen evolution and fresher iron surface condition (*12*). Zhang et al. (*5*) indicated that physically mixing of Pd and Fe particles produced no positive effects on trichloroethylene dechlorination. A close contact of the Fe with the catalyst metal in the bimetallic particle system is essential to obtain a positive effect, and through coating the catalyst on the iron surface or forming bimetal alloy can remarkly enhance reactivity of the system.

Synthesis of Pd/Fe

Typically, BNIPs are synthesized by co-reduction of the ionic precursors of the two metals (*6*) or post-deposition of the second metal on the surface of the fresh nZVI (*5, 13*). The latter is commonly adopted to synthesize Pd/Fe in which Pd is only a minute amount in the bimetallic particle, and it will ensure Pd deposition on the nZVI surface, forming discrete islets. In this synthesis route, the nZVI is first synthesized by reduction with $NaBH_4$, and the palladized nZVI is produced by soaking the freshly prepared nZVI particles in an acetone solution of palladium acetate. The synthesis processes are depicted in the following reactions:

$$Fe(H_2O)_4^{2+} + 2BH_4^- + 2H_2O \rightarrow Fe^0 \downarrow + 2B(OH)_3 + 7H_2 \uparrow \quad (1)$$

$$Pd^{2+} + Fe^0 \rightarrow Pd^0 + Fe^{2+} \quad (2)$$

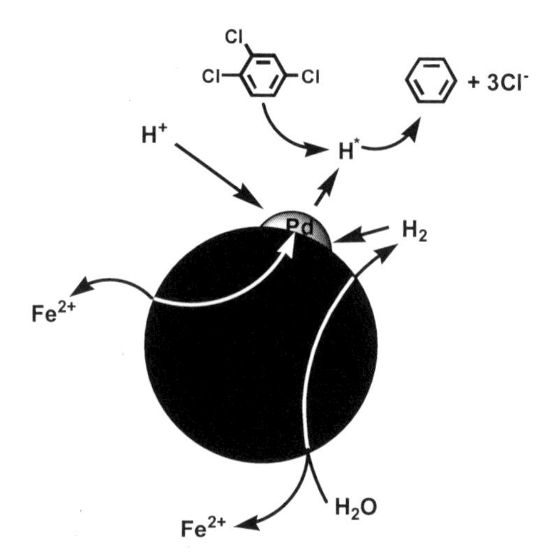

Figure 2. Hypothesized dechlorination mechanism for a chlorinated benzenes with Pd/Fe particle.

Samples with different Pd contents (1.0%, 0.5%, 0.2%, 0.1%, 0.05% and 0.01% w/w) have been synthesized by the authors (*9, 14*). The typical synthesis procedure is as follows. 20 mL of 12.41 g/L $FeSO_4 \cdot 7H_2O$ solution was added to a 70 mL bottle; 3.8 M NaOH solution was used to precipitate $Fe(OH)_2$. The $Fe(OH)_2$ formed was then reduced to ZVI when 20 mL 0.26 M $NaBH_4$ was added dropwise. The ZVI particles formed were isolated by centrifugation. To remove the remaining $NaBH_4$, the ZVI particles were rinsed with 50 mL water for each bottle. After discarding the rinsate, 10 mL acetone was added and mixed rigorously with the ZVI particles in each bottle, and then an appropriate amount of palladium (II) acetate was added. The Pd(II) would be reduced to Pd^o as illustrated by eq. 2, and deposited on the ZVI.

Characteristics

The actual Pd percents (w/w) in the synthesized Pd/Fe samples normally agreed with the stoichiometry illustrated in eqs. 1 and 2. The specific surface area of the various synthesized Pd/Fe particles with different Pd percentages was 27 ± 3 m^2/g. The point of zero charge of the Pd/Fe was around pH 8.1. The X-ray diffraction (XRD) patterns of the fresh sample corresponded to that of the body-centered cubic α-Fe^0 while the XRD patterns of aged samples showed characteristic peaks associated with iron oxides (*9*).

Through scanning electron microscope observation, the fresh Pd/Fe sample usually showed a homogenous chain-like texture, while the reacted sample exhibited platelet shaped crystals. Through transmission electron microscope (TEM), it could be confirmed that the fresh, spherical, Pd/Fe nanoparticles aggregated together forming chain-like structure (Figure 3a). The diameters of individual particles were typically 5 to 80 nm. In the aged Pd/Fe sample, authigenic iron (hydr)oxides formed on the surface of the particles (Figure 3b).

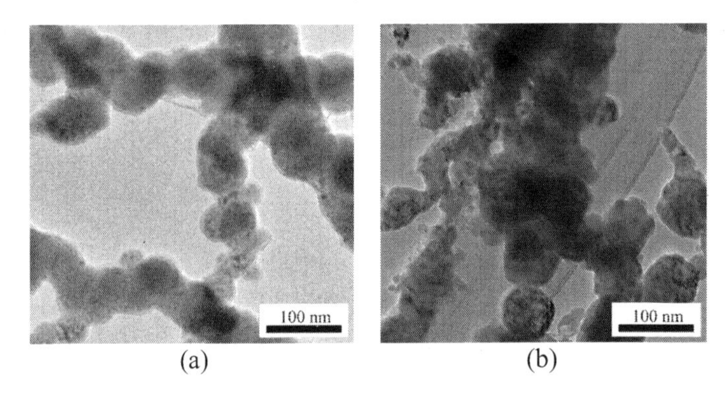

Figure 3. TEM images showing (a) the fresh Pd/Fe sample and (b) the aged Pd/Fe sample.

From X-ray photoelectron spectroscopy (XPS) analysis, the Fe 2p spectra of the fresh and reacted samples show a similar shape with binding energies of Fe $2p_{1/2}$ = 724.7 eV and Fe $2p_{3/2}$ = 710.9 eV (Figure 4a), corresponding to the oxidized iron. The two peaks at binding energies of 340.6 and 335.2 eV (Figure 4b) are associated with Pd^0 deposited on ZVI (*15*). The Ar^+ sputtering decreased the intensity of O and C peaks, while enhanced those for Fe and Pd. After Ar^+ sputtering, the peak at Fe $2p_{3/2}$ = 706.9 eV and Fe $2p_{1/2}$ = 719.9 eV emerged for the fresh and reacted sample. This confirms the core/shell structure of the Pd/Fe particles, and is consistent with the results found on ZVI particles described by the previous researchers (*16, 17*). The XPS analysis indicates that thickness of the iron oxides shell increased after reaction while the ZVI core shrank concomitantly. The process of oxide film formation on the ZVI and its evolution has been discussed by Huang and Zhang (*18*) and Noubactep (*19*). They suggested that a stratified ZVI corrosion coating could be formed after reaction in water, for which the outer and middle layers usually comprised both FeOOH and Fe_3O_4, while the inner layer mainly consisted of Fe_3O_4. Thus, the film is multi-layered with density increasing from the outer surface towards the core as the oxyhydroxides become more aged and less porous (*19*).

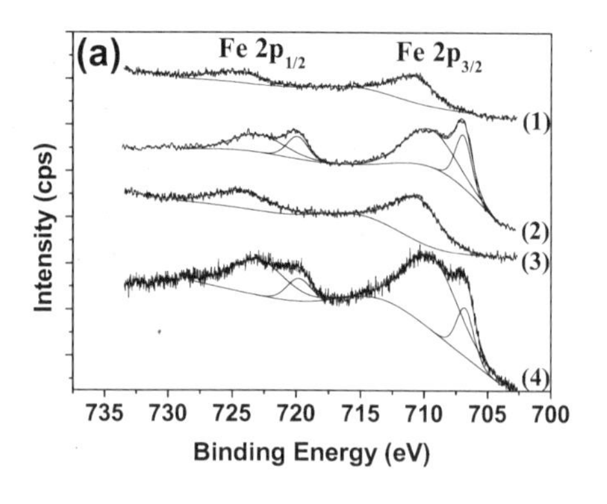

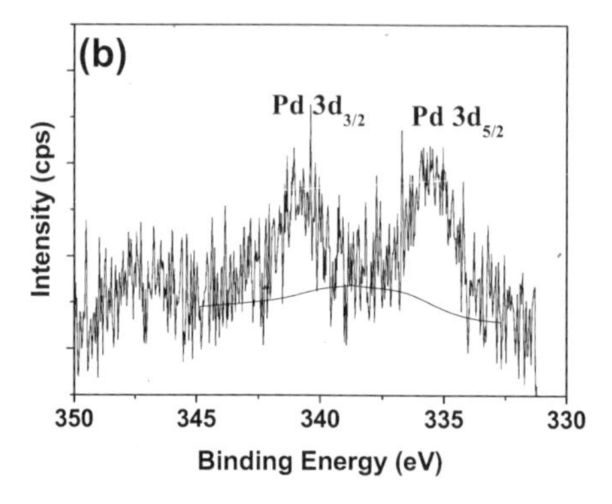

Figure 4. (a) Fe 2p, and (b) Pd 3d X-ray photoelectron spectra of the fresh Pd/Fe sample surface: (1) Fresh sample, (2) Fresh sample after 5 min of Ar^+ sputtering, (3) Aged sample, and (4) Aged sample after 5 min of Ar^+ sputtering (Reproduced with permission from reference 9. Copyright 2007 ACS).

Reactivity of Pd/Fe Nanoparticle

Reactivity

Monochlorobenzene (MCB), dichlorobenzenes (DCBs), and 1,2,4-trichlorobenzene (124TCB) dechlorination experiments in distilled water have been investigated with the freshly synthesized and aged Pd/Fe particles by the authors (*9,20*). The chlorinated benzenes could be completely reduced to benzene, following the pseudo-first-order kinetic model or Langmuir-Hinshelwood kinetic model. The reaction rates, when expressed with k_{SA} (specific reduction rate constant, $L \cdot min^{-1} \cdot m^{-2}$), followed the order TCB < DCBs < MCB (*9*). With the bare nZVI, there was no dechlorination process observed, and the dechlorination rate increase almost linearly with Pd content up to 0.5%

Pd (*9*). This finding suggests that in dechlorination of the chlorinated benzenes, Pd is the only reactive site in the Pd/Fe particles. Figure 5 shows the result of 12DCB dechlorination reaction by 0.1% Pd/Fe, together with their simulated degradation curves using the pseudo first-order kinetic model. The DCB was dechlorinated to half of its initial concentration within 5 min. The DCB might be dechlorinated to benzene following a stepwise (DCB→MCB→B) or concerted (DCB→B) pathway. The dechlorination rates among the isomeric DCBs followed the order 14DCB > 13DCB ≥ 12DCB (*9*), which is consistent with the finding on gas phase catalytic hydrodechlorination of DCBs over Ni/SiO_2 (*21*). The lower dechlorination rate constant of 12DCB compared with 14DCB was indicative of steric constraint effect.

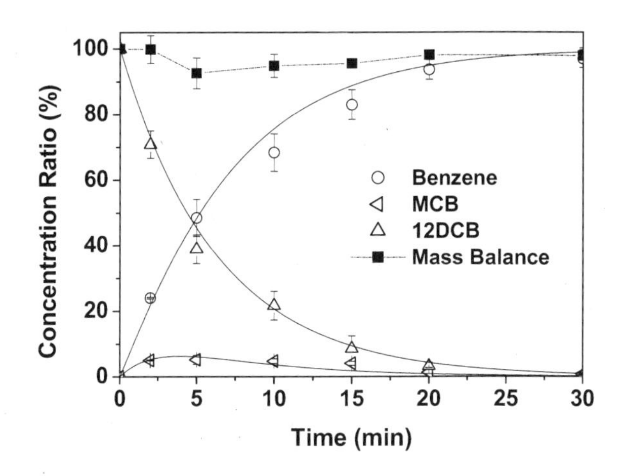

Figure 5. Dechlorination reaction time course for 12DCB with 0.1% Pd/Fe (Reproduced with permission from reference 9. Copyright 2007 ACS).

For dehalogenation of chlorinated and brominated methanes, adding Pd over ZVI could significantly enhance the reductive dehalogenation rates (*8,12*). Table I shows a comparison of reductive dehalogenation rates of carbon tetrachloride (CTC), chloroform (CF) and carbon tetrabromide (CTB), with palladized and bare nZVI. The reduction rate is higher for CTC than CF. With 0.2%Pd/Fe and bare nZVI, the rates were faster for CTB than CTC. It is worth noting that in this study, although the increase in Pd content could shorten the half life of the reduction reaction for the halogenated methanes, the intermediate byproduct distribution was not significantly affected (*12*). On the other hand, excessive Pd might not be beneficial to the dehalogenation reaction.

Table I. Values of reduction rate constants of halogenated methanes with nanoscale Pd/Fe and Fe particles

Particles	*Rate Parameters*	*CTC*	*CF*	*CTB*
	k_{obs} (h^{-1})	10.64	0.47	17.41
0.2% Pd/Fe	k_{SA} ($l\ h^{-1}\ m^{-2}$)	1.61E-1	7.11E-3	2.63E-1
	$t_{1/2}$ (h)	0.07	1.47	0.04
	k_{obs} (h^{-1})	56.04	0.89	>40
1% Pd/Fe	k_{SA} ($l\ h^{-1}\ m^{-2}$)	8.47E-1	1.35E-2	>6.05E-1
	$t_{1/2}$ (h)	0.01	0.78	<0.02
	k_{obs} (h^{-1})	3.96	3.39E-2	6.65
Fe	k_{SA} ($l\ h^{-1}\ m^{-2}$)	5.99E-2	5.13E-4	1.01E-1
	$t_{1/2}$ (h)	0.18	20.45	0.10

Matrix Effects

Since the reductive dehalogenation with Pd/Fe is a surface-mediated process, the reaction may be significantly influenced by the matrix species in the aqueous system because they may change the surface properties of the Pd/Fe particles or the physico-chemical properties of the interfacial region. The matrix species, either in ionic forms or as molecules can influence the dehalogenation reaction by interfering with process at either the reaction sites or the sorption sites on the iron surface (*14, 22, 23*). Successful implementation of BNIP treatment system in groundwater remediation therefore needs a thorough understanding of the effect of aqueous matrix species on the Pd/Fe particles and the reductive dehalogenation reaction.

Anions in groundwater may affect the haloorganics reduction with Pd/Fe by competing for the reactive sites. For example, several researchers (*24-26*) have found that nitrate can inhibit reductive reactions of contaminants with ZVI and itself be reduced to ammonia along with nitrite intermediate at low to neutral solution pH. Phosphate can easily form co-precipitation with the iron released from ZVI and the compounds deposited on the ZVI surface can subsequently inhibit removal of the target contaminants by the ZVI (*27-29*). Phosphate can be also strongly adsorbed to iron oxides (*30*). Silica has also been found to reduce ZVI activity in both batch experiment and column systems (*27, 31, 32*). Carbonate can accelerate corrosion of ZVI and enhance short-term reactivity at high concentration, but subsequent precipitation of $FeCO_3$ over time can result in ZVI surface passivation (*32*).

Table II shows the finding of a recent study by the authors that compared the effects of various anionic species commonly found in the surface water and groundwater on the reaction rates of 0.1%Pd/Fe and 124TCB. Also shown are the corresponding changes of solution pH, and the decreased Pd contents in various solution systems after 7 d of reaction. In the reaction system, the anions can be ranked in the order of control ≈ silica < perchlorate < carbonate < nitrate < phosphate < nitrite < sulfite < sulfide, based on the degrees of their influences on the Pd/Fe reactivity. The presence of nitrate or nitrite in the solution substantially suppressed the degradation reaction, due to competitive reductive reactions and also ZVI surface passivation by nitrate or nitrite. The decreased Pd/Fe reactivity towards 124TCB in phosphate and carbonate solutions could be due to the formation of inner-sphere complex on the particle surface. Sulfide and sulfite, which are usually present in groundwater as a result of sulfate reduction by sulfate reducing bacteria, are Pd poisons. They can completely halt the reaction of 124TCB and Pd/Fe. Based on the natures of their inhibitory effects on the dechlorination process, the anions can be classified as: (1) adsorption-precipitation passivating species (e.g., phosphate and carbonate), (2) redox-active species (e.g., nitrate, nitrite, and perchlorate), and (3) catalyst poisons (e.g., sulfide and sulfite) (*33*).

Besides ionic species, environmental amphiphiles such as natural organic matter (NOM) and surfactants have affinity for both HOCs and ZVI surface sites and thus may affect their interactions either favorably or adversely. Tratnyek et al. (*34*) postulated that amphiphiles may affect contaminant reductive transformation in the ZVI/water system through enhanced solubilization, enhanced sorption, competitive sorption, and electron transfer mediation.

NOM are the most omnipresent natural amphiphiles in aquatic environments (*35*). Therefore, NOM constitutes another important aqueous matrix species that can influence the contaminant dechlorination reaction by ZVI or BNIP. NOM can reduce the HOC dechlorination rate through competitive adsorption on reactive sites (*32*). Besides, NOM may also involve in redox reaction by acting as electron shuttles and accelerating the contaminant reduction rate (*34, 36, 37*). It has also been postulated that NOM can enhance dissolution of the iron oxides film, exposing more available reactive surface sites, and thereby accelerating the contaminant reduction (*37*).

Surfactants are synthetic amphiphiles that have been extensively used in household, agriculture, and many industrial applications. In groundwater and soil remediation, surfactants are introduced to the source zones to mobilize non-aqueous phase liquids. The influence of surfactant on dechlorination process depends on the surfactant type and concentration. Loraine (*38*) found that TX-100 (octylphenolpoly (ethyleneglycolether)$_x$) at concentrations below its critical micelle concentrations (CMC) enhanced PCE dechlorination rate but a reversed effect was observed at above its CMC. He also reported that below CMC, SDS (sodium dodecyl sulfate) showed a negligible effect on PCE and TCE reductive dechlorination. However, at a high SDS concentration far exceeding CMC, the dechlorination rate decreased due to partitioning of the PCE and TCE in the mobile micelles.

Table II. Pseudo-first-order reaction rates for 124TCB dechlorination under various conditions

		k_{obs} (min^{-1})	k_{SA} x 10^{-3} ($L\ m^{-2}\ min^{-1}$)[(a)]	pH Initial/Final	Pd content (w/w%)[(b)]
Ultrapure water		0.088	4.73	6.5/8.8	0.07
Nitrate	2 mM	0.029	1.55	4.9/9.8	-
	10 mM	0.009	0.48	5.0/10.1	0.057
Nitrite	2 mM	0.008	0.43	5.0/10.0	-
	10 mM	0.002	0.11	5.3/10.6	0.049
Perchlorate	0.2 mM	0.049	2.62	6.9/9.3	-
	2 mM	0.050	2.68	6.8/10.2	0.051
Phosphate	0.2 mM	0.021	1.12	6.1/10.6	-
	2 mM	0.015	0.80	5.9/11.1	0.077
Silica	0.2 mM	0.083	4.44	6.4/9.1	-
	2 mM	0.091	4.87	6.5/9.6	0.065
Carbonate	1 mM	0.047	2.52	8.1/10.5	-
	10 mM	0.046	3.00	8.9/11.1	0.043
Sulfite	2 mM	Negligible		8.7/9.0	-
	10 mM			9.2/9.7	0.045
Sulfide	2 mM			9.8/9.8	-
	10 mM			10.3/9.4	0.020

(a) The normalized surface reaction rate constant; the specific surface area of particles is 26.3 m^2/g and loading is 0.71 g/L.

(b) As determined after 7 d of reaction, with the initial content of 0.103%.

The influences of various amphiphiles, including NOM, cationic CTAB (cetyltrimethylammonium bromide) and DPC (dodecylpyridinium chloride), anionic SDS, and nonionic NPE (nonylphenol ethoxylate) and TX-100, on the dechlorination of 124TCB by the Pd/Fe nanoparticles have been comprehensively examined by the authors (*14*). The Langmuir-Hinshelwood model was used to elucidate the dechlorination kinetics, and it can provide indirect insight into the influence of the amphiphiles on 124TCB partitioning to the interfacial film and the resulting dechlorination rates. The Langmuir-Hinshelwood kinetic model is expressed as follow:

$$\frac{dC}{dt} = -\frac{K_A k_r S_t}{1 + K_A C} C \tag{3}$$

where K_A (μM^{-1}) is the Langmuir sorption coefficient of the substrate on reactive sites; k_r (min^{-1}) is the rate constant for the decay of the substrate at reactive sites; S_t (μM) is the abundance of reactive sites; and C (μM) is the substrate concentration in aqueous phase. Figure 6 shows the results of 124TCB dechlorination with the 0.1% Pd/Fe in the presence an amphiphile. Generally, the 124TCB concentration decreased exponentially with time, with concomitant appearances of intermediates (12DCB, 13DCB, 14DCB, and MCB) and end product (benzene) (*14*). Solid lines represent the simulated reaction kinetic curves with the Langmuir-Hinshelwood model, which appears to predict the reaction time course rater satisfactorily.

The findings from that study (*14*), which investigated surfactant concentrations over the ranges of below to above respective CMCs, showed that the pseudo first-order reaction rate constants increased by a factor of 1.5-2.5 with the presence of cationic CTAB. In the anionic SDS or nonionic NPE and TX-100 surfactant solutions, the 124TCB dechlorination rates were slightly increased over those observed in the ultrapure water at concentrations below their respective CMCs, but decreased drastically when their concentrations were above CMCs. It can be concluded that different amphiphiles manifest different effects, in both qualitative and quantitative manners, on chloroorganics dechorination by the nanoscale Pd/Fe, compared to their reported effects on the microsized ZVI. One main reason is that the dechlorination reactions associated with nZVI and nanoscale Pd/Fe involve different reactive sites, and different dechlorination mechanisms follows. Because the reaction between certain chloroorganics (e.g., chlorinated benzenes) and Pd/Fe occurred on Pd site only, steric congestion around the nanosized particles (due to the accumulation of macromolecules such as NOM) could be another reason for the dechlorination inhibition in the presence of high concentrations of amphiphiles. In addition, the NOM might be the competitive H_2 acceptors to 124TCB (*14*), and significantly retarded its catalytic dechlorination by the Pd/Fe particles. Thus, NOM appears to be a significant inhibitor to dechlorination of chlorinated benzenes by Pd/Fe. On the other hand, CTAB at concentration below CMC appeared to be the most benign to the 124TCB dechlorination.

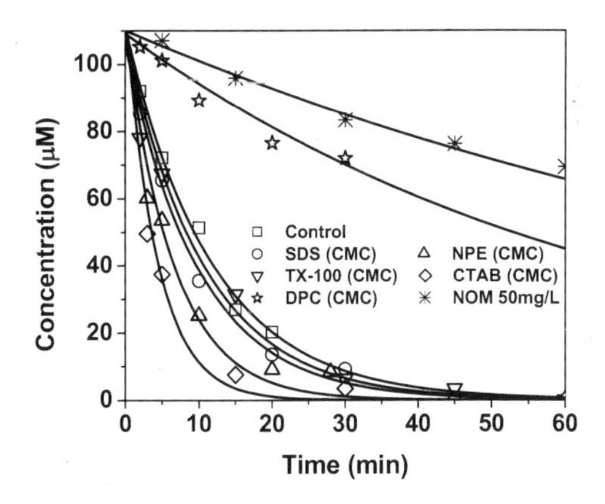

Figure 6. Dechlorination reaction time course for 124TCB with 0.1% Pd/Fe in water with a surfactant at CMC or 50 mg/L NOM.

Future Challenges

Deactivation due to Aging and Regeneration

ZVI corrodes in aqueous solution by forming corrosion products such as iron oxides or iron oxyhydroxides on its surface, thus affecting the metal-contaminant interaction. This phenomenon is amplified in the nZVI/water system since the nanoparticles have much larger specific surface area and are more reactive compared with the granular ZVI grains. In the BNIP/water system, the formation of iron oxide film may encapsulate the catalytic sites, resulting in deactivation of the catalyst. In the early study of Pd/Fe bimetallic particles, Muftikian et al. (*15*) observed deactivation of the Pd/Fe surface resulting from the formation of an iron oxide film. Similarly, Gui et al. (*39*) observed a decline in reactivity of a Ni/Fe bimetal and attributed it to encapsulation of the Ni islets by the iron oxide film. In summary, the decline of catalytic reactivity may be attributed to (as illustrated by Figure 7): (i) the loss of catalyst islet caused by the dissolution of underlying ZVI into the liquid solution, (ii) the formation of iron oxides or oxyhydroxides precipitation covering up the catalyst islets, and (iii) the extensive H_2 bubbles produced that form a mass transfer barrier (*11*). The formation of iron oxides layer and H_2 blanket will result in temporary catalyst passivation. If the used Pd/Fe is recoverable, e.g., if it is used in reactors, it can be reused again if its effective regeneration is possible.

At present, acid wash is a feasible method for the virgin ZVI to remove the passivating oxide layer or for the reacted ZVI to eliminate corrosion products, and to increase the ZVI surface area at the same time. The treatment of ZVI with dilute HCl has been reported to show a faster reduction rate compared to the untreated ZVI (*40*). The following explains the effects of HCl washing process that may be the reasons for the enhanced reaction rate of ZVI (*2, 41*): (i) dissolution of the oxide layer on the ZVI surface leaving cleaned ZVI surface free of nonreactive oxide or organic coatings; (ii) increasing surface area due to corrosion etching and pitting; (iii) increasing density of highly reactive sites resulting from the formations of steps, edges, and kinks on the ZVI surface; and (iv) acceleration of ZVI corrosion by the adsorbed H^+ and Cl^-. However, several drawbacks of this pretreatment method in practical applications have been also reported (*2, 41*): (i) production of strongly acidic waste stream with a high concentration of dissolved iron; (ii) loss of about 15% of the initial ZVI mass; and (iii) acceleration of corrosion by the adsorbed H^+ and Cl^-, and results in loss of activity.

Figure 8 shows that the Pd/Fe particles lost its 124TCB dechlorination reactivity significantly after ageing, i.e., $t_{1/2}$ increased from 5.3 min to 39 min. In the study, the aged Pd/Fe sample was also found losing its Pd content due to Pd dislodgment from the aged Pd/Fe particles, and the remaining Pd islets were encapsulated by the iron oxides film developed over the ageing period (*9*). The aged samples were then treated with HCl or $NaBH_4$. The results showed that the reactivity of the aged Pd/Fe could be only partially restored, with $t_{1/2}$ of 15 min, after HCl treatment. Treatment with $NaBH_4$ reduction method could not restore its activity, although the corroded iron was reinstated its zero-valent state. This indicated encapsulation of Pd islets by the regenerated ZVI shell which were originally the precipitated/deposited iron oxide on the aged Pd/Fe surface.

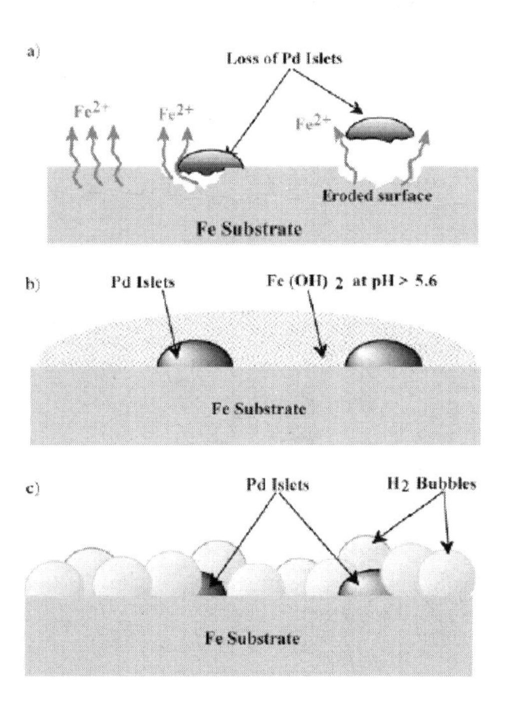

Figure 7. Illustrations of catalyst deactivation due to (a) loss of catalyst base because of dissolution of ZVI into the liquid solution; (b) ZVI hydroxide precipitation; and (c) extensive hydrogen gas formation (Adapted with permission from reference 42. Copyright 2005 ACS).

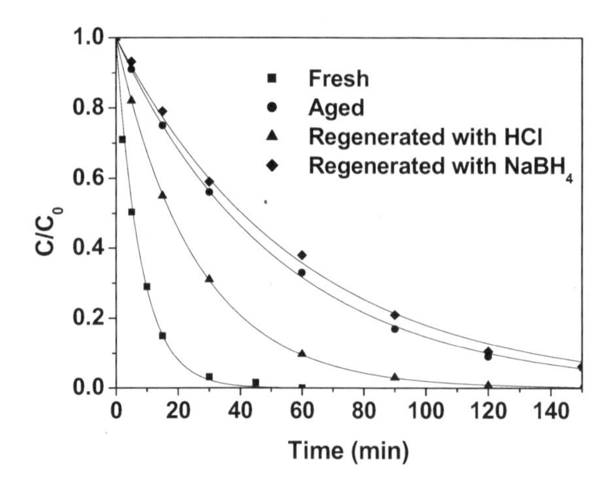

Figure 8. 124TCB dechlorination with fresh, aged and regenerated 0.1% Pd/Fe.

Field Remediation

With the recent advance in field applications of nZVI for remediation of contaminated groundwater (targeting the DNAPL source zone), similar application technique can be employed for injecting the BNIP in to the treatment zone. Table III lists several examples of sites which have been treated with BNIP. The cost of BNIP is much higher than that of the ZVI, due to the material cost of the noble metal and the addition synthesis steps involved.

Table III. Selected sites in the United States using bimetallic nanoscale iron particles for subsurface remediation*

Site Name & Location	*Nanoparticle Description*	*Remarks*
Naval Air Station, Jacksonville, FL	Nanoscale 0.1%Pd/Fe with polymer support	Groundwater contaminants: TCE, TCA, DCE, VC.
Naval Air Engineering Station, Lakehurst, NJ	Nanoscale Pd/Fe (PARS Environmental, Inc.)	Groundwater contaminants: PCE, TCE, TCA, cis-DCE, VC
BP – Prudhoe Bay Unit, North Slope, Alaska	BNIP (PARS Environmental Inc.)	TCA, Diesel fuel
Pharmaceutical Facility, Research Triangle park, NC.	BNIP (Lehigh University)	PCE, TCE, DCE, VC

* Main source of information: http://www.clu-in.org/download/remed/nano-site-list.pdf

Bare nZVI and similarly BNIP, once injected into groundwater may not travel far from the injection point. Because of the magnetism and high surface energy, BNIP nanoparticles tend to rapidly agglomerate in water into micro-scale or larger aggregates to achieve thermodynamic minimum, thereby losing their mobility and reactivity. In addition, the nanoparticles may also interact with the subsurface media and immobilized. Two techniques can be used to stabilize the nanoparticles, i.e., electrostatic stabilization and steric stabilization. Electrostatic stabilization is achieved by adsorption of ions to the metal surface which can form electrical double layer to repulse the individual particle. Steric stabilization can be carried out by surrounding the nanoscale iron particle with a layer of sterical bulk molecule which acts as protective shield to minimize particle-particle and particle-media interactions, and meanwhile possess high affinity for non-aqueous phase liquids. The main protective groups include polymers (or polyelectrolytes), copolymers, solvents, surfactants and oils (*43-47*). They act as molecular delivery carrier (or vehicle) to deliver the BNIP or nZVI to the targeted source zones, e.g., DNAPL.

While these surface modifications of the nZVI have been proven at field scale to gain certain degrees of success in source zone treatment as well as plume treatment, to the knowledge of the authors, such modifications have not been applied to BNIP. Though one may logically anticipate that the molecular delivery carriers tested on nZVI should show similar effects in terms of effective delivery to the source zone, their effects on the reactivity of BNIP with the target compound in the water system cannot be predicted based on the observations on nZVI. As revealed by the study carried by the authors (*14*), some molecular surface reactive agents can form physical barriers that inhibit HOC reduction by the BNIP. Also pointed in the preceding section, BNIP and nZVI exhibit different kinds of interactions with the contaminants in the presence of amphiphiles, and that may result in their contrary responses in a similar aqueous matrix system. More comprehensive and in-depth study targeted at BNIP are needed in the future to look into the potential effects of various molecular surface modifiers on BNIP, in terms of their effectiveness in delivering the BNIP to the contaminant source zone, ability to prolong longevity of the BNIP in groundwater, and interferences in the HOC reductive dehalogenation process. Potential of catalyst poisoning by sulfide or sulfite should not be overlooked.

References

1. Henderson, A. D.; Demond, A. H. *Envrion. Eng. Sci.* **2007**, *24*, 401-423.
2. Matheson, L. J.; Tratnyek, P. G. *Environ. Sci. Technol.* **1994**, *28*, 2045-2053.
3. Li, X. Q.; Elliott, D. W.; Zhang, W. X. *Crit. Rev. Solid State Mater. Sci.* **2006**, *31*, 111-122.

4. Scherer, M. M.; Balko, B. A.; Tratnyek, P. G. *Mineral-Water Interfacial Reaction: Kinetics and Mechanisms*; Sparks, D.L., Grundl, T.J., Eds; ACS Symp. Ser. 715, American Chemical Society: Washington, DC, **1998**, 301-322.
5. Zhang, W. X.; Wang, C. B.; Lien, H. L. *Catal. Today* **1998**, *40*, 387-395.
6. Schrick, B.; Blough, J. L.; Jones, A. D.; Mallouk, T. E. *Chem. Mater.* **2002**, 14, 5140-5147.
7. Lin, C. J.; Lo, S. L.; Liou, Y. H. *J. Hazard. Mater.* **2004**, *116*, 219-228.
8. Feng, J.; Lim, T.T. *Chemosphere* **2005**, *59*, 1267-1277.
9. Zhu, B. W.; Lim, T. T. *Environ. Sci. Technol.* **2007**, *41*, 7523-7529.
10. Wang, X.; Chen, C.; Chang, Y., Liu, H. *J. Hazard. Mater.* **2009**, *161*, 815-823.
11. Graham, L. J.; Jovanovic, G. *Chem. Eng. Sci.* **1999**, *54*, 3085-3093.
12. Feng, J.; Lim, T. T. *Chemosphere* **2007**, *66*, 1765-1774.
13. Wang, C. B.; Zhang, W. X. *Environ. Sci. Technol.* **1997**, *31*, 2154-2156.
14. Zhu, B. W.; Lim, T. T.; Feng, J. *Environ. Sci. Technol.* **2008**, *42*, 4513–4519.
15. Muftikian, R.; Nebesny, K.; Fernando, Q.; Korte, N. *Environ. Sci. Technol.* **1996**, *30*, 3593-3596.
16. Nurmi, J. T.; Tratnyek, P. G.; Sarathy, V.; Baer, D. R.; Amonette, J. E.; Pecher, K.; Wang, C.; Linehan, J. C.; Matson, D. W.; Leepenn, R.; Driessen, M. D. *Environ. Sci. Technol.* **2005**, *39*, 1221-1230.
17. Kanel, S. R.; Manning, B.; Charlet, L.; Choi, H. *Environ. Sci. Technol.* **2005**, *39*, 1291-1298.
18. Huang, Y. H.; Zhang, T. C. *Water Res.* **2006**, *40*, 3075-3082.
19. Noubactep, C. *Environ. Technol.* **2008**, *29*, 909-920.
20. Zhu, B. W.; Lim, T. T.; Feng, J. *Chemosphere* **2006**, *65*, 1137-1145
21. Keane, M. A.; Pina, G.; Tavoularis, G. *Appl. Catal. B* **2004**, *48*, 275-286.
22. Dries, J.; Bastiaens, L.; Springael, D.; Agathos, S. N.; Diels, L. *Environ. Sci. Technol.* **2004**, *38*, 2879-2884.
23. Nurmi, J. T.; Tratnyek, P. G. *Corros. Sci.* **2008**, *50*, 144-145.
24. Farrell, J.; Kason, M.; Melitas, N.; Li, T. *Environ. Sci. Technol.* **2000**, *34*, 514-521.
25. Alowitz, M. J.; Scherer, M. M. *Environ. Sci. Technol.* **2002**, *36*, 299-306.
26. Yang, G. C. C.; Lee, H. L. *Water Res.* **2005**, *39*, 884-894.
27. Su, C.; Puls, R. W. *Environ. Sci. Technol.* **2001**, *35*, 4562-4568.
28. Su, C.; Puls, R. W. *Environ. Sci. Technol.* **2003**, *37*, 2582-2587.
29. Su, C.; Puls, R. W. *Environ. Sci. Technol.* **2004**, *38*, 2715-2720.
30. Stumm, W.; Morgan, J. J. *Aquatic chemistry, chemical equilibria and rate in natural waters*, John Wiley & Sons, New York, **1996**.
31. Klausen, J.; Ranke, J.; Schwarzenbach, R. P. *Chemosphere* **2001**, *44*, 511-517.
32. Klausen, J.; Vikesland, P. J.; Kohn, T.; Ball, W. P.; Roberts, A. L. *Environ. Sci. Technol.* **2003**, *37*, 1208-1218.
33. Lim, T. T.; Zhu, B. W. *Chemosphere* **2008**, *73*, 1471–1477.
34. Tratnyek, P. G.; Scherer, M. M.; Deng, B.; Hu, S. *Water Res.* **2001**, *35*, 4435-4443.
35. Wandruszka, R. V. *Geochem. Trans.* **2000**, *1*, 10-15.

36. Lovley, D. R.; Coates, J. D.; Blunt-Harris, E. L.; Phillips, E. J. P.; Woodward, J. C. *Nature* **1996**, *382*, 445-448.
37. Xie, L.; Shang, C. *Environ. Sci. Technol.* **2005**, *39*, 1092-1100.
38. Loraine, G. A. *Water Res.* **2001**, *35*, 1453-1460.
39. Gui, L.; Gillham, R. W.; Odziemkowski, M. S. *Environ. Sci. Technol.* **2000**, *34*, 3489-3494.
40. Ponder, S. M.; Darab, J. G.; Bucher, J.; Caulder, D.; Craig, I.; Davis, L.; Edelstein, N.; Lukens, W.; Nitsche, H.; Mallouk, T. E. *Chem. Mater.* **2001**, *13*, 479-486.
41. Agrawal, A.; Tratnyek, P. G. *Environ. Sci. Technol.* **1996**, *30*, 153-160.
42. Jovanovic, G.; Znidarsic-Plazl, P.; Sakrittichai, P.; Al-Khaldi, K. *Ind. Eng. Chem. Res.* **2005**, *44*, 5099-5106.
43. Phenrat, T.; Saleh, N.; Sirk, K.; Tilton, R. D.; Lowry, G. V. *Environ. Sci. Technol.* **2007**, *41*, 284-290.
44. Phenrat, T.; Saleh, N.; Sirk, K.; Kim, H. J.; Tilton, R. D.; Lowry, G. V. *J. Nanopart. Res.* **2008**, *10*, 795-814.
45. He, F.; Zhao, D.; Liu, J.; Roberts, C. B. *Ind. Eng. Chem. Res.* **2007**, *46*, 29-34.
46. He, F.; Zhao, D. *Appl. Catal. B.* **2008**, *84*, 533-540.
47. Quinn, J.; Geiger, C.; Clausen, C.; Brooks, K.; Coon, C.; O'Hara, S.; Krug, T.; Major, D.; Yoon, W.S.; Gavaskar, A.; Holdsworth, T. *Environ. Sci. Technol.*, **2005**, *39*, 1309-1318.

Chapter 15

Use of Nanoscale Iron and Bimetallic Particles for Environmental Remediation: A Review of Field-scale Applications

Jacqueline Quinn[1], Daniel Elliott[2], Suzanne O'Hara[3], Alexa Billow[3]

[1]Mail Stop KT-D-3 (SLSL 308-2), Kennedy Space Center, Florida 32899
[2]Geosyntec Consultants, 3131 Princeton Pike Building 1B, Suite 205 Lawrenceville, New Jersey 08648;[3]130 Research Lane, Suite 2 Guelph, Ontario, Canada N1G5G3
[3]NASA Undergraduate Student Research Program, Harriet L. Wilkes Honors College Florida Atlantic University, 5353 Parkside Drive, Jupiter, Fl. 33458

A number of field-scale remediation applications using nanoscale reactive metals have taken place since the start of the new millennium. This paper reviews the chemistry behind the use of nanoscale metals for environmental site remediation and offers a synopsis of some of the field –scale applications that have been published to date. Site reviews include those applications of nanoscale iron conducted at Trenton, NJ, Cape Canaveral Air Force Station, FL, Naval Air Station, Jacksonville, FL., and at Research Triangle Park, NC.

Introduction

Over the past decade, significant advances have occurred with respect to the utilization of nanotechnology in environmental processes. While nanotechnology encompasses a wide variety of materials, devices, and techniques from many different disciplines that function, operate, or are otherwise focused on dimensions measured in nanometers (*1*). Nano-sized zero-valent metal particles, including monometallic and bimetallic systems, are of great interest to environmental remediation researchers and practitioners due to their reactivity with a wide variety of key contaminant classes including the ubiquitous chlorinated solvents. Since 1996, the emerging nZVI technology has been extensively studied in both bench and field-scale tests with some success (*2*, *3*, *4*, *5*).

Background

Nanoscale zero-valent iron particles (nZVI) hold promise for their ability to degrade chlorinated organic contaminants in groundwater, including chlorinated methanes and ethenes such as carbon tetrachloride (CT), tetrachloroethene (PCE) and trichloroethene (TCE) (*6*). While the majority of research in peer-reviewed publications has focused on structurally simple chlorinated aliphatic hydrocarbons (CAHs), nZVI has likewise shown promise in the treatment of alicyclic compounds such as the hexachlorohexanes (HCHs), which were commonly utilized pesticides (*7*) and the destruction of certain congeners of polychlorinated biphenyls (PCBs) (*8*) and chlorinated benzenes (*9*,*10*,*37*). nZVI has also been applied in the destruction of other types of contaminants including nitroaromatics (*11*), hexavalent chromium (*12*), perchlorate (*13*), and has been used to sequester heavy metal ions (*14*). In general, the contaminant classes most amenable to treatment by nZVI are those which are relatively oxidized. As an example, the CAHs are oxidized compounds by virtue of the one or more highly electronegative chlorine substituents on the carbon backbone.

The nZVI technology represents an extension of more conventional applications of zero valent iron (ZVI) in environmental remediation, specifically the widespread use of permeable reactor barriers to treat groundwater impacted by chloroethenes like PCE and TCE (15,16). PRBs entail the emplacement of granular ZVI filings, turnings, or other macroscale particles to form a permeable and reactive "iron wall" perpendicular to the flow of contaminated groundwater. Since the mid-1990s, the PRB technology have been demonstrated to be effective in treating plumes of dissolved chlorinated solvents (*17*, *18*) and have been installed at over a hundred sites around the world.

While both the nZVI and PRB technologies revolve around the reducing capabilities of zero valent iron, particle size represents a major and differentiating factor. The small particle size (e.g. < 100 nm) of nZVI means that there is a corresponding larger specific surface area (SSA) in comparison to microscale or granular irons. For example, the SSA for nZVI tested was 27 ± 3 m^2/g in one study (*10*) and 33.5 m^2/g in another (*8*), versus approximately 1 m^2/g or less for microscale iron powders (e.g. particle size ~10 micrometers) or granular (e.g. sizes measured in millimeters) material. Previous research has demonstrated the direct link between increasing SSA and reactivity towards the target contaminant. (2,36). Nanoparticles can be highly reactive due to the large SSA and the presence of a greater number of reactive sites, which is important because reactions between chlorinated contaminants and ZVI particles are surface-mediated (*11*); these two factors can improve nZVI's reactivity over iron powders by 10 to 1000 times (*19*).

Nanoscale iron particles can also be emplaced more easily in the subsurface than macroscale or microscale iron and have a greater likelihood of being mobile in the subsurface once introduced and therefore traveling under prevailing flow conditions (2). The nZVI particles can be suspended as an aqueous slurry, generally using groundwater from the targeted formation, and introduced to the subsurface by simple gravity-driven injection (*7*) into wells. In less permeable zones, the nanoparticle slurry may need to be emplaced using direct push Geoprobe pressure injection, or potentially pneumatic fracturing or other aggressive delivery techniques.

Modifications to the nanoscale particle surfaces can include the addition of catalysts and coatings such as polymers or surfactants (*20*). To increase the electron-donating capacity of iron and improve its reactivity for chemical reduction, nZVI particles may be coated with a small amount of a less active metal, commonly palladium (Pd) (*21*), but also silver (Ag) (*9*) and nickel (Ni) (*22*). These bimetallic nanoparticles typically exhibit more rapid reactivity towards the target contaminant(s) and, as is the case with chlorinated aromatic hydrocarbons, the noble metal catalyst (usually Pd, Pt, or Ag) is essential for the reduction of the parent compound. Whereas the chlorinated benzenes are generally poorly to modestly reactive with most microscale and nanoscale irons, recent literature indicates that nZVI/Pd can effectively degrade mono-, di-, and trichlorobenzenes (10, 37).

Other modifications of the nZVI technology include the addition of polyelectrolyte polymers to alter the surface architecture and help increase mobility in the subsurface by lessening the likelihood of aggregation of the particles. Aggregation, which can be attributed nZVI's strong magnetic properties and low surface charge, among other factors, adversely affects subsurface mobility potential due to the increased likelihood of interception by the soil matrix of the larger iron particle assemblages. The nZVI aggregates often form chain-like or dendritic structures, as shown in Figure 1, and nZVI reactivity by decreasing effective surface area and preventing access by contaminants to reactive sites. Lowry and his colleagues demonstrated that freshly prepared reactive nanoscale iron particles (RNIP) aggregated into the microscale (i.e. > 1 micron) within approximately 10 minutes (*23*). Moreover, in addition to helping to lessen the effects of aggregation, surface-modified iron can be used to help target areas of dense non-aqueous phase liquids (DNAPL).

One surface modification technique entails anchoring the particles on polymeric or zeolite (or other) support structure (23). In addition, surface coatings such as polymers or surfactants added to the nZVI slurry or co-injected with the iron are also being evaluated to determine their potential impacts regarding subsurface mobility. Biodegradable polymers and/or surfactants also represent potential additional electron donors to enhance secondary anaerobic degradation of the contaminants once the iron-mediated abiotic reactions are essentially completed.

In addition to aggregation-related challenges, the mobility of injected nZVI is also controlled by the lithologic and hydrogeologic properties. .nZVI adheres to soil particles or is filtered out (i.e. interception by natural geologic media "collectors"), such that a proportion of what is injected remains at or near the point of injection (*19*). The fraction of nZVI that is filtered out by surrounding geologic material at the point of injection varies depending on parameters such as the type of nZVI (e.g. whether it is surface modified or not), storage and handling of the particles prior to injection, use of surfactants, the method of injection, and subsurface conditions such as sediment particle size and groundwater geochemistry. For nZVI particle sizes on the order of 1 micrometer or smaller, Brownian motion dominates over gravitational settling and interception with respect to governing subsurface transport (*14*).

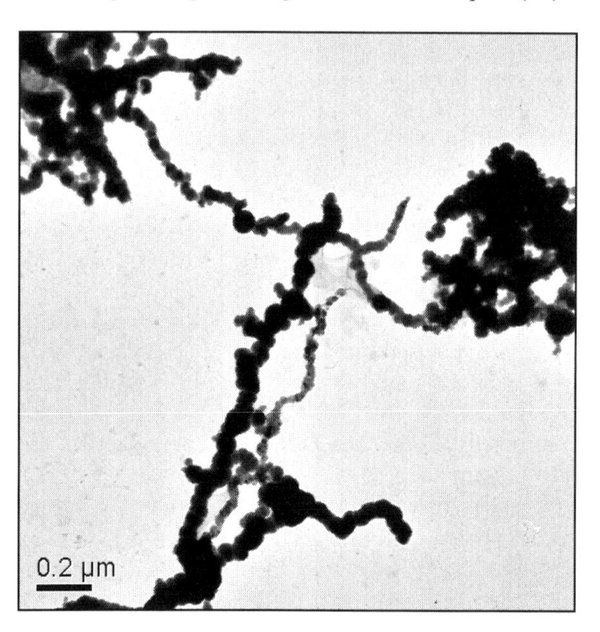

Figure 1. TEM Image of nZVI. Aggregation into chains is evident in this image

Another ZVI application in environmental remediation features the use of microscale or nanoscale iron in an emulsified form specifically for the treatment of chlorinated solvent DNAPLs. This material, referred to as emulsified zero-valent iron (EZVI), is a surfactant-stabilized, biodegradable emulsion that forms emulsion droplets consisting of an oil-liquid membrane surrounding ZVI particles in water. Chlorinated solvents DNAPLs provide long term sources of groundwater contamination as the DNAPL dissolves. Although nZVI particles can be injected into can be injected directly into DNAPL source areas, the ZVI particles require water for the degradation reactions to occur. Therefore, injecting the particles into a DNAPL source zone will still require the dissolution of the DNAPL into the surrounding water before degradation can occur. Figure 2 shows a schematic and a magnified image of an EZVI droplet. Since the exterior oil membrane of the emulsion particles has hydrophobic properties similar to that of DNAPL, the emulsion is miscible with the DNAPL. EZVI therefore increases the contact between the nZVI particles and the contaminant and is also capable of penetrating DNAPL pools and sequestering DNAPL prior to degrading it (*5*, *24*).

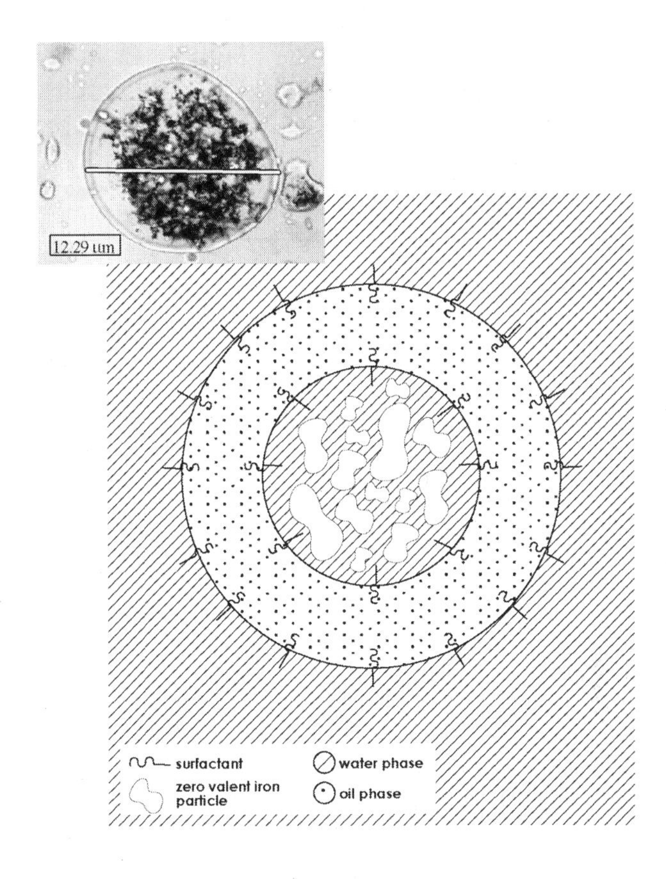

Figure2. Schematic and photograph of EZVI droplet showing the oil-liquid membrane surrounding particles of nZVI in water. After Quinn et al (5).

Chemistry of nZVI and Bimetallic Particles

Iron is an excellent electron donor in aqueous solution regardless of particle size (*25*, *15*, *2*, *7*). The degradation of chlorinated organic compounds such as PCE and TCE by ZVI, regardless of particle size, is believed to occur via both β-elimination and reductive dechlorination at the iron surface and require excess electrons produced from the corrosion of the ZVI in water (*26*). In β-elimination, the principal degradation-related intermediate for TCE is chloroacetylene whereas in reductive dechlorination, cis-DCE and VC are often observed. Using palladized iron synthesized using the borohydride method, Lien and Zhang (1998) observed principally ethene and ethane with no chlorinated intermediates as is shown in Figure 3. In this experiment, 20 mg/L aqueous TCE was treated by 5 g/L of nZVI with an average particle size of 100-200 nm (38, 39). As depicted in Figure 3, the data demonstrate that the half-life of TCE is on the order of 15 minutes and that ethane is the dominant end product with lesser amounts of ethene being observed. Although a chlorine balance was not performed, the experimental data allowed a reasonable carbon balance (normalized concentrations of at least 0.80) to be determined. Minor potential losses of ethene and ethane at 0.17 hours of elapsed can be rationalized in terms of the fact that both analytes are light hydrocarbon gases. The reduction of chlorinated organic compounds is coupled with the oxidation of Fe^0 (*11*):

$$RCl + M^0 + H^+ \rightarrow RH + M^{2+} + Cl^-$$

where "R" represents the molecular group to which the chlorine atom is attached (e.g., for TCE, "R" corresponds to C_2HCl_2-) and M is the metal (e.g., iron). Under aerobic conditions, water and dissolved oxygen may compete for electrons with the intended substrate (*11*). Ferrous iron (Fe^{2+}) is readily oxidized further to ferric iron (Fe^{3+}) (*7*). Hydrogen gas is evolved as a result of the competing reduction of water at the iron surface, which is reflected by the precipitous decline of solution oxidation-reduction potential (ORP). The reduction or "splitting" of water also produces hydroxide (OH^-) which causes an increases solution pH and contributes to lasting particle stability (*25, 2, 14*). Shown in Figure 4, the characteristic ORP decline and pH increase are useful field indicators of iron activity but they do not necessarily indicate that satisfactory contaminant reduction is also occurring.

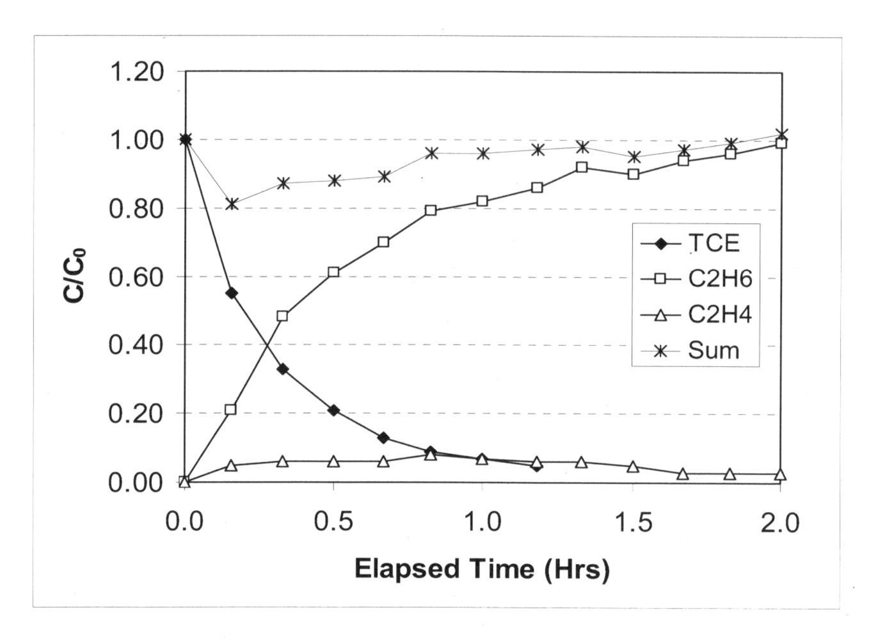

Figure 3. 5 g/L 100-200 nm nZVI with 20 mg/L aqueous TCE in batch system

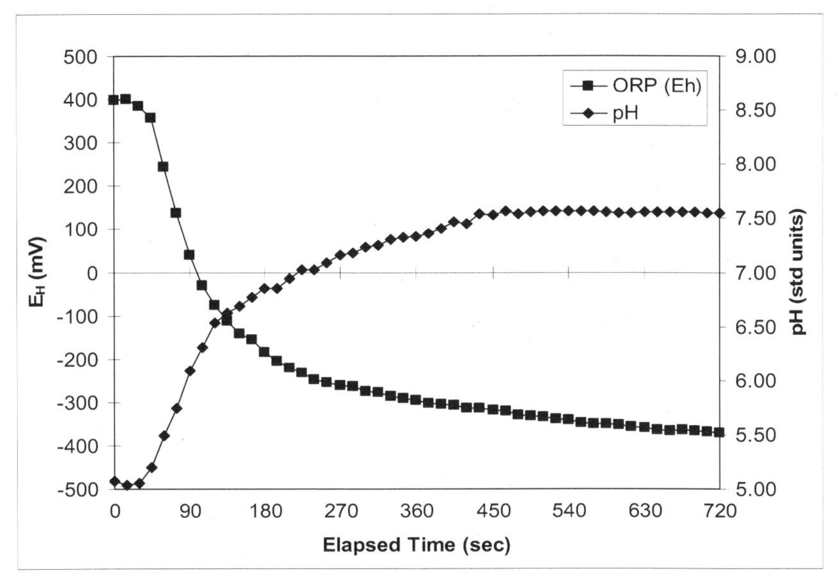

Figure 4. Batch reactions of 0.10 g/L nZVI in DI water at 25°C and 550 rpm

The surface-mediated mechanism of reductive dehalogenation is sufficient for dechlorination of chlorinated aliphatics by nZVI alone (*27*). However, in the case of aromatics, iron must be activated by another, less active metal for complete dechlorination (*8*). A bimetallic system forms a galvanic cell, improving the redox capacity of the iron; the noble metal is cathodically protected, whereas iron is oxidized preferentially as the anode (*2*). Addition of a noble metal also accelerates electron transfer, promotes the dechlorination reactions, and has significant positive effects on the reaction rates (*28*, *29*). Studies have indicated that the dehalogenation mechanism by iron alone is different from that in a bimetallic system. For example, Xu and Bhattacharyya (*22*) report that when TCE is dechlorinated by nZVI alone, significant concentrations of chlorinated byproducts such as *cis*-dichloroethene (cDCE) and vinyl chloride (VC) are detected. This differs from findings by Lien and Zhang in which 20 mg/L of aqueous TCE was degraded by nZVI yielding only ethene and ethene as observed intermediates (38). However, Fe/Ni nanoparticles degraded aqueous TCE with ethene and ethane being the only hydrocarbon products observed. This data appears to suggest that the presence of the nickel acts in a catalytic fashion yielding only the fully-dechlorinated, light hydrocarbon gases ethene and ethane (22). This mechanism may be highly preferable, as certain intermediates of TCE degradation (i.e. VC) are quite toxic themselves, and the process of degradation should proceed as near as possible to completion. However, field results indicate that even bimetallic degradation is not always complete, and multiple mechanisms including reductive dechlorination have been observed (*19*). Other metals including Pd (*8*), Ag (*9*), and Ni (22) have been shown to serve a similar catalytic function. In a system with Fe/Pd as the reductant(s), Pd provides the only possible reactive site for reduction of chlorobenzenes (*10, 37*).

Synthesis of Nanoparticles

Nanoparticles can be synthesized in the laboratory via a "bottom up" approach by various methods including the reduction of iron salts and oxides (*8*) and thermal decomposition of iron organometallics (*30*). Precipitation of Fe^{3+} with sodium borohydride (*8*), the so-called "borohydride method", was one of the first methods used to develop nZVI for environmental remediation applications.

$$Fe(H_2O)_6^{3+} + 3BH_4^- + 3H_2O \rightarrow Fe^0\downarrow + 3B(OH)_3 + 10.5\ H_2$$

Metal catalyst (Pd in the following example) may then be deposited by oxidation of Fe (*8*):

$$Pd^{2+} + Fe^0 \rightarrow Pd^0\downarrow + Fe^{2+}$$

Palladium and iron bimetallic particles are the most common and are commercially available.

In addition to the "bottom up" chemical synthetic procedures, a newer "top down" approach entails the use of a ball mill to grind microscale ZVI into the desired nanoscale size range (31). This "size reduction" process can result in significantly larger nZVI batch sizes and while the ball mill is energy intensive, the overall risk management profile of the manufacturing process is more favorable than many of the chemical syntheses, especially the borohydride method which requires management of significant evolution of flammable and explosive hydrogen gas. nZVI can then be transformed into bimetallic particles as described or surface-modified to achieve specific field performance characteristics (i.e. addition of biodegradable tri-block polymeric surface architectures to impart greater resistance to aggregation, DNAPL affinity characteristics, etc.)

Since nanoscale iron and bimetallic particles have only been commercially available for less than a decade, there often exist significant differences with respect to size, reactivity, mobility and longevity depending on the method of manufacture and by the vendor providing the particle (*32*). The intrinsic reactivity of iron-based nanoparticle products indicates that material performance properties may change significantly over time. Therefore, it is important that remediation practitioners obtain key process and quality assurance/quality control (QA/QC) data from the manufacturer prior to field utilization. Key process-related data include the batch and/or lot numbers of the nZVI products, the date of manufacture, and details associated with any stabilization techniques used (e.g. presence of any polymers or other surface modifying agents or storage under water, ethanol, glycols, or other liquids). Key QA/QC parameters of interest include a particle size distribution (PSD), determination of specific surface area (SSA), measurement of zeta potential or isoelectric point (defined as the pH at which the surface potential is zero), and aqueous batch pH/Eh profile (see Figure 4). Common field practice for full-scale sites includes a fixed-duration (e.g. 4 to 6 hours) reactivity batch test of the nZVI material using a standard reductate like TCE. Hold times for nZVI products vary based upon the manufacture and repeated reactivity tests may be warranted if the materials experience extended hold times prior to subsurface injection.

Application of Nanoparticles in the Field

As a result of their high reactivity, a passivating layer quickly forms on the nZVI particle surface – such that a core-shell structure is formed with the core being zero valent iron and the shell being composed of mixed valent iron oxides. The composition of this oxide shell varies depending on the water chemistry but it can effectively limit the ZVI corrosion rate (*33*). Dilute acid-washing may restore reactivity of Fe0 by removing the oxide layer (*10*), though this may be troublesome to apply in the field. The oxide shell is capable of shuttling electrons from the ZVI core so the particles remain reactive essentially until the ZVI core is consumed. In the case of bimetallic particles because the ZVI particle forms the support for any noble metal that is coupled with it, deterioration of the particle means loss of catalyst and a corresponding drop in reactivity, since the catalyst provides reactive sites (9,*10,37*). However, it is important to note that these highly reactive nanoscale particles change over time, with handling, during storage (in a slurry of water) and with exposure to natural environments where constituents in groundwater will decrease the reactivity of the particle surface.

Iron powders and monometallic nZVI show little reactivity to chlorinated aromatics (*8, 9*). In the presence of a noble metal catalyst, significant degradation of chlorobenzenes (*9, 10, 37*) and PCBs (*8*, *22*, *31*) can be achieved. Nanoscale Fe/Pd has been shown to completely degrade Aroclor 1254, a commercial mixture of certain PCBs (*8*). Other results suggest that degradation is more selective. Fe/Ag has been applied in the degradation of hexachlorobenzene, with very little final product (benzene) detected in the end; instead there was a pronounced accumulation of 1,2,4,5-tetrachlorobenzene and 1,2,4-trichlorobenzene. Completeness of degradation was correlated with Ag concentration, suggesting that Ag was necessary for providing reactive sites (*9*). In contrast to the nZVI-mediated degradation of chlorinated aliphatic compounds, the reaction rate of Fe/Pd with chlorobenzenes decreases as degree of chlorine substitution increases. For tri- and greater substituted benzenes, 1,2-substituted intermediates tended to accumulate and it seemed unlikely that steric effects hindered substitution. The rate was also observed to increase with Pd concentration (*10,37*).

Rather than degrade certain contaminants, nZVI may also be used to sequester them. nZVI has a potential for metal sorption that exceeds that of conventional carbon, zeolite, and others (*14*). Mechanisms for metal reaction and removal by the complex FeOOH/Fe0 structure of oxidized nZVI particles are being explored. Excellent removal efficiency for metals such as Cu, Ag, Pb, Cr, Zn, Ni, and Cd (*14*) and Tc (*34)* has been reported.

The mechanisms for degradation of chlorinated solvents using EZVI made with nZVI are the same as that of nZVI. However, since the nZVI is contained within an oil/water emulsion droplet it is miscible with the DNAPL. It is believed that as the oil emulsion droplets combine with pure phase DNAPL (TCE for example), the TCE dissolves and diffuses into the aqueous droplet containing ZVI at the center of the oil emulsion droplet. While the nZVI particles in the aqueous emulsion droplet remain reactive, the concentration gradient across the oil membrane is maintained by the fact that the chlorinated compounds are continually degraded within the aqueous emulsion droplets, which establishes a driving force for additional TCE migration into the aqueous emulsion droplet where additional degradation can occur (*24*).

Toxicity and safety concerns

Significant research is being conducted by academia and industry into the potential toxicity of nanoscale particles. There is concern that, due to the increased surface area and reactivity, nanoparticles may have greater biological activity than micro- or macroscale particles of the same composition. Due to their very small size they may be able to reach and accumulate in areas where larger particles may not. Nanoparticles may have negative impacts to human health if inhaled, absorbed through the skin or ingested that their micro- or macro-scale counterparts do not (*35*). nZVI is being used in site remediation even though little toxicological research has been done on nanoparticles. Research has shown that nZVI applied in the environment will not move very far from the point of injection due to aggregate formation and filtering by soil particles. However, while still largely focused in the laboratory, efforts are being made to coat particles with various substances including surfactants or polymers to increase their mobility in the subsurface (23). Increased mobility could result in unintentional impacts to areas outside of the treatment zone such as release of nanoparticles to surface water bodies or drinking water supplies.

Field demonstrations summary

nZVI technology has been sparsely demonstrated for *in situ* site remediation, however the handful of publications documenting the use of nZVI in field-scale applications shows great promise for the remediation of chlorinated organic contaminants in groundwater. Following is a discussion of applications of various nZVI technologies at several contaminated sites.

Trenton, NJ (*2*):

One of the first proof-of-concept studies on the use of NZVI was performed in the summer of 2000 using bimetallic nanoscale particles at an active manufacturing facility in Trenton, NJ. The demonstration area was approximately 4.5m by 3.0m in footprint, and extended to a depth of approximately 6.0m. Within the test area, existing well DGC-15 (screened over the depth interval of 3.0 to 4.5 meters bgs) was used for nZVI slurry injection and a series of nested piezometer couplets (denoted PZ-1S and 1D, 2S and 2D, and 3S and 3D) were situated at intervals of 5 feet downgradient of the injection well. The S-series piezometers were screened over the shallower interval of 2.4

to 3.6 meters below ground surface (bgs) while the D-series piezometers were screened over 4.7 to 5.9 meters bgs. The 100-200 nm bimetallic nanoscale particles used for the demonstration were synthesized using 0.50 M $NaBH_4$ and 0.09 M $FeCl_3$ solutions and were precipitated using previously published procedures (*8*). The palladium was deposited onto the nanoscale iron particles by reductive deposition of $(Pd(C_2H_3O_2)_2)_3$. Palladium was present in a ratio controlled at 1:300 w/w. The resulting bimetal was gravity fed into a screened well where the hydraulic conductivity of the aquifer was estimated to be 0.2 cm/sec.

Contaminants at the site were well-characterized, with supporting data collected for over a decade. Chlorinated hydrocarbons dominate the species of concern and include tetrachlorethene (PCE), TCE, cis-dichloroethene (c-DCE), vinyl chloride (VC), chloroform (CF), carbon tetrachloride (CT), and 1,1-dichloroethene. Contaminant concentrations range from the detection limit of 1 μg/L to 4600 μg/L. TCE was the predominant constituent of concern, with persistent recorded values of 445 μg/L to 800 μg/L. Field measurements for ORP and pH were measured prior to, during and after initiating the field test.

Injection of the bimetal occurred in two phases. During phase I, approximately 490 L of a 1.6 g/L nanoparticle suspension was gravity fed over a four hour period into an established recirculation loop. In phase II, two injections of iron nanoparticles were accomplished under gravity feed conditions. In the first injection, 890 L of 1.5 g/L bimetallic nanoparticles were gravity fed into the injection well over a period of 6.3 hours. The second injection of phase II occurred one day later in which 450 L of 0.75 g/L nanoiron slurry was gravity fed into the injection well. Six weeks after the injection, a pump test was performed in the injection area with no significant impact to hydraulic conductivity noted.

For this demonstration, both dissolved iron (Fe^{2+}) and total iron (Fe_T) were monitored in lieu of efforts to directly detect nanoparticles. Background levels of total and dissolved iron concentrations were 1.0 mg/L or less. Total iron levels rose to 135 mg/L in the injection well prior to declining to 32.2 mg/L. At down-gradient piezometers (4.5 m from the injection point), Fe_T peak concentrations realized were in the 10-20 mg/L range. Using Fe_T as an indicator, the nanoparticle plume traveled at velocities of up to 0.8 m/d, which exceeded the site's calculated seepage velocity of 0.3 m/d.

ORP and pH were also evaluated at the site. Historic ORP values for the site groundwater ranged from +150 to +250 mV. After iron injection, the ORP values rapidly declined to indicate strongly reducing conditions within four days at the injection well head (-360 mV). ORP values at the down-gradient monitoring points showed only slightly reducing conditions. (-20 to -75 mV). Groundwater pH increased from between 4.6 and 5.2 to between 5.1 and 7.7 due to H_2 evolution and depletion of DO.

Overall, the project measured successful TCE removal efficiencies. Results varied considerably by time and location, with TCE concentrations declining by 1.5% to 96.5%. As shown in Figure 5, the results from the phase I injection indicated that the extent of TCE concentration declines were transient and tended to be most significant at monitoring locations closest to the point of injection, DGC-15. PZ-1S/1D, PZ-2S/2D, and PZ-3S/3D are located 5, 10, and

15 feet downgradient, along the generally westerly direction of flow from injection well DGC-15. It also merits mention that these reductions were attained despite the very small iron dose and the fact that the injections were accomplished within a significant chloroethenes plume. Interestingly, as evident in Figure 5, the magnitude of the observed decline in % TCE concentration rose during the course of phase I monitoring in two of the three "deep" piezometer locations, PZ-1D and PZ-2D. This suggests a possible downward migration of the injected nZVI mass through the test area and concomitant reduction in TCE concentration. Over a period of several months, TCE concentrations gradually recovered to near pre-injection conditions.

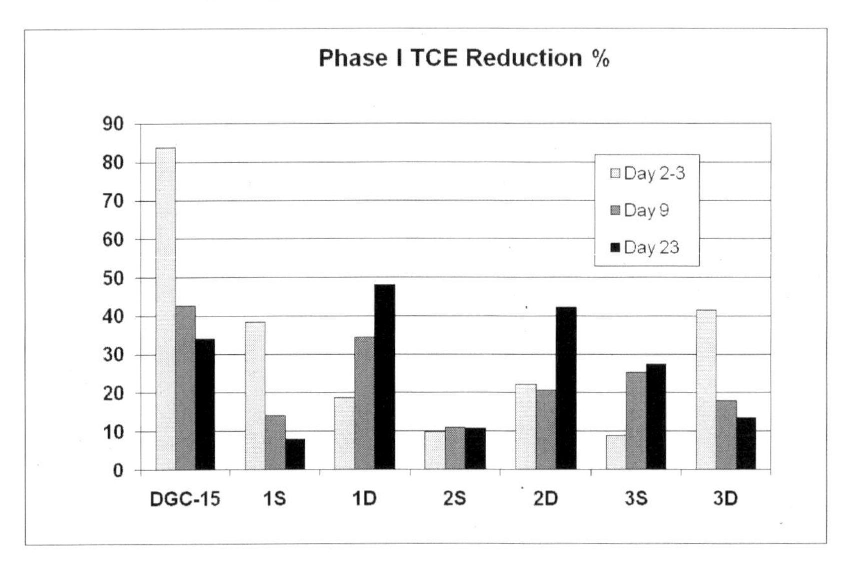

Figure 5. Results from Phase I injection of nZVI injection at the Trenton site. In addition to injection well DGC-15, the results from piezometer couplets PZ-1S and 1D, PZ-2S and 2D, and PZ-3S and 3D are shown. The piezometer couplets were 5, 10, and 15 feet, respectively, downgradient from DGC-15.

Naval Air Station, Jacksonville, FL (*19*):

In January 2004, a proprietary formula of nanoscale iron/palladium-doped particles mixed with polymers was injected into Hangar 1000's source area on the Naval Air Station in Jacksonville, FL. The contaminants of concern at the site were primarily VOCs and include PCE, TCE, 1,1-dichloroethene, 1,1,1-trichloroethane (TCA) and cis-1,2-dichloroethene. Background concentrations for TCA, PCE and TCE were as high as 25,300 μg/kg, 4,360 μg/kg and 60,100 μg/kg respectively.

Previous remediation efforts by hydrogen peroxide at this site had shown little success due to the hypothesized presence of DNAPL. In order to minimize the aggregation of Fe/Pd particles and their adhesion to soil particles, NZVI/Pd supported by a biodegradable food-grade (i.e. edible) polymer formulation was selected for the full-scale remedial effort. Groundwater was withdrawn from the subsurface, mixed with the NZVI, and reintroduced into injection wells using only a gravity feed-system. Loading rates for the 150 kgs of iron/palladium suspension ranged from 2.0 g/L to 4.5 g/L. No clogging of the injection wells or recirculation wells was noted during the deployment. A total of two pore volumes were recirculated during the first injection period. A second injection took place approximately two weeks after the first. The loading rate for the entire second injection was set at 4.5 g/L of iron/palladium suspension. After approximately 24 hours of recirculation and injection, approximately 2.5 pore volumes were recirculated within the target zone.

Standard field parameters were monitored as indicators of system performance. These included ORP, pH, DO and contaminant concentration levels. Immediately after injection, ORP values decreased rapidly from background levels ranging from 100 to -100 mV to strong reducing conditions in the -200 to -550 mV range. Dissolved oxygen levels also declined from 1.0 to 0.2 mg/L. Also noted was visual staining of the groundwater collected from the wells. The groundwater was either a cloudy gray color, or dark black stains were noticed when the wells were sampled.

Reductive dechlorination followed the expected sequence of biologically-mediated degradation products including cis-DCE, VC, ethene, and ethane. Significant variability in the data collected over the long-term monitoring period exists. The largest decrease in contaminants of concern was noted during the first quarter following injection. All subsequent monitoring efforts did not repeat the first quarter's performance level. However, significant decreases in the parent compounds TCE and TCA were observed. In general, the demonstration showed good groundwater contaminant removal efficiencies, with reductions of individual compounds ranging from 65 to 99 percent. Furthermore, the production of innocuous fully dechlorinated end products such as ethane, ethene acetylene was noted.

Slug tests and soil cores were collected prior to iron/palladium injection and approximately one year after the second injection/recirculation effort. Ten pre-demonstration soil cores and ten post-demonstration soil cores were collected and analyzed for contaminant concentration. Between 8 and 92 percent cumulative contaminant reductions were observed in the soil data. Slug test data revealed a 45 percent reduction in soil permeability as a result of the NZVI addition.

Although some rebounding in groundwater contaminant concentrations occurred, the Florida Department of Environmental Protection (FDEP) and the US EPA concluded that the contamination was sufficiently reduced and the site would transition to monitored natural attenuation with institutional controls as the long-term site remedy.

Research Triangle Park, NC (*21*)

In September 2002, nanoscale iron was injected into shallow bedrock approximately 40m from a former waste disposal area. Previously, the disposal area was excavated; however residual impacts remained a persistent and significant obstacle for the site owner. Radial groundwater flow predominated the disposal area, with the subsurface lithology yielding a hydraulic conductivity of approximately 10^{-3} cm/sec. Prior to the injection sequence and for a period after the injection, a number of field parameters were monitored using multi-parameter, in-well data loggers. Monitored parameters included groundwater flow rate, water levels, ORP, DO, pH, specific conductance and temperature. Additionally, both pre and post-injection concentration levels were monitored for PCE, TCE and lesser chlorinated degradation products including TCE, c-DCE, and VC.

The iron particles were injected as a slurry suspension using potable water. The average injection rate of 0.6 gpm introduced 11.2 kg of nanoscale iron into the target aquifer over a two day period. The results of this field deployment agree with those of the aforementioned site deployments using nanoscale iron slurries. The site noted over 90% reduction of total VOC concentration, dropping from 14,000 μg/l total VOCs within only a few days. PCE, TCE and DCE concentrations were lowered to near groundwater standards within a six week period, with no increases to the background concentration of vinyl chloride. ORP values also dropped considerably from iron-reducing conditions (+50 to -100 mV) to methanogenic conditions (-700 mV) during and after injection.

A radius of influence for the gravity feed injection system was projected from contaminant reduction profiles. The area of impact was estimated to be in the six to ten meter range, radial from the injection well. The research speculate that impact distances of up to 20 meters could be achieved with engineering aids such as pressurized injection or cross-gradient pumping.

Launch Complex 34, Cape Canaveral Air Force Station, FL (*5*):

The first field demonstration for EZVI was conducted at NASA's Launch Complex 34 on Cape Canaveral Air Force Station in Florida. Pressure pulsing technology (PPT) was used to distribute EZVI, made with Toda nZVI, beneath the site's Engineering Support Building within a 9 X 15 ft plot at a depth of 16-26 ft below the foundation. DNAPL is present in the subsurface at the Site as a result of historical releases from Site operations. The DNAPL consists primarily of TCE, although cis-1,2-dichloroethene (cDCE) and vinyl chloride (VC) also are present in groundwater as a result of intrinsic TCE biodegradation processes. The remediation pilot test focused on the upper aquifer unit which is composed of medium to coarse-grained sand and crushed shells and extends from ground surface to approximately 18-25 ft bgs.

Six continuous soil core samples were collected prior to the injection of EZVI. Figure 6 illustrates the location of the soil cores collected and their relationship to other monitoring and injection points in the demonstration test area. A groundwater control system was used to create a closed-loop recirculation cell and forced gradient conditions across the target treatment zone, which would allow for a comparison of the flux to groundwater from the DNAPL source zone before and after treatment with the EZVI. A series of four multilevel monitoring wells with five separate sample intervals and one fully screened well were installed in the pilot test area (PTA) to provide groundwater chemistry data to evaluate the changes in concentrations and mass flux before and after EZVI injection.

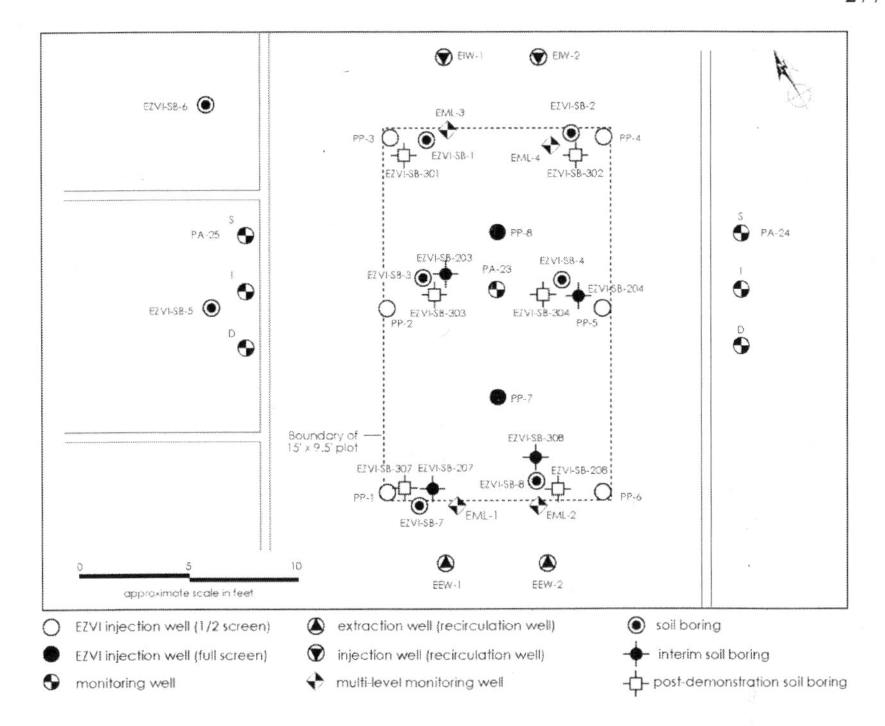

Figure 6. Sampling and monitoring locations within the demonstration test area.

The EZVI was injected into eight wells in the demonstration test area at two injection intervals per well (16 to 20.5 ft bgs and 20.5 to 24 ft bgs) using an injection method called pressure pulse technology (PPT). PPT applies large-amplitude pulses of pressure to porous media increasing fluid flow and minimizing the "fingering" effect that occurs when a fluid is injected into a saturated media. Approximately 670 gallons of EZVI were injected into the PTA over a 10-foot depth interval. EZVI was injected in August of 2002 over a four day period.

A set of ten post-demonstration soil cores were obtained and soil samples collected and analyzed following the same procedures used for the pre and interim demo soil samples. Figure 6 shows the locations of all soil cores and the injection and monitoring wells. Figure 7 shows a droplet of EZVI distributed within the sand grains of a core. Four months after the injection of EZVI post-injection groundwater samples were collected to evaluate changes in concentrations and mass flux. In addition, groundwater samples were collected 19 months after injection of EZVI to evaluate the "Long Term" impacts of the treatment on VOC concentrations in groundwater.

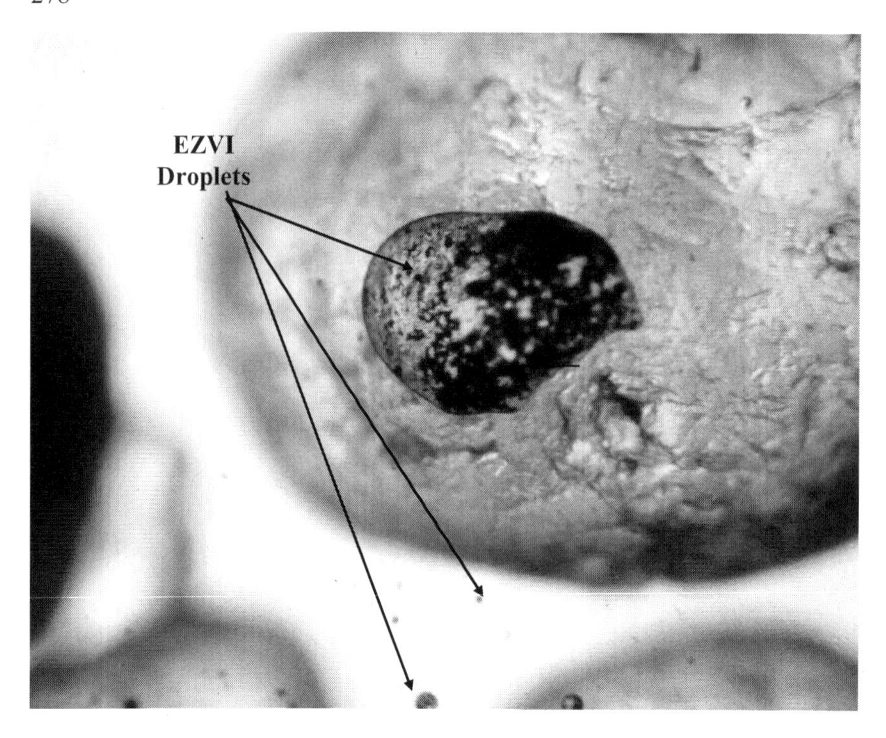

Figure 7: Soil Core Sample Post-EZVI Injection Showing EZVI on an individual sand grain

The EZVI had a tendency to migrate up from the injection depth interval to more shallow intervals when injected using PPT. It is believed that this upward migration of the EZVI resulted in less effective degradation of TCE in the target treatment intervals in some of the locations within the PTA. However, significant reductions in TCE concentrations (> 80%) were observed at all soil boring locations with the exception of SB-3 and SB-8 where visual observations suggest that most, if not all, of the EZVI migrated up above the target treatment depth. Significant reductions in TCE concentrations (i.e., 57% to 100%) were observed at all depths targeted with EZVI (16 to 24 feet bgs). The average reduction in concentration for the downgradient transect (E-ML-1 and E-ML-2) is 68% and the mass flux measured for this multilevel well transect decreased by approximately 56% from 19.2 mmoles/day/ft2 to 8.5 mmoles/day/ft2 over a period of 6 months. Groundwater samples from these wells also showed significant increases in the concentrations of cDCE, VC, and ethene.

Nineteen months after injection, a series of additional groundwater samples were collected and the results show that significant additional reductions in TCE concentrations occurred after the initial set of post-injection groundwater samples were collected. The molar concentrations of TCE, cDCE, VC and ethene in selected monitoring wells in the downgradient transect and PA-23 is presented in Figure 8. The additional decreases in TCE observed in the long-term groundwater samples suggest that a portion of the removal of TCE may be due biodegradation enhanced by the presence of the oil and surfactant in the EZVI emulsion.

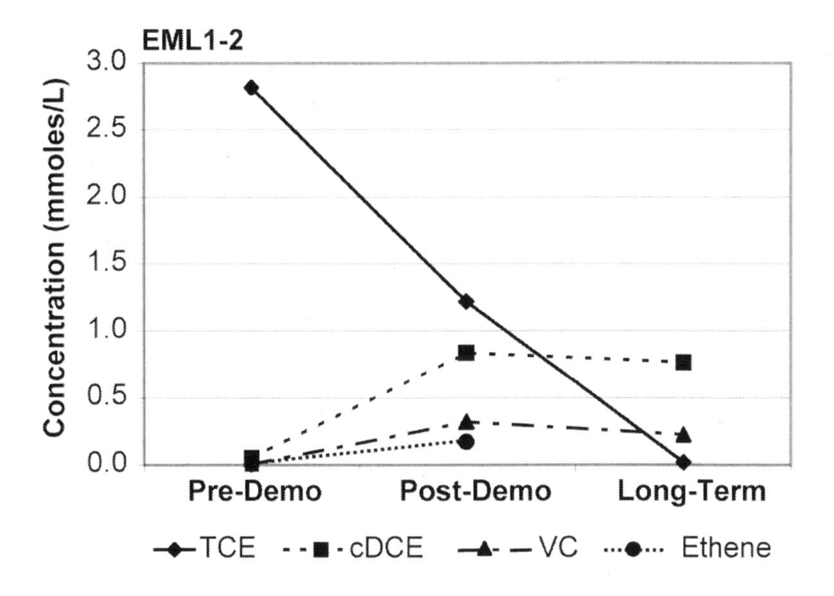

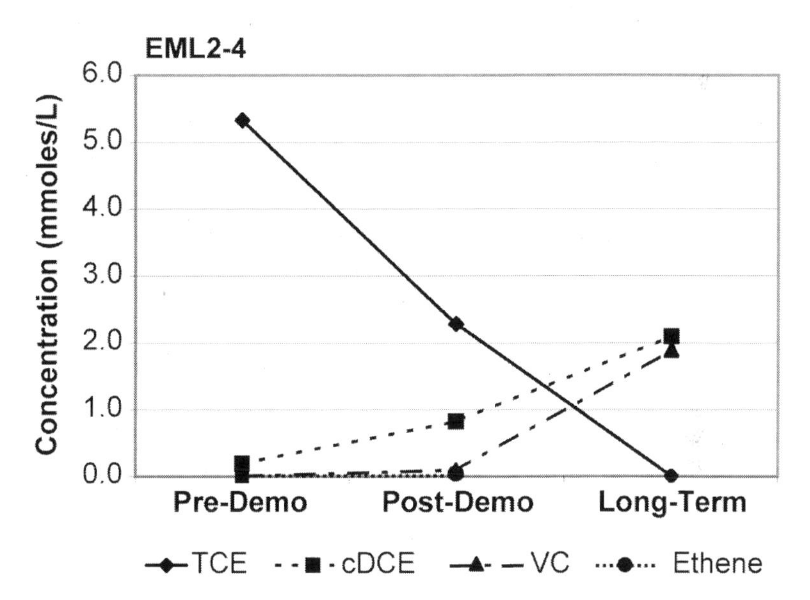

Figure 8 TCE, cDCE, VC and Ethene concentrations in selected monitoring wells over time.

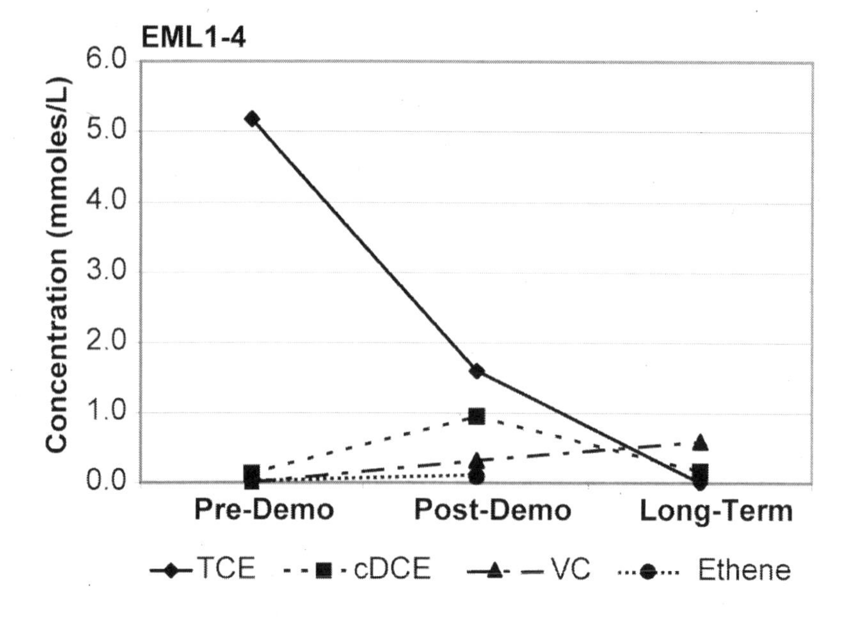

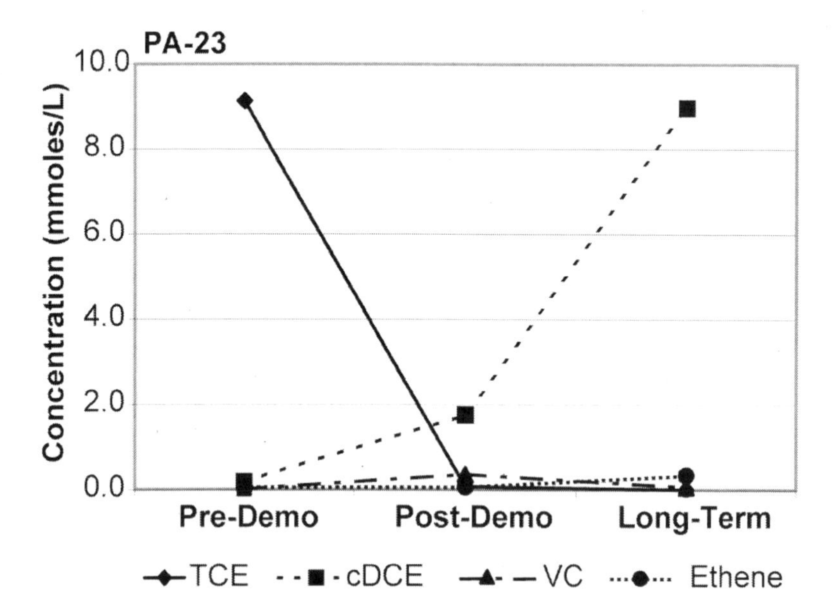

Figure 8 (con't) TCE, cDCE, VC and Ethene concentrations in selected monitoring wells over time.

Summary

Nanoscale iron holds promise for its ability to degrade a wide variety of key contaminant classes including chlorinated organic contaminants in groundwater such as CT, PCE and TCE. The nZVI technology is an extension of more conventional applications of ZVI in environmental remediation, such as reactive barriers to treat groundwater impacted by chloroethenes. The small particle size (e.g. < 100 nm) of nZVI indicates there is a corresponding larger specific surface area in comparison to microscale or granular irons. Since degradation is a surface-mediated degradation reaction, the rates of reaction are significantly enhanced. nZVI can be applied in a variety of ways in the field for environmental remediation. The nZVI particles can be suspended as an aqueous slurry and injected into the subsurface, it can be surface-modified through the addition of substances like polymers for improved mobility or emulsified with vegetable oil to help target contaminants such as non- aqueous phase liquids. The use of nZVI particles for *in situ* site remediation has been sparsely demonstrated to date, however the publications documenting the use of nZVI in field-scale applications summarized in this paper show great promise for the remediation of chlorinated organic contaminants in groundwater.

References

1. Masciangioli, T., and Zhang, W.-X. Environ. Sci. Technol. 37 *(2003)*, 102A-108A.
2. Elliott, D. and Zhang, W.-X. Environ. Sci. Technol. 35 *(2001)*, 4922-4926
3. Li, X.Q., Elliott D.W., and W.Z. Zhang. *Crit. Reviews in Solid State Physics*. 2006. 31. 111-122.
4. Gavaskar, A., L. Tatar, and W. Condit. 2005. Cost and Performance Report: Nanoscale Zero-valent Iron Technologies for Source Reduction. Contract Report CR-05-007-ENV. Naval Facilities Engineering Service Center (Port Hueneme, CA).
5. Quinn, J., Geiger, C., Clausen, C., Brooks, K., Coon, C., O'Hara, S., Krug, T., Major, D.; Yoon, W., Gavaskar, A, Holdsworth, T. "Field Demonstration of DNAPL Dehalogenation Using Emulsified Zero-Valent Iron." *Environ. Sci. and Technol.* March 2005 Vol 39, pp 1309-1318.
6. Lien, H.-L., and Zhang, W.-X. Colloids and Surfaces 191 *(2001)*, 97-105.
7. Zhang, W.-X., and Elliott, D. Remediation 16 *(2006,)* 7-21.
8. Wang, C.-B., and Zhang, W.-X. Environ. Sci. Technol. 31 *(1997)*, 2154-2156.
9. Xu, Y., and Zhang, W.-X. Ind. Eng. Chem. Res. 39 *(2000)*, 2238-2244.
10. Zhu, B.-W., and Lim, T.-T. Environ. Sci. Technol. 41 *(2007)*, 7523-7529.
11. Choe, S., Lee, S.H., Chang, Y.Y., Hwang, K.Y., Khim, J.,2001. Rapid reductive destruction of hazardous organic compounds by nanoscale Fe0. *Chemosphere*, 42(4):367-372.
12. Cao, J. and W.X. Zhang. *J. Hazard. Mater*. 2006. 132, 213-219.
13. Cao, J., Elliott, D.W., and W.X. Zhang. "Perchlorate Reduction by Nanoscale Iron Particles." (*2005*). 7, 499-505.

14. Li, X.-Q., and Zhang, W.-X. J. Phys. Chem. C 111 *(2007)*, 6939-6946.
15. Gillham, R.W. and S.F. O'Hannesin. 1994. Groundwater, v. 32, p. 958-967.
16. Powell, R.M., D.W. Blowes, R.W. Gillham, D. Schultz, T. Sivavec, R.W. Puls, J.L. Vogan, P.D. Powell and R. Landis. 1998. *Permeable Reactive Barrier Technologies for Contaminant Remediation*, EPA/600/R-98/125, September 1998.
17. O'Hannesin, S.F., and R.W. Gillham. 1998. *Ground Water* 36, 164-170.
18. Vogan, J.L., R.M. Focht, D.K. Clark, and S.L. Graham. 1999. *Journal of Hazardous Materials* 68, 97-108.
19. Henn, K., and Waddill, D. Remediation 16 *(2006),* 57-77.
20. Saleh, N., H.-J. Kim, T. Phenrat, K. Matyjaszewski, R.D. Tilton, and G. Lowry. 2008. *Environmental Science and Technology*. 42:3349-3355.
21. Zhang, W.-X. 2003. *Journal of Nanoparticle Research*. 5:323-332.
22. Xu, J., and Bhattacharyya, D. Environmental Progress 24 *(2005)*, 358-366.
23. Phenrat, T., Saleh, N., Sirk, K., Kim, H.J., Tilton, R.D., and G.V. Lowry. 2007. *J. Nanoparticle Research*. 10: 795-814.
24. O'Hara, S., Krug, T., Quinn, J., Clausen, C., and Geiger, C. *Remediation. (Spring, 2006)* 35-56.
25. Matheson L. J., and P.G. Tratnyek. 1994. *Environ. Sci. Technol*. 28: 2045-2053.
26. Arnold, W. A, A.L. Roberts. 2000. Environ. Sci. Technol. 2000, vol 34 pp 1794-1805. 2000.
27. Liu, Y., Phenrat, T., and Lowry, G. Environ. Sci. Technol. 41 *(2007)*, 7881-7887.
28. Lim, T.T. and B.W. Zhu. **2007**. *Environ. Sci. Technol*. 41 (21): 7523-7529.
29. Zheng, T., Zhan, J., He, J., Day, C., Lu, Y., McPherson, G., Piringer, G., and John, V. Environ. Sci. Technol. 42 *(2008)*, 4494-4499.
30. Park, S., Kim, S., Lee, S., Khim, Z.G., Char, K. and Hyeon, T., 2000. J. Am. Chem. Soc., 122 (35), 8581 -8582, 2000. 10.1021/ja001628c S0002-7863(00)01628-0
31. U.S. EPA. Science Policy Council. 2007. "Nanotechnology White Paper" *U.S. Environmental Protection Agency*. February. http://es.epa.gov/ncer/nano/publications/whitepaper12022005.pdf
32. Miehr, R., P. G. Tratnyek, J.Z. Bandstra, M.M. Scherer, M.J. Alowitz, E.J. Bylaska. 2004. *Environmental Science and Technology* 38(1):139-147.
33. Nurmi,, J.T, P. G. Tratnyek, V. Sarathy, D. R. Baer, J. E AMonette, K. Pecher, C. Wang, J C. Linehan, D. W. Matson, R. L. Renn, and M. D. Driessen. 2005. Environ. Sci. Technol. 2005, vol 39, No. 5, pp 1221-1230. March 2005.
34. Mallouk, T., Ponder, M., and Darab, J., for US DoE. "Removal of Technetium, Carbon Tetrachloride, and Metals from DoE Properties." Final report, September 2000.
35. Kreyling, W.G., M. Semmler-Behnke, and W. Möller. 2006. *Journal of Nanoparticle Research*. 8:543-562.
36. Johnson, T.L., Scherer, M.M., and P.G. Tratnyek. 1996. 30(8): 2634-2640.
37. Lim, T.T., Feng. J. and B.Z. Zhu. 2008. *Environ. Sci. Technol*. 42(12): 4513-4519.

38. Zhang, W.X., Wang, C.B., and H.L. Lien. 1998. *Catalysis Today*. 40: 387-395.
39. Lien, H.L., and W.X. Zhang. 1999. *J. Environ, Eng*. 125(11): 1042-1047.

Indexes

Author Index

Subject Index

E

F

G

H

I

J

K

L

M

O

P

R

S

T

V

W

X

Z